MEMBRANE
RESEARCH

ACADEMIC PRESS RAPID MANUSCRIPT REPRODUCTION

Proceedings of a Conference
Held at Squaw Valley, California
March 13-17, 1972

First ICN – UCLA Symposium on Molecular Biology

MEMBRANE RESEARCH

edited by

C. FRED FOX

Department of Bacteriology
University of California
Los Angeles, California

Academic Press　　*New York and London*　　*1972*

ACADEMIC PRESS, INC.
111 Fifth Avenue, New York, New York 10003

United Kingdom Edition published by
ACADEMIC PRESS, INC. (LONDON) LTD.
24/28 Oval Road, London NW1

LIBRARY OF CONGRESS CATALOG CARD NUMBER: 72-77341

PRINTED IN THE UNITED STATES OF AMERICA

CONTENTS

IV. STRUCTURE AND ASSEMBLY
OF VIRAL MEMBRANES

V. MEMBRANES OF TRANSFORMED CELLS

VI. MEMBRANE-DNA INTERACTIONS

PARTICIPANTS

Abrams, Adolph, Department of Biochemistry, University of Colorado School of Medicine, Denver, Colorado

Aithal, H. N., Department of Biochemistry, University of Southern California School of Medicine, Los Angeles, California

Allen, C. Freeman, Department of Chemistry, Pomona College, Claremont, California

Altendorf, Karlheinz, Division of Research, National Jewish Hospital and Research Center, Denver, Colorado

Ames, Giovanna F.-L., Department of Biochemistry, University of California, Berkeley, California

Andreoli, A. J., Department of Chemistry, California State College, Los Angeles, California

Aswad, Dana W., Department of Biochemistry, University of California, Berkeley, California

Atkinson, Paul H., Department of Pathology, Albert Einstein College of Medicine, Bronx, New York

Baker, Nome, Radioisotope Research, Veterans Administration Hospital, Los Angeles, California

Banker, Gary, Department of Psychobiology, University of California, Irvine, California

Barber, Albert A., Department of Zoology, University of California, Los Angeles, California

Barber, Mary Lee, Biology Department, San Fernando Valley State College, Northridge, California

Barnes, Eugene M., Jr., Department of Biochemistry, Baylor College of Medicine, Houston, Texas

Bartholomew, James C., The Salk Institute for Biological Studies, San Diego, California

Barton, Peter G., Department of Biochemistry, University of Alberta, Edmonton, Canada

Bayer, Manfred E., Department of Molecular Biology, Institute for Cancer Research, Philadelphia, Pennsylvania

Berg, Howard C., Department MCD-Biology, University of Colorado, Boulder, Colorado

Bernard, George W., Department of Anatomy, University of California School of Medicine, Los Angeles, California

Birdwell, Charles R., Division of Biology, California Institute of Technology, Pasadena, California

Black, Douglas, R., Division of Infectious Diseases, Stanford University School of Medicine, Stanford, California

Blasie, J. K., Johnson Research Foundation, Department of Biophysics and Physical Chemistry, University of Pennsylvania, Philadelphia, Pennsylvania

Boedeker, Edgar C., Department of Medicine, Boston University Medical School, Boston, Massachusetts

Bose, Subir K., Department of Microbiology, St. Louis University School of Medicine, St. Louis, Missouri

Boyer, P. D., Molecular Biology Institute, University of California, Los Angeles, California

Branton, Daniel, Department of Botany, University of California, Berkeley, California

Bratt, Michael A., Department of Bacteriology, Harvard University Medical School, Boston, Massachusetts

Brewer, Gregory J., Biology Department, University of California, San Diego, California

Brodie, A. F., Department of Biochemistry, University of Southern California School of Medicine, Los Angeles, California

Brown, William E., Department of Molecular Biophysics and Biochemistry, Yale University, New Haven, Connecticut

Burge, Boyce, Massachusetts Institute of Technology, Cambridge, Massachusetts

Burger, Max M., Department of Biochemical Science, Princeton University, Princeton, New Jersey

Burgeson, Robert, Department of Zoology, University of California, Los Angeles, California

Cadenhead, David A., Department of Chemistry, State University of New York, Buffalo, New York

Carraway, Kermit L., Department of Biochemistry, Oklahoma State University, Stillwater, Oklahoma

Carter, James R., Jr., Presbyterian-University of Pennsylvania Medical Center, Philadelphia, Pennsylvania

Caspar, Donald L. D., Children's Cancer Research Foundation, Boston, Massachusetts

Chan, Sunney I., Noyes Laboratory of Chemical Physics, California Institute of Technology, Pasadena, California

Choppin, Purnell W., The Rockefeller University, New York, New York

Chun, Paul W., Department of Biochemistry, J. Hillis Miller Health Center, University of Florida, Gainesville, Florida

Collier, R. John, Department of Bacteriology, University of California, Los Angeles, California

Conner, R. L., Biology Department, Bryn Mawr College, Bryn Mawr, Pennsylvania

Cooperman, Barry S., Department of Chemistry, University of Pennsylvania, Philadelphia, Pennsylvania

Cotman, Carl W., Department of Psychobiology, University of California, Irvine, California

Criddle, Richard S., Department of Biochemistry and Biophysics, University of California, Davis, California

Darlington, Robert W., St. Jude Children's Hospital, Memphis, Tennessee

Deamer, D. W., Department of Zoology, University of California, Davis, California

Dennis, Edward A., Department of Chemistry, University of California, San Diego, California

Devaux, Philippe, Department of Chemistry, Stanford University, Stanford, California

Dratz, Edward A., Division of Natural Sciences, University of California, Santa Cruz, California

Earhart, Charles F., Jr., Department of Microbiology, University of Texas, Austin, Texas

Ebner, Kurt E., Department of Biochemistry, Oklahoma State University, Stillwater, Oklahoma

Eckhart, Walter, The Salk Institute for Biological Studies, San Diego, California

Edelman, Gerald M., The Rockefeller University, New York, New York

Eisenberg, Moishe, California Institute of Technology, Pasadena, California

Elovson, John, Department of Biology, University of California, San Diego, California

Elsbach, Peter, Department of Medicine, New York University School of Medicine, New York, New York

Epstein, Ruth, The Salk Institute for Biological Studies, San Diego, California

Epstein, Wolfgang, Department of Biochemistry, University of Chicago, Chicago, Illinois

Esfahani, Mojtaba, Department of Biochemistry, Baylor College of Medicine, Houston, Texas

Fessler, John H., Department of Zoology, University of California, Los Angeles, California

Fisher, Knute A., Department of Botany, University of California, Berkeley, California

Fitz-James, Philip, Department of Bacteriology, University of Western Ontario, London, Canada

Fong, Susie W., Department of Pediatrics, University of California, Irvine, California

Fox, C. Fred, Department of Bacteriology, University of California, Los Angeles, California

Francke, Bertold, The Salk Institute for Biological Studies, San Diego, California

Fried, Victor A., Institute of Molecular Biology, University of Oregon, Eugene, Oregon

Fuller, Gerald M., Department of Human Biological Chemistry and Genetics, University of Texas Medical Branch, Galveston, Texas

Furlong, Clement E., Department of Biochemistry, University of California, Riverside, California

Gallaher, William R., Department of Ophthalmology, Hospital of the University of Pennsylvania, Philadelphia, Pennsylvania

Galsworthy, Peter R., Department of Biochemistry, University of Western Ontario, London, Canada

Ganser, Allen L., Department of Physiology-Anatomy, University of California, Berkeley, California

Gibson, Wade, Department of Microbiology, University of Chicago, Chicago, Illinois

Gomatos, Peter John, Sloan-Kettering Cancer Research Institute, New York, New York

Goodenough, Daniel A., Department of Anatomy, Harvard University Medical School, Boston, Massachusetts

Gray, Gary R., Department of Biochemistry, University of California, Berkeley, California

Griffith, Jack, Department of Biochemistry, Stanford University School of Medicine, Stanford, California

Grinna, Lynn, Department of Zoology, University of California, Los Angeles, California

Hakomori, Sen-Itiroh, Department of Preventive Medicine, University of Washington School of Medicine, Seattle, Washington

Hanawalt, Philip C., Herrin Biology Laboratory, Stanford University, Stanford, California

Hargrave, Paul A., Biology Division, California Institute of Technology, Pasadena, California

Heine, Jochen W., Department of Microbiology, University of Chicago, Chicago, Illinois

Heller, Joram, Jules Stein Eye Institute, University of California School of Medicine, Los Angeles, California

Hendler, Sheldon, Department of Biology, University of California, San Diego, California

Hirata, H., Department of Biochemistry, University of Southern California School of Medicine, Los Angeles, California

Hokin, L. E., Department of Pharmacology, University of Wisconsin Medical School, Madison, Wisconsin

Holden, Joseph T., Neurosciences Division, City of Hope Medical Center, Duarte, California

Horwitz, Alan F., Laboratory of Chemical Biodynamics, University of California, Berkeley, California

Hourani, Benjamin, Section on Chemical Pathology, National Institute of Arthritis and Metabolic Diseases, Bethesda, Maryland

Huang, Josephine, Department of Surgery, University of California School of Medicine, San Francisco, California

Hubbell, Wayne L., Department of Chemistry, University of California, Berkeley, California

Hung, Paul P., Department of Virology, Abbott Laboratories, North Chicago, Illinois

Hunter, A. R., The Salk Institute for Biological Studies, San Diego, California

Ito, Junetso, Scripps Clinic and Research Foundation, La Jolla, California

James, Robert B., Department of Botany, University of California, Berkeley, California

Jan, Lily, Division of Biology, California Institute of Technology, Pasadena, California

Jan, Y. N., Division of Biology, California Institute of Technology, Pasadena, California

Jewett, Susan, Department of Bacteriology, University of California, Los Angeles, California

John, Philip, Department of Physiology-Anatomy, University of California, Berkeley, California

Jordan, John Maxwell, Department of Chemistry, University of California, Los Angeles, California

Jost, Patricia, Institute of Molecular Biology, University of Oregon, Eugene, Oregon

Kaback, H. R., Roche Institute of Molecular Biology, Nutley, New Jersey

Kahane, I., Section on Chemical Pathology, National Institute of Arthritis and Metabolic Diseases, Bethesda, Maryland

Kalira, V. J., Department of Biochemistry, University of Southern California School of Medicine, Los Angeles, California

Kalnins, V. I., Department of Anatomy, University of Toronto, Toronto, Canada

Kasper, Charles B., McArdle Laboratory, University of Wisconsin, Madison, Wisconsin

Keller, John M., Department of Biochemistry, University of Washington School of Medicine, Seattle, Washington

Kelly, Françoise, Cold Spring Harbor Laboratory, Cold Spring Harbor, New York

Khare, Gyan P., ICN Nucleic Acid Research Institute, Irvine, California

Khwaja, Tasneem, ICN Nucleic Acid Research Institute, Irvine, California

Kijimoto, Shigeko, Department of Pathobiology, University of Washington, Seattle, Washington

Kim, Young S., G. I. Research Laboratory, Veterans Administration Hospital, San Francisco, California

King, Glen I., Cardiovascular Research Institute, University of California School of Medicine, San Francisco, California

Kirk, Betty I., Department of Biochemistry, University of California, Berkeley, California

Klein, William L., Molecular Biology Institute, University of California, Los Angeles, California

Kornberg, Arthur, Department of Biochemistry, Stanford University School of Medicine, Stanford, California

Kreishman, George P., ICN Nucleic Acid Research Institute, Irvine, California

Lansman, Robert A., Department of Biological Sciences, Stanford University, Stanford, California

Law, John H., Department of Biochemistry, University of Chicago, Chicago, Illinois

Lazdunski, Andree, Department of Biochemistry, University of Washington, Seattle, Washington

Lazdunski, Claude J., Department of Biochemistry, University of Washington, Seattle, Washington

Leive, Loretta, Laboratory of Biochemical Pharmacology, National Institute of Arthritis and Metabolic Diseases, Bethesda, Maryland

Lewis, Lowell N., Citrus Research Center and Agricultural Experiment Station, University of California, Riverside, California

Li, Joseph K. K., Department of Bacteriology, University of California, Los Angeles, California

Linden, Carol, Department of Bacteriology, University of California, Los Angeles, California

Lux, Samuel E., National Institutes of Health, Bethesda, Maryland

Machtiger, Neal A., Department of Bacteriology, University of California, Los Angeles, California

Magilen, Gilbert, Department of Nutritional Sciences, Agricultural Experiment Station, University of California, Berkeley, California

Magnuson, James Andrew, Department of Chemistry, Washington State University, Pullman, Washington

Marchesi, V. T., Section of Chemical Pathobiology, National Institute of Arthritis and Metabolic Diseases, Bethesda, Maryland

McConnell, H. M., Department of Chemistry, Stanford University, Stanford, California

McElhaney, Ronald N., Department of Biochemistry, University of Alberta, Edmonton, Canada

McFarland, Betty G., Department of Chemistry, Stanford University, Stanford, California

McGuire, Edward John, Department of Biology, Johns Hopkins University, Baltimore, Maryland

McMurray, W. C., Department of Biochemistry, University of Western Ontario, London, Canada

McNamee, Mark G., Department of Chemistry, Stanford University, Stanford, California

Melnick, Ronald L., Department of Physiology-Anatomy, University of California, Berkeley, California

Menninger, John R., Institute of Molecular Biology, University of Oregon, Eugene, Oregon

Meryman, Harold T., American National Red Cross Blood Research Laboratory, Bethesda, Maryland

Messer, Anne, Institute of Molecular Biology, University of Oregon, Eugene, Oregon

Meyer, David, Department of Zoology, University of California, Los Angeles, California

Moyer, Richard W., Department of Biochemistry, Columbia University, New York, New York

Miller, Jon P., ICN Nucleic Acid Research Institute, Irvine, California

Miyamoto, Vernon K., Cardiovascular Research Institute, University of California School of Medicine, San Francisco, California

Mokrasch, Lewis C., Department of Biochemistry, Louisiana State University Medical Center, New Orleans, Louisiana

Mommaerts, Wilfried, Department of Physiology, University of California, Los Angeles, California

Montal, Maurice, Department of Biochemistry, National Polytechnic Institute, Mexico City, Mexico

Moyer, Sue A., Department of Microbiology and Immunology, Albert Einstein College of Medicine, Bronx, New York

Mukherjee, Barid B., Department of Biology, McGill University, Montreal, Canada

Nicolson, Garth L., The Salk Institute for Biological Studies, San Diego, California

Nojima, Shoshichi, Department of Chemistry, National Institute of Health, Tokyo, Japan

Okada, Yoshio, Department of Preventive Medicine, Research Institute of Microbial Diseases, Osaka University, Osaka, Japan

Oppenheimer, Steven B., Department of Biology, San Fernando Valley State College, Northridge, California

Osborn, Mary J., Department of Microbiology, University of Connecticut School of Medicine, Farmington, Connecticut

Oseroff, Allan, Department of Biology, Massachusetts Institute of Technology, Cambridge, Massachusetts

Oxender, Dale L., Department of Biological Science, Stanford University, Stanford, California

Ozanne, Brad, Cold Spring Harbor Laboratory, Cold Spring Harbor, New York

Parks, Leo W., Department of Microbiology, Oregon State University, Corvallis, Oregon

Perdue, James F., McArdle Laboratory for Cancer Research, University of Wisconsin Medical Center, Madison, Wisconsin

Peter, James B., Institute of Rehabilitation and Chronic Diseases, Los Angeles, California

Racker, E., Division of Biological Sciences, Cornell University, Ithaca, New York

Ray, Dan S., Department of Zoology, University of California, Los Angeles, California

Reinert, Joe C., Department of Environmental Health Sciences, Oregon State University, Corvallis, Oregon

Revel, Jean-Paul, Division of Biology, California Institute of Technology, Pasadena, California

Richmond, Jonas E., Department of Nutritional Science, Agricultural Experiment Station, University of California, Berkeley, California

Rittenhouse, Harry G., Department of Chemistry, Washington State University, Pullman, Washington

Rittenhouse, Judy, Department of Chemistry, Washington State University, Pullman, Washington

Robbins, April, Division of Biological and Medical Sciences, Brown University, Providence, Rhode Island

Robertson, J. David, Department of Anatomy, Duke University Medical Center, Durham, North Carolina

Robinson, William S., Department of Medicine, Stanford University School of Medicine, Stanford, California

Roizman, Bernard, Committee on Virology, University of Chicago, Chicago, Illinois

Romano, Antonio H., Microbiology Section, University of Connecticut, Storrs, Connecticut

Rose, Steven P., Department of Biological Chemistry, University of California, Los Angeles, California

Rosenberg, Murray D., Department of Genetics and Cell Biology, University of Minnesota, St. Paul, Minnesota

Rothfield, Lawrence, Department of Microbiology, University of Connecticut School of Medicine, Farmington, Connecticut

Rothstein, Aser, Research Institute, Hospital for Sick Children, Toronto, Canada

Rousseau, Guy, Department of Biochemistry, University of California School of Medicine, San Francisco, California

Rousseau, Robert J., ICN Nucleic Acid Research Institute, Irvine, California

Ruth, Royal F., Department of Zoology, University of Alberta, Edmonton, Canada

Ryser, Hugues J.-P., Department of Cell Biology and Pharmacology, University of Maryland School of Medicine, Baltimore, Maryland

Sanwal, B. D., Department of Medical Cell Biology, University of Toronto, Toronto, Canada

Sato, Gordon, Department of Biology, University of California, San Diego, California

Scandella, Carl, Department of Chemistry, Stanford University, Stanford, California

Scarborough, Gene A., Department of Biochemistry, University of Colorado School of Medicine, Denver, Colorado

Schimke, Robert T., Department of Pharmacology, Stanford University, Stanford, California

Schneider, Allan, National Institute of Mental Health, Bethesda, Maryland

Segrest, J. P., National Institute of Arthritis and Metabolic Diseases, Bethesda, Maryland

Seifert, Willi, The Salk Institute for Biological Studies, San Diego, California

Shapiro, Bennett M., Department of Biochemistry, University of Washington School of Medicine, Seattle, Washington

Shapiro, David, Department of Pharmacology, Stanford University School of Medicine, Stanford, California

Sheppard, J. R., University of Colorado Medical Center, Denver, Colorado

Siddiqui, Bader, Department of Pathobiology, University of Washington, Seattle, Washington

Simon, Lee, Institute for Cancer Research, Philadelphia, Pennsylvania

Simon, Melvin, Department of Biology, University of California, San Diego, California

Simoni, Robert D., Department of Biological Sciences, Stanford University, Stanford, California

Smart, John E., Division of Biology, California Institute of Technology, Pasadena, California

Smith, Marvin A., Graduate Section of Biochemistry, Brigham Young University, Provo, Utah

Smith, Roberts A., Department of Chemistry, University of California, Los Angeles, California

Smith, Robbin Peggy Piety, Department of Biochemistry, University of Washington, Seattle, Washington

Spear, Patricia G., The Rockefeller University, New York, New York

Staehelin, Andrew, Department of Molecular, Cellular and Developmental Biology, University of Colorado, Boulder, Colorado

Steck, Theodore L., Department of Medicine, University of Chicago, Chicago, Illinois

Stellner, Klaus, Department of Pathobiology, University of Washington, Seattle, Washington

Stoeckenius, Walther, Cardiovascular Research Institute, University of California School of Medicine, San Francisco, California

Strauss, James H., Division of Biology, California Institute of Technology, Pasadena, California

Szabo, Gabor, Department of Physiology, University of California, Los Angeles, California

Takemoto, Jon, Department of Bacteriology, University of California,
Los Angeles, California

Taylor, John M., Department of Pharmacology, Stanford University School of
Medicine, Stanford, California

Teng, Nelson, Laboratory of Chemical Biodynamics, University of California,
Berkeley, California

Terry, Thomas, Department of Microbiology, University of Connecticut, Storrs,
Connecticut

Thomas, David, Department of Biological Science, Stanford University, Stanford,
California

Thompson, Edward D., Department of Microbiology, Oregon State University,
Corvallis, Oregon

Tiffany, John J., Department of Ophthalmology, Hospital of the University of
Pennsylvania, Philadelphia, Pennsylvania

Tinberg, Harold M., Department of Physiology-Anatomy, University of California,
Berkeley, California

Tomasz, Alexander, Laboratory of Genetics, The Rockefeller University, New York,
New York

Trudell, James R., Department of Anesthesia, Stanford University School of Medicine,
Stanford, California

Tsukagoshi, Norihiro, Department of Bacteriology, University of California,
Los Angeles, California

Tustanoff, Eugene R., Department of Pathological Chemistry, University of Western
Ontario, London, Canada

Umbreit, Jay, Biological Laboratories, Harvard University, Cambridge, Massachusetts

Vernon, Leo P., Brigham Young University, Provo, Utah

Victoria, Edward J., Laboratory of Biochemistry, National Heart and Lung
Institute, Bethesda, Maryland

Volcani, B. E., Department of Marine Biology, University of California, San Diego,
California

Wakil, Salih J., Department of Biochemistry, Baylor College of Medicine, Houston,
Texas

Walter, Gernot, The Salk Institute for Biological Studies, San Diego, California

Werchau, Hermann, The Salk Institute for Biological Studies, San Diego, California

Whitehead, James S., Veterans Administration Hospital, San Francisco, California

Whiteley, Norman, Department of Molecular, Cellular and Developmental Biology,
University of Colorado, Boulder, Colorado

Wickner, Bill, Department of Biochemistry, Stanford University School of Medicine,
Stanford, California

Williams, Mary Ann, Department of Nutritional Sciences, University of California, Berkeley, California

Wisnieski, Bernadine, Department of Bacteriology, University of California, Los Angeles, California

Wlodawer, Alexander, Department of Chemistry, University of California, Los Angeles, California

Woodward, Dow, Department of Biology, Stanford University, Stanford, California

Yu, Kam-Yee, Cardiovascular Research Institute, University of California School of Medicine, San Francisco, California

PREFACE

The general catalog of the University of California at Los Angeles states: "The Molecular Biology Institute was established to serve interested departments of the biological, medical, and physical sciences in the coordination, support, and enhancement of research and training in molecular biology. Interests and activities of the Institute encompass all approaches which aim to explain biology at a molecular level, with particular emphasis on correlation of structure and function." It is within this framework of responsibility that the Institute has organized various activities, including small conferences, on the UCLA campus. This conference, the first in a series of annual conferences on selected topics in molecular biology, represents a broadening of the scope of the Institute to serve a wider academic community and need, while at the same time enhancing research and training in molecular biology at UCLA. The selection of membrane research as the topic for the first of these conferences is appropriate both from the standpoint of its timeliness and its interdisciplinary nature.

We are deeply indebted to the perspective shown by the International Chemical and Nuclear Corporation in providing the financial support that made the conference possible. We are also grateful to Drs. Walther Stoeckenius, Walter Eckhart, Jon Singer, William Robinson, Harden McConnell, Arthur Kornberg, and Gordon Sato for their expert advice and direct participation in the selection of the speakers, to the speakers themselves who individually and collectively deserve appreciation for the success of the conference and for making the preparation of this volume possible, and finally to Dr. Roberts Smith for his indispensible contributions.

C. Fred Fox, *Conference Chairman*

Paul D. Boyer, *Director*
Molecular Biology Institute

MEMBRANE
RESEARCH

I

CELL SURFACE FLUIDITY

MOLECULAR MOTION IN MEMBRANES AS INDICATED
BY X-RAY DIFFRACTION

J. Cain, G. Santillan and J.K. Blasie

Johnson Research Foundation
University of Pennsylvania
Philadelphia, Pennsylvania 19104

ABSTRACT. Photopigment molecules in oriented frog retinal
receptor disk membranes may present a model for protein
molecular motion in the plane of biological membranes.
Changes in the local arrangement or clustering of photopigment molecules over the surface of the disk membrane may be
induced by variation in temperature, pH, ionic strength
and/or bleaching. In addition, bleaching causes a reduction in net electric charge on the photopigment molecule
and induces a movement of the molecule in the transmembrane
direction.
      Oriented model membranes based on the lecithin bilayer
may present a model for lipid molecular motion in membranes.
The fatty-acid chain packing and electron density profile
for the dipalmitoyl lecithin bilayer depends critically on
temperature in the region of 40-45° C and on a specific
interaction of H_2O with the ester linkages of the molecule.
Information relating to molecular motion in the model membrane may be obtained from a correlation of the x-ray
structure with spectroscopic data obtained from probe molecules which are sensitive to molecular motion incorporated
into the host membrane, especially if they induce only
minimal perturbation of the host structure. Changes in
bonding between an incorporated molecule and the lecithin
in the polar headgroup region of the bilayer may strongly
affect chain packing as shown by the lecithin:ubiquinone Q_3
membranes on oxidation-reduction of Q_3.

INTRODUCTION

In general, x-ray diffraction cannot distinguish bet-

ween the ensemble-average of a frozen statistically disordered structure and a dynamical structure whose time-aver-
age is identical to the ensemble-average of the frozen
structure (1). In order to gain information about molecular motion within a structure, one may correlate structure deduced by x-ray diffraction with data from probes of
structure sensitive to molecular motion such as spectroscopy. On the other hand, molecular motion may be inferred
from structural changes between initial and final states
deduced by x-ray diffraction produced by artificial or
natural perturbation of the structure.

Photopigment molecules in oriented pellets of isolated
disk membranes provide a system for the detection of the
molecular motion of an intrinsic membrane protein since
x-ray diffraction can be obtained which arises from the
planar arrangement of photopigment molecules over the membrane surface (2). Changes in the local arrangement or
clustering of photopigment molecules produced by artificial
or natural perturbation as well as changes in the embedding
of the molecules in the hydrocarbon core (3-4) of the disk
membrane are detected by x-ray diffraction (5-7). The
mobility of these molecules in the plane of the disk membrane and the transmembrane direction may then be inferred.

X-ray diffraction from oriented multilayers of model
membranes whose basic structural element is a lecithin bilayer may provide structural information relating to the
mode of packing of the fatty-acid chains and molecular
packing in the plane of the membrane as well as the distribution of molecular components through the cross-section of
the membrane at near-atomic resolution if the phase-problem
for the lamellar reflections from the oriented multilayer
can be solved. We have developed several methods for the
solution of the phase problem for these structures (8-11).
Information relating to molecular motion in these model
membranes may be obtained by a correlation of the x-ray
structure with data from intrinsic probes of molecular
motion such as ^{13}C or proton nuclear magnetic resonance
spectroscopy (12-14) or with data from extrinsic probes incorporated into the model membrane such as fluorescence
spectroscopy of 12-(9-anthroyl)-stearate (AS) or N-octade-
cylnaphthyl-2-amine-6-sulfonate (ONS), (15), or electron
paramagnetic resonance spectroscopy of "spin-labelled"
probe molecules (20). Changes in the x-ray structure of
model membranes induced by temperature or chemical pertur-

bation, e.g. the oxidation-reduction of Q_3 in lecithin-ubi-quinone Q_3 membrane, may indicate modes of molecular motion in these membranes.

METHODS

X-ray diffraction from photopigment molecules in oriented pellets of isolated disk membranes and that dependence of this diffraction data on temperature, pH, ionic strength and photopigment bleaching together with an analysis of these data which provide a description of the local planar arrangement of photopigment molecules in the surface of the membrane has been described (2,5-6). In addition, the bleaching-dependent location of the photopigment molecules in the cross-section of the disk membrane has been determined (7).

X-ray diffraction from oriented multilayers of certain model membranes has been described (8-11,16-17) and interpretations of diffraction data arising from fatty-acid chain packing in the bilayer membrane have been presented (16). Several methods for solving the phase-problem for the lamellar reflections have been developed by us (8-11) and provide the cross-sectional electron density profile [the projection of electron density in the membrane along a direction parallel to the plane of the membrane onto an axis normal to the plane in the absence of molecular motion (9)] for the model membrane to a few Ångströms resolution. The incorporation of cholesterol (16), chlorophyll (17), and the fluorescent probes AS and ONS (11) into phospholipid bilayer membranes has been studied previously. The structures of dipalmitoyl lecithin bilayer membranes at various water contents and temperatures have been determined in a similar fashion (10-11) as were the structures of dipalmitoyl lecithin: oxidized/reduced ubiquinone Q_3 model membranes. The latter structure was determined independently at each stage of the cycle oxidized $Q_3 \rightarrow$ reduced $Q_3 \rightarrow$ reoxidized Q_3 in order to provide the proper control experiments.

RESULTS

The local planar arrangement of photopigment molecules derived from x-ray diffraction data provides average nearest neighbor numbers and average nearest neighbor

separations (5). With decreasing temperature from 42.5° C
to 4.5° C, average nearest neighbor separations increase
from 52 Å to 64 Å while average nearest neighbor numbers
vary from 2.5 to 3.8 in a manner which preserves the average membrane area per photopigment molecule (5). The diameter of the photopigment molecule in the plane of the disk
membrane is ∿ 42 Å (5). Increasing pH (6.0 to 8.0) or decreasing ionic strength increases the average nearest
neighbor separation (6). The average nearest neighbor
separation at a given pH in the range 6.0-8.0 is greater
for unbleached photopigment compared with bleached photopigment. Diffraction arising from the lipid fatty-acid
chain packing in the plane of the membrane at ∿ 4.5 Å was
invariant under these perturbations.

The reduction in the scaled total diffracted x-ray
intensity from the planar arrangement of photopigment molecules on increasing the average electron density of the
sedimentation medium surrounding the disk membranes indicates that the photopigment molecule is embedded ∿ 14 Å into the lipid hydrocarbon core of the disk membrane with ∿
28 Å protruding into the aqueous surface-layer of the disk
membrane (7). The molecule sinks ∿ 7 Å deeper into the
hydrocarbon core on bleaching.

The electron density profiles for the dipalmitoyl
lecithin bilayer at 25° C and 49° C at similar water contents for the multilayers are shown in figure 1. The thickness of the bilayer is decreased at the higher temperature
while both the polar headgroup region and hydrocarbon core
of the bilayer are considerably modified. The dipalmitoyl
lecithin bilayer structure at 5 Å resolution at lower temperatures has been derived previously by us (11) while the
high-temperature profile is a preliminary result. Defining
the axis of the lamellar reflections as the meridian in the
diffraction pattern from the lecithin multilayers, a sharp
Bragg reflection at 4.15 Å occurs on the equator with relatively little off-equatorial spreading for the lower
temperature bilayer while a 4.5 Å broad maximum with considerable off-equatorial spreading (nearly ring-like)
occurs at 49° C. The degree of orientation of the lamellar
reflections is similar at the two temperatures.

The electron density profiles for dipalmitoyl lecithin
bilayers at 25° C and two different water contents are
shown in figure 2a and 2b. Some alterations in the polar
headgroup region are apparent while the hydrocarbon core is

is relatively invariant. With increasing water content, the 4.15 Å equatorial Bragg reflection becomes relatively less intense with an increase in off-equatorial spreading of this reflection into a considerably broader line-profile with a maximum at nearly the same diffraction angle as judged by microdensitometry.

The electron density profiles for dipalmitoyl lecithin and dipalmitoyl lecithin:palmitate = 1:1 bilayers are shown in figure 3. Large differences in the polar headgroup regions are apparent while the hydrocarbon cores are nearly identical as are the near-equatorial Bragg reflections at $\sim$ 4.15 Å.

The electron density profiles for dipalmitoyl lecithin: AS = 4:1 molar ratio and dipalmitoyl lecithin: ONS = 2:1 bilayers are shown in figure 2c and 2d. Considerable alteration of the polar headgroup region and hydrocarbon core occurs on the addition of AS while ONS appears only to increase the width of the hydrocarbon core by some 4 Å, all other aspects of the structure remaining relatively invariant. Incorporation of AS causes the 4.15 Å equatorial reflection to become relatively less intense with considerable increase in the broad component spreading well off-equatorial at equivalent diffraction angle. ONS slightly broadens the 4.15 Å equatorial reflection and induces some additional off-equatorial spreading of this sharp reflection.

Electron density profiles for dipalmitoyl lecithin and dipalmitoyl lecithin:oxidized ubiquinone Q_3 = 2:1 are shown in figure 4. Differences in the polar headgroup region are apparent with negligible differences in the hydrocarbon cores of the two bilayers. Incorporation of oxidized Q_3 results in a dramatic reduction of the relative intensity of the 4.15 Å equatorial reflection with a considerable increase in the intensity of the broad component at equivalent diffraction angle spreading well off-equatorial. Stoichiometric reduction of the Q_3 results in the return of a equatorial 4.15 Å reflection which is somewhat broader but shows less off-equatorial spreading than the lecithin alone at comparable humidity. Partial re-oxidation of the Q_3 induces a reversion toward that described for fully oxidized Q_3. During the reduction and re-oxidation of Q_3, small changes occur in the polar headgroup region of the electron density profile near the inner peak facing the hydrocarbon core while the core-region of the bilayer is relatively invariant.

DISCUSSION

The ability of photopigment molecules to alter their local arrangement or clustering over the surface of the disk membranes as induced by alteration in temperature, pH, etc. while the average lipid fatty-acid chain spacing in the plane of the disk membrane remains unchanged implies a lateral mobility for these protein molecules in the plane of the membrane. Such lateral mobility would be expected to depend primarily on the viscosity of the lipid hydrocarbon core of the disk membrane due to the substantial embedding of the photopigment molecule in the core and on the net electric charge occurring on that portion of the molecule protruding into the aqueous surface layer of the membrane, as well as shielding of this electric charge by counter-ions in this layer, due to nearest-neighbor interactions. The nature of the embedding of the molecules in an aqueous-lipid hydrocarbon interface prevents tumbling of the molecule about an axis lying in the plane of the membrane but allows rapid rotation about a axis normal to the plane of the disk membrane as verified spectroscopically (18-19). In addition, a small transmembrane mobility for the photopigment molecules exists depending on bleaching.

The electron density profile in the hydrocarbon core for dipalmitoyl lecithin below 40° C together with the nature of the near-equatorial diffraction indicate that the fatty-acid chains occur predominately in an extended form normal to the plane of the bilayer and are most likely arranged in a hexagonal lattice in the plane of the bilayer with an inter-chain separation of 4.8 Å. The 4.15 Å Bragg reflection for this hexagonal lattice may be a higher order reflection of a "super-lattice" of twice the dimensions since several lower-angle equatorial reflections including the 4.15 Å reflection from such a lattice appear at very low water contents. This "super-lattice" may represent the lecithin molecular lattice. The interior methylene groups of the chains form the region of constant electron density in the core-region of the profile while the terminal methyl group and possibly 1-2 adjacent methylene groups lie in the trough of relatively lower electron density at the center of the bilayer. Higher order lamellar reflections (>10) which could narrow this feature of the profile are

not observed except at very low water contents when more
than 22 orders appear. These 2-3 carbon atoms at the terminal methyl end of the chains may then be statistically
or motionally disordered in the unit cell structure inorder
to account for the width of this central trough. Above 45°
C, the hydrocarbon core of the bilayer thins dramatically
and the fatty acid chains are highly disordered in a state
similar to that of liquid long-chain normal hydrocarbons as
indicated by the nature of the broad intensity maximum at
$\sim$ 4.5 Å as described. However, the terminal methyl ends of
the chains appear to still be predominantly localized near
the center of the bilayer as indicated by the fact that a
somewhat broader trough of lower relative electron density
is still distinguishable in the center of the bilayer. The
general features of the high-temperature profile for the
hydrocarbon core of the bilayer could be indicative of
increased statistical or motional disorder along the chains
toward their terminal methyl groups particularly in the
last 3-4 carbon atoms. The latter interpretation would be
consistant with results obtained by several spectroscopic
methods (12-14,20) as is the reorganization of the polar
headgroup region above the transition temperature.

The outer peak of the two peaks in the polar headgroup
region of the dipalmitoyl lecithin electron density profile
(below 40° C) can be considered to contain the phosphate
and tri-methyl ammonium groups while the inner peak contains
the ester groups with one water molecule coordinated on
each ester group, as indicated by model calculations (21).
This assignment is supported by the invariance of the hydrocarbon cores and the differences in the polar headgroup
regions of the dipalmitoyl lecithin profile as compared with
the dipalmitoyl lecithin:palmitate = 1:1 profile, as indicated by model calculations.

Increasing water content in the lecithin multilayers
renders the ester region of the polar headgroup region of
the profile less structured. In addition, a slightly
greater portion of segments of fatty acid chains are packed
in a statistically disordered amorphous state and a lesser
portion of segments occur in an extended-chain hexagonal
packing, as indicated by changes in the profile in the
hydrocarbon core and near-equatorial diffraction.

A model for the location of AS in the dipalmitoyl
lecithin bilayer consistant with the electron density profile and near-equatorial diffraction places the AS molecule

in an extended form within the bilayer placing its anthra-
cene chromophore deep within the hydrocarbon core of the
bilayer penetrating to within $\sim$ 5 Å of the center of the
bilayer and its carboxyl group in the polar headgroup region (11). This penetration of the chromophore causes the
chains to become packed predominately in a statistically
disordered amorphous state.[1] This is verified by the high
and nearly constant value of the polarization of fluorescence from AS between 10° and 42° C (22). Above 42° C,
this polarization of fluorescence becomes markedly reduced
indicative of considerable motion of the chromophore during
its excited state lifetime consistant with a motionally
disordered chromophore environment as described above for
the fatty acid chain packing for dipalmitoyl lecithin alone
above 42° C. The incorporation of the saturated normal
hydrocarbon chain of ONS into the lecithin bilayer in an
extended form induces little perturbation of chain packing
in the hydrocarbon core (11); however, its naphthalene
chromophore is similarly sensitive to this phase transition
for the fatty-acid chains even though it is located outside
the hydrocarbon core region of the mixed bilayer (22). It
appears that both modes of chain packing, amorphous and
well-ordered hexagonal lattice, have similar phase transitions for chain melting.

A model for the location of oxidized ubiquinone Q_3 in
the dipalmitoyl lecithin bilayer consistant with the profile and near-equatorial diffraction places the quinone
moiety in the ester layer of the polar headgroup region
with its branched unsaturated hydrocarbon chain penetrating
in an extended form into the hydrocarbon core of the bilayer. This penetration destroys the fatty acid chain
lattice and induces a statistically disordered amorphous
packing of the chains[1] even more disordered than the amorphous chain packing induced by AS incorporation. Stoichiometric reduction of Q_3 appears to induce the formation of
a "co-lattice" or mixed-lattice of the chains in which
chain segments occur predominately in a somewhat disordered
hexagonal lattice oriented normal to the plane of bilayer.
Partial re-oxidation of the Q_3 causes a reversion to the
amorphous chain packing. The hydroquinone protons might
be expected to participate in hydrogen-bonding in the ester
layer of the polar headgroup region of the lecithin and
produce a specific reduced Q_3-lecithin interaction in that
region of the bilayer. Such a specific interaction in the

polar headgroup region could induce a more regular chain packing as suggested by the dependence of the lecithin chain packing on hydration. The changes occurring in the polar headgroup region indicate some structural rearrangement in that region with oxidation-reduction of the quinone moiety of Q_3. More detailed experiments on this point are in progress. However, these results suggest that changes in hydrogen bonding interactions in the polar headgroup region of this model membrane may induce alterations in the mode of chain packing in its hydrocarbon core.

ACKNOWLEDGMENTS

This work was supported by National Institutes of Health grants GM 12202 and EY 00673. We should like to thank Drs. A. Waggoner and L. Stryer for samples of AS and ONS.

REFERENCES

1. Hosemann, R. and Bagchi, D.N., Direct Analysis of Diffraction by Matter, North-Holland Publishing Co., Amsterdam (1962).
2. Blasie, J.K., Worthington, C.R. and Dewey, M.M., J. Mol. Biol., 39, 407 (1969).
3. Gras, W.J. and Worthington, C.R., Proc. Nat'l Acad. Sci. U.S., 63,233 (1969), and Biophys. J., 12, 255a (1972).
4. Blaurock, A.E. and Wilkins, M.H.F., Nature, 223, 906 (1969).
5. Blasie, J.K. and Worthington, C.R., J. Mol. Biol., 39, 417 (1969).
6. Blasie, J.K., Biophys. J., 12, 2, 205 (1972).
7. Blasie, J.K., Biophys. J., 12, 2, 191 (1972).
8. Lesslauer, W. and Blasie, J.K., Biophys. J., 12, 2, 175 (1972).
9. Lesslauer, W. and Blasie, J.K., Acta Cryst., A27, 456 (1971).
10. Lesslauer, W., Cain, J. and Blasie, J.K., Biochim. Biophys. Acta, 241, 2, 547 (1971).
11. Lesslauer, W., Cain, J. and Blasie, J.K., Proc. Nat'l. Acad. Sci. U.S., (in press).
12. Metcalfe, J.C., et al., Nature, 233, 199 (1971).
13. Lee, A.G., et al., Biochim. Biophys. Acta, 255, 43

(1972).
14. Horowitz, A.F., Horsley, W.J. and Klein, M.P., Proc. Nat'l. Acad. Sci. U.S., (in press).
15a. Waggoner, A. and Stryer, L., Proc. Nat'l Acad. Sci. U.S., 67, 579 (1970).
15b. Yguerabide, J. and Stryer, L., Proc. Nat'l. Acad. Sci. U.S., 68, 6, 1217 (1971).
16. Levine, Y.K. and Wilkins, M.H.F., Nature, 230, 69 (1971).
17. Cain, J.E. and Blasie, J.K., Biophys. J., 12, 45a (1972).
18. Liebman, P.A., Biophys. J., 2, 161 (1962).
19. Brown, P.K., Biophys. J., 11, 248a (1971).
20. Hubbell, W.L. and McConnell, H.M., J. Amer. Chem. Soc. 93, 314 (1971).
21. Wilkins, M.H.F., "X-Ray Diffraction Studies on Membranes and Model Systems", Meeting N.Y. Acad. Sci., New York, 1971, June, Annal N.Y. Acad. Sci., (in press).
22. Vanderkooi, J. and Chance, B., FEBS Letters, (in press).

[1] The statistically disordered amorphous packing refers to chain packing in the plane of the bilayer membrane. It results from a combination of lattice disorder coupled with substitution disorder, i.e. although the chains are relatively extended, the conformation and/or composition of neighboring chains differ. Hence, diffraction from the chains arises from most probable inter-chain vectors instead of a Bragg plane for a well-ordered lattice (1).

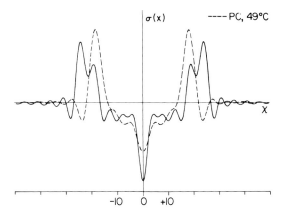

Figure 1. Electron density profiles for dipalmitoyl
lecithin at 25° C and 49° C at 5 Å and 7 Å resolution and
∿ 5% and ∿ 8% water contents, respectively.

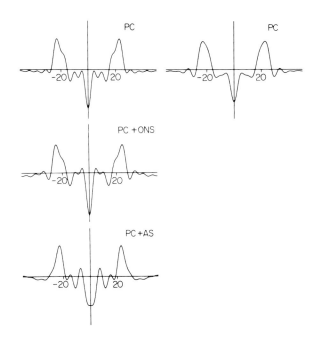

Figure 2. Electron density profiles for various bilayers
at comparable resolution (∿ 7 Å)
  (a) dipalmitoyl lecithin at ∿ 3% water content.
  (b) dipalmitoyl lecithin at ∿ 16% water content.
  (c) dipalmitoyl lecithin: AS = 4:1 molar ratio at ∿ 3%
      water content.
  (d) dipalmitoyl lecithin: ONS = 2:1 molar ratio at ∿ 5%
      water content.

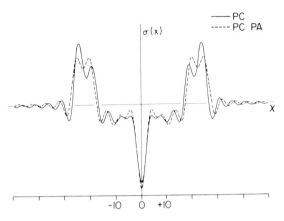

Figure 3. Electron density profiles for dipalmitoyl lecithin and dipalmitoyl lecithin:palmitate = 1:1 molar ratio at 5 Å resolution and similar water contents.

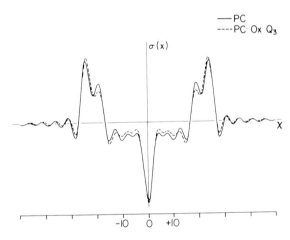

Figure 4. Electron density profiles for dipalmitoyl lecithin and dipalmitoyl lecithin:oxidized ubiquinone Q_3 = 2:1 molar ratio at 5 Å resolution and similar water contents.

ASPECTS OF PLASMA MEMBRANE FLUIDITY

Michael Edidin

Department of Biology
The Johns Hopkins University
Baltimore, Maryland 21218

ABSTRACT. Antibodies may be used to detect rearrangements of plasma membranes induced by other agents, or
may themselves be used to remodel the surface of cells.
Both sorts of experiments suggest that the animal cell
plasma membrane is fluid-like and this raises the problem of
restriction of surface receptors and enzymes in functionally
differentiated tissue cells.

INTRODUCTION

Recently, several probes of plasma membranes and
phospholipid model systems have indicated that the cell
surface may consist of a fluid lipid continuum in which are
embedded protein molecules free to rotate and to move laterally in the plane of the lipid (1, 2, 3, 4).

In this paper I want to present further experimental
evidence consistent with a fluid plasma membrane, and to
consider experiments that deal with some objections that may
be raised to our earlier work in the area. Finally I want
to consider ways of restricting protein mobility in the
membrane, and some of the evidence that such restrictions
may operate in intact tissues.

The experiments to be discussed all employ antibodies
as probes of the state of the cell surface. Antibodies may
be used either to measure the state of a membrane after it
has been perturbed by addition of new and antigenically
different membrane (by cell fusion) or may be used themselves as perturbing agents.

15

FUSION EXPERIMENTS

In initial experiments in our laboratory (4) it was found that when mouse and human cultured cells were fused using inactivated Sendai virus, their surface antigens were intermingled within minutes after initiating fusion. The marker antigens used for mouse cell membrane were Histo-compatibility-2, transplantation antigens, which appear to be intimately bound to the membrane and which strongly resist extraction in aqueous phase (5, 6). In short, these appear to be "integral" membrane proteins as defined by Singer and Nicholson (1), and hence adequate cell surface markers. Human cell surface proteins were defined by an heterologous rabbit anti-human cell antiserum; this seemed not to detect lipids, since it could not be absorbed by chloroform-methanol extracts of human cells, but it was not otherwise characterized.

Surface antigens were visualized by reacting newly-fused cells with the antisera described and then, after washing, with fluorescein or tetramethyl rhodamine-labelled anti-globulin reagents. Initially, red and green stains were localized in separate hemispheres of a given doubly stained cell. However, within one hour, all doubly-stained cells showed complete intermixing of red and green. The formation of these completely intermixed or "mosaic" cells was not inhibited by puromycin or cyclohexi-mide, even when parent cells were pre-treated for six hours before fusion, and was also not affected by inhibitors of ATP generation. It was, however, blocked to a greater or lesser extent by temperature. Table 1 shows the proportion of all heterokaryons that were completely intermixed after forty minutes at a given temperature. A graph of these data yields a sigmoidal curve with an abrupt increase in percent of mosaic cells between 15 and 20°C. Originally, we interpreted this sharp break in the curve as indicating a melting, a true phase transition, in the membrane lipids. Though such transitions may occur around 15-20°C (7), it has been pointed out (8, 9) that the curve may also be that of a linear process with a Q_{10} of around 5. This Q_{10} is about that observed for rotation of rhodopsin molecules in rod outer segments (9). It may be that by choosing to measure percent of mosaic cells at a fixed time after initiating fusion we exaggerated the difference between the 15° and the 30° experiments. If the time required to reach a 50% level of heterokaryons had been measured over a temperature range, we might have been able to distinguish between the two possibilities.

The data on heterokaryons indicate that membrane antigen mixing probably occurs by passive diffusion of surface proteins rather than by some synthetic process or by addition of preformed subunits to the membrane. Now, such diffusion could either occur in the plane of the membrane, or, with intermediate exit and entrance steps, in the cytoplasm, with marker antigens leaving the membrane and reentering after some steps of diffusion. The first possibility seems the more likely since:

1) It is the most direct and simplest model for the phenomenon observed. Cytoplasm diffusion would not necessarily show the steps in mixing observed and continuous spread of antigen from one hemisphere to the other. There is no prediction in this model that an antigen molecule entering the cytoplasm and diffusion would reappear in the membrane at the edge of the existing patch of antigen, rather than at the opposite pole of the cell.

2) It would not require protein molecules facing the outside of the membrane to move through the interior of the membrane to the cytoplasm. Experiments on phospholipid micelles indicate that such transitions of probes are rare (10).

3) Migration occurs more slowly than it should if antigen movement is through the cytoplasm. The best measurements for cytoplasmic viscosity indicate that it is a fraction of a poise (11), while, given an average size and shape for the marker antigens, their time of spread is consistent with movement in a medium of viscosity 1-2 poise, about that of many oils, and similar to that found by Cone for the environment of rhodopsin in retinal rods (9).

In summary, we feel that the data are best explained in terms of antigens diffusing in the plane of a fluid-like membrane.

MARKING EXPERIMENTS

A labelling experiment has been devised that eliminates the possibility of native surface membrane molecules
entering the cytoplasm and migrating through it to re-
emerge on the surface; the experiment also eliminates the
use of Sendai virus. It involves putting a spot of labelled antibody on the surface of a cell and watching the
change in breadth and breadth and intensity of this mark
with time. This was done in collaboration with Dr. Douglas
Fambrough, of the Carnegie Institution of Washington, using
cultured muscle fibers, formed from explanted myoblasts
(12), as the cells to be marked. These fibers may attain
lengths of 1 mm or greater within a few weeks of culture,
and their great size, plus the fact that they are elongate
cells, make them eminently suitable for marking with a spot
of anti-surface membrane antibody applied with a micro-
pipette. Thus, heterogeneity may be introduced into an
otherwise homogenous membrane by labelling molecules in a
small area. The label is seen as a bright line at the edge
of the cell, and if the labelled molecules are free to
diffuse in the membrane the line ought to increase in length,
and then appear to dim as the marker is spread over a greater and greater surface area.

In the first series of experiments, unlabelled antibody
to purified adult rat muscle membranes was locally applied,
and the plate was flooded with fluorescent antiglobulin.
This produced bright discreet patches, which remained
essentially static for up to 24 hours of culture. In view
of the electron microscope observations on patches on
lymphocytes stained with multi-layer antibody systems (13,
14) this is not surprising, for the patches formed by such
double antibody layers appear to be thousands of angstroms
long. The fastest observed rate of spread was around 24
microns in 24 hours. This rate is indeed that expected for
a patch of protein about 10,000Å in diameter floating in a
fluid matrix. However, observations of small changes in
patch size over very long times were not satisfactory. In
order to escape these problems by not forming large patches
of stained molecules at the cell surface, we split the
intact anti-muscle antibody into monovalent fragments, Fab,
by digestion with papain (15). The purified fragments were
directly coupled with tetramethyl rhodamine, and used to

make patches of stain on cultured muscle fibers. Each Fab has only one antigen binding site, and therefore the fluorescent patches made should consist of collections of single membrane antigenic sites bound to single Fab fragments, with no cross-links formed between the sites.

When fragments were marked with fluorescent Fab reagent, the patches were seen to fade rapidly with time. Part of the fading was temperature dependent, and could be reversed by cooling fibers from room temperature, about 25°C, to 0°C. The remaining component of fading was associated in a number of instances with measurable increases in the observed length of the marked area. In a typical experiment, a marked fiber was photographed while bathed in ice cold medium. This medium was then exchanged for warm medium, the fiber remaining in place on the microscope stage. At intervals of five to ten minutes the fiber was shifted back to cold medium and re-photographed.

When the photographs obtained were enlarged to final magnifications of around 300x, the patch lengths seen at various times after warming the fiber could be measured and compared. Typical data, showing actual measurements made, and the rates of spread calculated from them are shown in Table 2. It appears that the patch of stain spreads rapidly and that this spread reflects a fluid state in the muscle fiber membrane. Lowered temperature inhibits this spread, as can be seen for fiber   3. A mechanism involving entrance or exit to and from the surface is here rather improbable, since the surface antigens bear antibody fragments which ought to hinder any usual cytoplasmic migration of vesicles or molecules.

The single alternative to spread of the patches by diffusion of the marked molecules that readily suggests itself is that the Fab fragments progressively dissociate and reassociate with new surface molecules, "walking" along the membrane and thus showing a spreading pattern that does not reflect the movement of the underlying membrane antigens. We controlled for this possibility by flooding newly marked fibers with an excess of unlabelled intact antibody. In this situation, if a fluorescent Fab molecule leaves the antigen to which it was bound, it should be replaced by an unlabelled intact antibody molecule. If replacement occurs

we ought to see a fading of the patch without any increase
in its size. On the other hand, if the Fab marker remains
bound to the antigen sites with which it reacts initially,
then either the usual increase in patch dimensions, followed by fading should be seen, or a constant spot, neither
fading nor spreading should remain. In fact the latter two
situations seem to obtain. The spots do not fade rapidly
in the presence of excess unlabelled antibody, and several
of them appear to have increased in length with time of
incubation. Thus it seems that the spread of spots of
fluorescent Fab applied to part of a muscle fiber surface
is a diffusion-like process, and is not an artifact of
antigen-antibody equilibrium.

CAP FORMATION AS AN INDICATOR OF MEMBRANE FLUIDITY

Observations on the patterns of staining of lymphocytes with fluorescent antibody to immunoglobulin have
shown that the antiglobulin reagents cause aggregation of
the membrane-bound molecules, followed by collection of the
aggregates into a large cap at one pole of the cell, and by
the pinocytosis of this cap (3, 16). The aggregation of
the stain and its collection into a cap appear to be separate
events since the former occurs when the latter is inhibited,
in cells treated with sodium azide or cytochalasin B. Both
processes appear to be inhibited by temperature. The two
events involved appear to be first, aggregation of immuno-
globulins by the anti-immunoglobulin reagent, and second,
movement of the aggregate to one pole of the cell to form a
cap. While the nature of the process or processes causing
movement is obscure, it appears that the first step in cap
formation, aggregation or as it has been termed "patch"
formation, is a reflection of immunoglobulin molecules'
mobility in the plane of the membrane.

The experiments thus far reported have dealt only with
caps on lymphocytes and primarily with immunoglobulin molecules as surface markers. We have attempted to broaden the
study of aggregates of antigen and formation of caps, by
examining the H-2 antigen of L-cells. Mr. Arthur Weis
has been able to show that C11d growing on plastic can be
made to "cap" by the addition of anti-H-2 alloantibody in a
concentration somewhat lower than usually used for immuno-
fluorescent staining, followed by fluorescent anti-mouse
immunoglobulin. Since the cultured cells may round up

20

somewhat during their staining, two main types of stained cells are seen, and caps on these cells vary accordingly. The rounded cells show frank, polarized caps of the sort described and pictured for Ig caps. Spread cells, on the other hand, show segregation of fluorescence towards the middle of the cell, leaving its processes unstained. In the second case, the fluorescence comes to lie internally, as fluorescent perinuclear vacuoles. The formation of both sorts of caps or large patches seem to involve parameters similar to those defined for immunoglobulin caps, since cyanide, fluoride, dinitrophenol lowered temperature cytochalasin B all inhibit cap formation.

The appearance of an overt cap should not be taken as the first sign of surface antigen mobility, but, rather, as a consequence of antigen aggregation. Indeed if human epithelial cells of the VA-2 line, which are highly pinocytotic, are treated with antiserum to the cell surface and antiglobulin, though fluorescent spots, probably reflecting antigen aggregates, do form, caps are not seen, and instead the aggregates are locally pinocytosed, quickly filling a cell's cytoplasm with fluorescent vacuoles. While caps of one sort or another may prove to be an easily-scored indicator of membrane fluidity their absence cannot be construed as indicating the presence of a rigid membrane, or a membrane in which proteins are anchored. Perhaps, formation of discontinuities in a ring of fluorescent stain is a more accurate reflection of membrane molecule mobility than is accumulation of a cap.

RESTRICTION OF MOLECULAR MOBILITY
IN THE PLANE OF THE CELL MEMBRANE

While the experiments described and others on both animal and bacterial cells, indicate that the plasma membrane may exist in a fluid state in many cells, a consideration of tissue architecture and cytochemistry quickly suggests that there must be mechanisms acting to restrict mobility of membrane molecules when cells are organized into tissues. Thus, liver parenchymal surface enzymes such as 5'-nucleotidase adenyl cyclase and Ca, Mg-dependent ATPase, also show a restriction, at least of areas of high enzyme concentration, to portions of the parenchymal cell membrane (17). In intestine too, saccharidases, seem to be present

at the luminal end of epithelial cells, while the sodium
pump enzymes are localized at the basal end of these cells.
In muscle, acetylcholine and cholinesterase receptors are
localized in the endplate, and in normal innervated muscle
are not found elsewhere. Further examples of this sort of
restriction, will readily suggest themselves from histology
and embryology.

The observed restrictions suggest that the marker molecules so restricted are somehow inhibited in their diffusion
in the plane of the membrane. Though it is possible that
sites of high enzyme and receptor concentration represent
peaks of concentration from which molecules rapidly diffuse,
diluting to levels not detectable with present techniques,
such a constant pumping and loss of molecules into the
surface seems energetically and metabolically intolerable.
How then, might protein molecules, antigens or enzymes be
fixed in the surface of tissue cells? This fixation could
occur in three ways:

1. By alteration of the lipid phase of the membrane
so that it either became very viscous, or actually changed
to a more organized phase than that which exists in the
membrane of free or cultured cells. This would be true
alteration of membrane fluidity.

2. By aggregation of surface protein molecules to
form patches so large (10,000Å or so) that their diffusion
rates would be greatly reduced from those of single molecules. This involves alteration of protein mobility.

3. By anchoring of proteins extending completely
through the membrane through attachment to microfilaments
or other specialized organelles within the cell. Like 2,
this is a change in mobility of membrane components and not
in membrane matrix fluidity. Arguments can be made for all
three possibilities, based on experimental evidence. In
favor of 1 is the fact that cells can indeed remodel their
membrane lipids (18). Indeed, data on mycoplasma membranes
indicate that cells may be viable with membrane lipids
having melting points between -20°C to around 50°C (7).
But then highly selective mechanisms for differential
remodelling of membrane lipids might be required, since we
assume that only some cells in some tissues would restrict
the fluidity of their membranes.

Of the two approaches to altering membrane protein mobility, only 3 seems reasonable. Aggregates of protein would diffuse more slowly than unaggregated proteins only by a linear factor. Thus since it seems that a site a few 100 Å in diameter can move about 1 μm^2 per minute (4, 19), a site 10,000 Å in diameter could move the same distance in under an hour, far less than the life of the cell bearing it. Restriction of protein mobility by aggregation with other surface proteins seems an inefficient, short-term way of stabilizing a membrane configuration.

This leaves the possibility of anchoring membrane proteins in the cell cytoplasm. Such anchoring requires that a portion, not necessarily the antigenic portion, of a surface molecule extend through the bi-layer. It would not be inconsistent with the observations that ferritin-conjugated Concanavalin A stains only the external and not the internal face of the erythrocyte membrane (20). Anchoring by this method would yield stable specific sites for enzymes or antigens on a cell. It should be noted further that all of the local surface differentiations that I have cited are associated with morphological specializations in the membrane, in particular with microvilli or membrane infloodings. Thus some combination of membrane folding, stabilization, perhaps by microfilaments could anchor surface proteins in place on the cell.

Finally, one ought to mention that, whatever the mechanism or restriction of molecular motion on cell surfaces, it may well be dependent upon contact of one cell with another. While the catalog of fluid membranes is still only sketchy it seems that we might now, in parallel with further probes of fluidity, turn attention to systems of interacting cells that offer the possibility of surprising the mechanism of membrane locking and stabilization.

ACKNOWLEDGMENTS

The experiments described were supported by NIH grant AM-11202. I am grateful to Drs. Richard Cone and Douglas Fambrough for stimulating discussion, and to Dr. H. M. McConnell for sending me a preprint of his latest paper. This is contribution number 677 from the Department of Biology.

23

REFERENCES

1. Singer, S. J. and Nicholson, G. L., Science 175, 720 (1972).
2. Hubbell, W. L. and McConnell, H. M., J. Am. Chem. Soc. 93, 314 (1971).
3. Taylor, R. B., Duffus, W. P. H., Raff, M. C. and dePetris, S., Nature (NB) 233, 225 (1971).
4. Frye, L. D., and Edidin, M., J. Cell Sci. 7, 319 (1970).
5. Nathenson, S. G., Ann. Rev. Genetics 4, 69 (1970).
6. Mann, D. L. and Fahey, J. L. Ann. Rev. Microbiol. 25, 679 (1971).
7. Reinert, J. C. and Steim, J. M., Science 168, 1580 (1970).
8. Brown, P. K., Nature (NB) 236, 35 (1972).
9. Cone, R. A., Nature (NB) 236, 39 (1972).
10. Kornberg, R. D. and McConnell, H. M., Biochem. 10, 1111 (1971).
11. Casley-smith, J. R. and Chin, J. C., J. Microscop. 93, 167 (1971).
12. Yaffe, D. Curr. Topics Devel. Biol. 4, 37 (1969).
13. Stackpole, C. W., Aoki, T., Boyse, E. A., Old, L. J., Lumley-Frank, J. and deHarven, E., Science 172, 472 (1971).
14. Davis, W. C., Science 175, 1006 (1972).
15. Porter, R. R. Biochem. J. 73, 119 (1959).
16. Yahara, I. and Edelman, G. M., Proc. Nat. Acad. Sci. 69, 608 (1972).
17. Reik, L., Petzold, G. L., Higgins, J. A., Greengard, P. and Barnett, R. J., Science 168, 382 (1970).
18. van Deenen, L. L. M., Ann. N. Y. Acad. Sci. 137, 717 (1966).
19. Kornberg, R. D. and McConnell, H. M., Proc. Nat. Acad. Sci. 68, 2564 (1971).
   ; Devaux, P. and McConnell, H. M. manuscript in press.
20. Nicholson, G. and Singer, S. J. Proc. Nat. Acad. Sci. 68, 942 (1971)

Table 1

Effect of Temperature on the
Formation of "Mosaic" Heterokaryons

Incubation temperature (°C)	Percent of all heterokaryons whose antigens were completely intermingled
0	0
0	7
0	0
15	8
15	0
20	42
20	49
26	77
26	76

Table 2

Elongation of Fluorescent Fab Spots on Muscle Fibers

Fiber	Time after warming (min)	Spot length x330 (mm)	Spot length actual (μm)
1	0	65	197
	10	70	212
2	0	26	79
	5	31	94
	12	38	115
3	0*	40	121
	30	41	124

*
Kept on ice for 30 minutes.

25

LATERAL DIFFUSION AND PHASE SEPARATIONS

IN BIOLOGICAL MEMBRANES

Harden M. McConnell, Philippe Devaux and Carl Scandella

Stauffer Laboratory for Physical Chemistry
Stanford, California 94305

Introduction

We have recently studied the rates of lateral diffusion of phospholipids in pure phospholipid model membranes and in sarcoplasmic reticulum from rabbit muscle [1][2] [3]. These studies, which have utilized certain spin-labeled lecithins, are described in detail elsewhere [1][2] [3]. The purpose of this Chapter is to summarize briefly the results of these studies, and to discuss their relevance to the problem of phase separations in biological membranes.

Rates of Lateral Diffusion of Phospholipids in Pure Phospholipid Bilayers.

In two-dimensional diffusion, a molecule travels a root mean square distance s in a time t where

$$s^2 = 4Dt \qquad (1)$$

This equation may be regarded as the definition of the diffusion constant D. In discussing the lateral diffusion of phospholipids in a two-dimensional bilayer, it is plausible to assume that the lipid molecules form a quasi-crystalline two-dimensional lattice, and that the translation step for diffusion involves an interchange of neighboring molecules. An interchange of this type is sketched in Figure 1, where for simplicity the lipids are regarded as cylindrical objects that form a hexagonal array. In each step, the molecule moves a distance λ in an (average) time τ. Thus

$$\lambda^2 = 4D\tau \qquad (2)$$

If it is further assumed that the quasi-crystalline array is hexagonal, then λ is the distance between neighboring molecules and the area A per lipid molecule is $\sqrt{3}/2\ \lambda^2$. Equation (2) may then be written in the following convenient form.

$$\tau^{-1} = \frac{2\sqrt{3}\ D}{A} \qquad (3)$$

which gives the jumping rate τ^{-1} in terms of the diffusion constant D.

Three different methods have been used to estimate D and/or τ^{-1} in egg lecithin bilayers. Kornberg and McConnell studied the proton nuclear resonance line widths of the choline methyl groups in vesicles of egg lecithin containing low concentrations ($\sim 1\%$) of spin labeled lecithin (I) molecules [3]. This study showed that the jump rate

$$CH_3\!-\!(CH_2)_{\overline{14}}\ CO\!-\!\overset{\displaystyle H\ \ H\ \ O}{\underset{\displaystyle H\ \ H\ \ O^-}{\text{(I)}}}$$

(I)

τ^{-1} is at least 3000 sec^{-1}. (With A = 65 Å^2, $D \geq 10^{-12}$)

In a second study, Devaux and McConnell determined D for spin label I in planar phosphatidyl choline multilayers [2]. A value of $D = 1.8 \times 10^{-8}$ cm^2/sec was obtained.

This second study involved a measurement of the time-evolution of paramagnetic resonance spectra due to the spreading of small concentrated patches of I in non-labeled host phosphatidyl choline multilayers. The resonance spectra were analyzed in terms of empirical reference spectra obtained by recording the spectra of known uniform concentrations in multilayers.

Using a third method. Scandella, Devaux and McConnell determined D for II in egg lecithin dispersed in water (liposomes, no sonication) from a theoretical analysis of the (time-independent) spectra of II in terms of the collision rates between molecules of II[1].

$$CH_3-CH_2-\overset{\overset{\displaystyle O \quad N \rightarrow O}{\diagdown \diagup}}{C}-(CH_2)_{14}-\overset{\overset{\displaystyle O}{\|}}{C}-O-\overset{\overset{\displaystyle \overset{\displaystyle H \quad H \quad O}{\diagdown \diagup \quad \|}}{C-O-C-(CH_2)_n-CH_3}}{\underset{\underset{\displaystyle \overset{\diagup \diagdown}{H \quad H} \quad O^-}{C-O-\overset{\|}{P}-C-(CH_2)_2-\overset{+}{N}(CH_3)_3}}{\underset{\displaystyle O}{\overset{\displaystyle |}{C}-H}}} \qquad (II)$$

This analysis yielded a value of D equal to $5 \times 10^{-8} cm^2/$sec at $25°C$. In an earlier, independent study, Sackmann and Träuble [4] used this method of analysis to determine the diffusion constant of an androstan spin label in phosphatidyl choline bilayers, and obtained a value for $D \simeq 10^{-8} cm^2/sec$ for this steroid molecule.

These results lead to some interesting conclusions regarding lateral diffusion in phosphatidyl choline bilayers. Lateral diffusion in these bilayers is fast in the sense that a lipid molecule can travel thousands of molecular diameters in a second. For example, if a lipid molecule in an E. coli bacterium were characterized by a diffusion constant $D \simeq 1.8 \times 10^{-8} cm^2/sec$, then the lipid molecule could travel from the end of the bacterium to the other in a time of the order of one second. A further discussion of lateral diffusion in bacteria is given later.

The results of the three methods are in remarkably good agreement with one another, particularly in view of the fact that three quite different labels have been used, different types of lipid preparations (vesicles derived by sonication, planar multilayers, and non-sonicated liposomes) have been employed, and entirely different methods of analysis have been applied to the resonance line shapes. The last two methods in fact yield diffusion constants that are in order-of-magnitude agreement with one another. Some of the derived diffusion constants are given in Table I. (The first method only gave a lower limit to the diffusion constant.) This agreement between three various methods gives us confidence in the results; the second method also provides direct evidence that the lipid motion does in fact follow a diffusion equation.

Rates of Lateral Diffusion of Phospholipids in Sarcoplasmic Reticulum.

The sarcoplasmic reticulum of rabbit muscle can be isolated in the form of closed membrane vesicles which accumulate calcium in the presence of ATP[5]. By introducing low concentrations (0.4 - 5 mole%) of spin label II in this membrane, and analyzing the resonance spectra in terms of spin-exchange collisions at various temperatures and label concentrations, the diffusion constants shown in Table I were obtained. Note that the diffusion constants so obtained are, at the physiological temperature, comparable to those obtained in 4:1 egg lecithin:cholesterol bilayers. The spin-labeled vesicles retained the calcium pumping activity. These results demonstrate that the lateral diffusion of II in the intact, functional sarcoplasmic reticulum is indeed rapid.

Phase Separation in Sarcoplasmic Reticulum.

Before discussing the subject of phase separations in the sarcoplasmic reticulum, and in other biological membranes, let us distinguish carefully between "phase

separations", and "phase transitions." The latter terminology has often been used in discussing the properties
of biological membranes whereas the former is doubt-
less more appropriate in many cases. We shall limit
the term "phase transition" to a transformation in the
physical state of a pure chemical compound. Such transitions take place at a unique temperature (e.g., the
melting temperature), under given external conditions
such as pressure. We use the term "phase separation"
to describe the separation of one phase from another,
such as the separation of a solid phase from a liquid
phase. Thus, the freezing out of ice from a salt-water
solution is a phase separation, and only the freezing of
pure water is a phase transition. A phase separation
may involve the separation of a pure solid substance
from a liquid solution of more than one component, as
well as the separation of a solid solution.

At $40°C$ and above the paramagnetic resonance
spectrum of spin label II in the sarcoplasmic reticulum
shows clearly that the label is in fact in "dilute solution"
in the membrane lipids, and that the label undergoes
rapid lateral diffusion [1]. When the mole fraction of
label in the membrane is high (~5%) a phase separation
of pure II, or at least very concentrated II, is observed
when the temperature is reduced below 20°C [1]. This
is a readily reversible change. Not all of the label II
is present in this concentrated phase; the resonance
spectrum shows that a substantial fraction of II remains
in dilute solution in the remainder of the membrane
lipids.

The tendence of II to separate in a pure, or concentrated phase in the sarcoplasmic reticulum might
arise from at least two different effects. (a) If pure
label II has a high melting point, then the tendency of
II to separate as a concentrated phase would be understandable. (b) If the lipids of the sarcoplasmic reticulum tend to freeze themselves, then these frozen lipids
might tend to exclude (sequester) II, and thus lead to a

concentrated phase II.

Spin label II has a phase transition at ~15° C,as can be seen from the paramagnetic resonance spectrum of II itself. A plot of this line width vs. temperature (Figure 2) shows clear evidence for a phase transition in the 10° -20° C range. The rather large apparent width of the transition is probably due to heterogeneity in the composition of the α - fatty acid chains, since label II was derived from egg lecithin. Thus, the separation of II in the sarcoplasmic reticulum may be due in part to the tendency of II itself to freeze. On the other hand, this can hardly be the sole effect. No evidence for a comparable degree of phase separation of concentrated II is observed when II is dissolved in egg lecithin, egg lecithin/cholesterol, or in lipids estracted from the sarcoplasmic reticulum. In fact, the paramagnetic resonance spectrum of II in a 50:50 mixture of dipalmitoyl lecithin and egg lecithin shows no evidence for the sequestering of II at low temperatures, since at low temperatures quite distinct spectra are visible in the (solid) phase that is rich in DPL, and in the liquid phase of egg lecithin. Illustrative spectra are shown in Figure 3.

Even though we have not yet succeeded in demonstrating the strong sequestering of the label II in pure phospholipid model systems due to freezing of the solvent lipids there can be little doubt that this is a likely mechanism. Clear evidence for this sequestering effect has been seen in pure phospholipid systems containing several other spin labels [4][6]. For example, a fatty acid label III

$$CH_3-(CH_2)_{10}-C-(CH_2)_3-C \overset{O}{\underset{OH}{\diagdown}} \qquad (III)$$

is sequestered from pure DPL below the freezing temperature of pure DPL (40° C) when the fatty acid DPL mole ratio is 1/100 or higher [6]. Also, there is good evidence that the phospholipid label III is retained in the fluid phase in mixtures of dioleyl lecithin and diplamitoyl lecithin at low temperatures where the solid phase of dipalmitoyl lecithin begins to separate [7].

On the other hand, our failure to detect the separation of II in extracted lipids from sarcoplasmic reticulum indicates that the protein does pay an important role in this particular experiment. This role may be relatively simple; perhaps the protein merely decreases the fluidity of the sarcoplasmic lipids, and reduces the solubility of II in these lipids.

Phase Separations in Biological Membranes.

Our studies of spin label II in the sarcoplasmic reticulum have shown that a lipid molecule diffuses rapidly in this membrane at the physiological temperature, and undergoes a phase separation at lower temperatures. There is now further evidence that lipids, as well as other membrane components, undergo lateral diffusion and phase separations in other biological membranes. One particularly interesting example of this is the work of Fox and co-workers [8] and Overath and co-workers [9] who have studied the temperature dependencies of lactose uptake in E. coli fatty acid auxotrophs supplied with various exogeneous unsaturated fatty acids. At the present time, it appears that the data obtained by these investigators can be understood in terms of phase separations of the higher melting lipid components in these membranes, and in terms of the lateral diffusion of the lipid components that are not "frozen." Temperature breaks in the Arrhenius plots of lactose uptake of E. coli fatty acid auxotrophs supplied with a single exogeneous unsaturated fatty acid apparently can be understood in terms of a "phase transition" of a relatively pure lipid, whereas such temperature breaks from auxotrophs supplied with more than one unsaturated fatty acid can be understood in terms of phase separations of the higher melting lipids [8][9]. Quite recently Fox has shown that different sequences of lac inducation, growth, and composition of fatty acid supplement can lead to more than one break in the uptake vs. temperature curves, as explained in detail elsewhere [10]. Again, all of these results are understandable if there is complete mixing by lateral diffusion of all the lipids that are above their phase separation temperatures.

It appears likely that a number of important properties of biological membranes depend on lateral diffusion and phase separation of membrane components. From the point of view of the physical chemist, the challenge in this area is to measure these rates of lateral diffusion, and to establish appropriate phase diagrams to describe the phase separations. From the point of view of the biologist, it is of interest to establish what biological functions depend on these phenomena.

Acknowledgements

We are greatly indebted to Dr. C. F. Fox for many interesting discussions concerning bacterial membranes. This work has been supported by the National Science Foundation under grant No. NSF GB-19638 and has benefited from facilities made available to Stanford University by the Advanced Research Projects Agency through the Center for Materials Research.

References

1. C. J. Scandella, P. Devaux and H. M. McConnell, submitted to Nature.
2. P. Devaux and H. M. McConnell, J. Am. Chem. Soc., in press.
3. R. D. Kornberg and H. M. McConnell, Proc. Nat. Acad. Sci., US, 68, 2564 (1971).
4. E. Sackmann and H. Träuble, J. Am. Chem. Soc., in press.

5. W. Hasselbach, Prog. Biophys. Mol. Biol., 14, 169 (1964).
6. W. L. Hubbell and H. M. McConnell, J. Am. Chem. Soc., 93, 314 (1971).
7. B. G. McFarland, in Biomembranes (a volume of Methods in Enzymology, Eds. Fleischer, Packer and Estabrook, Academic Press, N.Y.).
8. N. Tsukagoshi, P. Fielding and C. F. Fox, Biochem. Biophys. Res. Commun. 44, 497 (1971).
9. P. Overath, F. F. Hill and I. Lamnek-Hirsch, Nature New Biology 234, 264 (1971).
10. C. F. Fox, private communication.

TABLE I Diffusion Constants (cm^2/sec x 10^8)

t °C	$\frac{Egg\ PC}{Cholesterol}\ \frac{4}{1}$	Sarcoplasmic Reticulum	Lipid Extract
70	20 ± 1.5	12 ± 1.5	
60	17 ± 1.5	9.5 ± 1.5	10 ± 3
50	15 ± 1.5	7.5 ± 1.5	
40	(12 ± 2)	(6 ± 2)	

These values were calculated from the spin label collision rates using a value of 60$Å^2$ for a^2. The values given in parenthesis at 40° were obtained by extrapolating the values at higher temperatures. The stated error limits reflect the uncertainty in the values obtained by comparing experimental and calculated spectra. These diffusion constants are subject to an uncertainty of 2-3 fold because of assumptions introduced in the calculations [1].

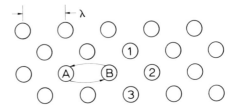

Figure 1. Schematic representation of the lateral diffusion of phosphatidyl choline molecules in a bilayer, as seen from above the plane of the bilayer. For simplicity the lipids are regarded as cylindrical objects that form a hexagonal array. The diffusion is assumed to take place by the interchange of neighboring pairs of molecules, such as A and B. After the interchange, one molecule of the pair, A, has three new neighbors, numbered 1, 2, 3. The distance between neighboring molecules is λ.

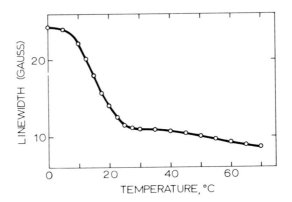

Figure 2. The paramagnetic resonance line width of an aqueous dispersion of neat phospholipid spin label II.

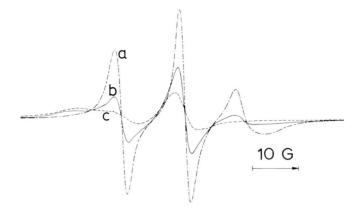

Figure 3. Spectra of spin label II (1 mole %) in liposomes of a) egg lecithin (— - — - —), b) egg lecithin:
dipalmitoyl lecithin, 1:1 (————), c) dipalmitoyl lecithin (-----). These spectra are recorded at $0°$ C. At
$45°$ the spectra of these three samples are identical.

II

LOCALIZATION OF PROTEINS IN MEMBRANES

MOLECULAR FEATURES OF HUMAN ERYTHROCYTE GLYCOPHORIN

V.T. Marchesi, J.P. Segrest, and I. Kahane

Laboratory of Experimental Pathology, NIAMD, NIH
Bethesda, Maryland

Abstract

The major glycoprotein of the human red blood cell membrane has been isolated, purified and its overall molecular anatomy partially characterized, and some ideas are suggested as to how this molecule might be oriented in the intact cell membrane. This molecule, named glycophorin, is a single polypeptide chain composed of $\sim$ 200 amino acids and $\sim$125 sugar residues which carries many blood group antigens and lectin receptors. The carbohydrates are covalently linked to the N-terminal one-half of the polypeptide chain and are exposed to the external environment of the cell. The C-terminal one-third of the polypeptide chain appears to be located internal to the lipid barrier of the membrane and part of it may extend into the cytoplasm of the cell. The segment connecting these two portions is composed predominantly of non-polar and hydrophobic amino acids ($\sim$35 residues) which span the lipid region of the membrane. Glycophorin molecules are distributed uniformly over the red cell surface and appear to be associated with the intramembranous particles.

Introduction

Most of the carbohydrate residues associated with the human red cell membrane are covalently linked to a membrane-bound glycoprotein. This glycoprotein, named glycophorin (1), is a single polypeptide chain of approximately 200 amino acids. Extraction procedures which employ neutral salt solutions fail to solubilize this molecule from the membrane, but lithium diiodosalicylate (2) and other solvents (3,4) cause its dissociation into a

water-soluble form. Solubilized glycophorin retains antigens and receptors previously shown to be present on the
glycoprotein of the intact membrane.

This report summarizes the results of our studies
concerning the physical and chemical properties of the
isolated glycoprotein and of its peptide components which
are produced by controlled enzymatic (trypsin) and chemical (cyanogen bromide) cleavages. Studies have also been
carried out on the distribution of glycophorin molecules
on the surfaces of intact red cells using the technique
of freeze-etching and electron microscopy to localize
appropriate lectin-ferritin conjugates.

The results of these studies provide us with a
general idea of the molecular configuration of glycophorin,
and from this we have deduced a provisional model for the
organization of this molecule in the membrane.

General Properties of Glycophorin

Glycophorin can be isolated in water-soluble form
from red cell ghosts using lithium diiodosalicylate (LIS)
as described previously (2). The purified molecule
appears to be a single molecular species when analysed by
polyacrylamide gel electrophoresis and it has leucine as
its N-terminal amino acid. The molecular weight of the
monomeric unit of glycophorin appears to be ~50,000. This
value is based on mobility in SDS-acrylamide gels (5) and
is consistent with the number of peptides obtained when
the molecule is cleaved by cyanogen bromide (described
below). Since the molecule is 60% carbohydrate and 40%
protein by weight, the molecular weight of the polypeptide
portion is ~20,000 or 200 amino acids. This value is
also very close to the sum of the amino acids of the
individual cyanogen bromide fragments (6). A value of
50,000 for the molecular weight of this glycoprotein
differs from the values reported by others (these range
from 31,000 to 160,000); the basis for this discrepancy
remains unexplained. Preliminary amino acid sequence
analyses have been carried out on some of the tryptic and
cyanogen bromide peptides (7) and the results suggest that
glycophorin is composed of a single polypeptide chain
rather than a family of closely related molecules as was

suggested earlier (8).

Glycophorin carries a variety of blood group antigens
(AB, MN, I, and probably others) and has multiple copies
of the receptors for kidney bean phytohemagglutinin (PHA),
wheat germ agglutinin (WGA), and influenza viruses. These
activities have all been measured on glycophorin preparations which were extracted repeatedly with chloroform-
methanol to eliminate the possibility that some of the
activities might be due to contaminating glycolipids.

The receptors for the plant agglutinins (PHA, WGA)
and for influenza viruses are specific carbohydrate or
oligosaccharide residues which are attached to the
polypeptide chain by covalent linkages to either
threonine/serine or to asparagine (8). One of the PHA
receptors is a complex oligosaccharide with galactose and
mannose residues as its determinant (9). N-acetyl
glucosamine seems to be an important component of the WGA
site (10) and sialic acid residues are clearly involved in
influenza binding (11). Since this glycoprotein is 60%
carbohydrate and 25% sialic acid by weight it should contain approximately 130 sugar residues 40 of which are
sialic acids. We do not have any reliable estimate as to
how many different oligosaccharide chains are attached to
a single glycoprotein molecule, but based on the studies
of Winzler and others (8) and on the results of alkaline
reduction carried out by Jackson in our laboratory (12)
we estimate roughly that 20-30 oligosaccharide units might
be attached to the polypeptide backbone of each molecule.
Many of these oligosaccharides are probably incomplete
forms or variations of the tetrasaccharide complex
described by Thomas and Winzler (13), but some of these
oligosaccharides are also complex units comprising 10 or
more sugar residues, such as the oligosaccharide chain
bearing the PHA receptor described by the Kornfelds (9).
Although the exact number of oligosaccharide units per
polypeptide chain is not known, we do know that
essentially all the sugar residues are attached to points
along the N-terminal one-half of the polypeptide chain.
This conclusion is based on the results of tryptic
digestion described below.

Linear Arrangement of Peptide Components

Many investigators have isolated glycopeptides from red cell membranes by treating intact cells with trypsin or other proteases (14). Trypsin releases approximately 50-60% of the total protein-bound carbohydrate from intact red cells, while pronase reportedly can release close to 100% of the sialic acid and presumably most of the neutral sugars and hexosamines as well if the digestion is prolonged.

Four unique glycopeptides are generated when isolated glycophorin preparations are trypsinized, and these account for over 90% of the total sugar content of the molecule. These peptides have been purified and partially characterized in terms of amino acid and sugar composition and their biologic activities (12). The evidence suggests that all four peptides can be derived from a single glycophorin molecule. However, if the glycoprotein is trypsinized while it is still bound to the intact membrane, only three of the four glycopeptides are produced. The fourth glycopeptide (labelled β) is approximately 25 amino acids long and is only accessible to tryptic digestion after the molecule is isolated from the membrane. We have taken advantage of this property to determine how the glycopeptides are arranged in the molecule. Intact red cells were trypsinized and the released glycopeptides isolated, then the remaining glycoprotein fragments were extracted from the membrane and subjected to a second trypsin digestion. By determining which glyco-peptides are released at the different steps we were able to establish the order of tryptic glycopeptides in the molecule as shown in fig. 1. This arrangement was also confirmed by an analysis of peptides derived from cyanogen bromide cleavage described below.

In addition to the four water-soluble glycopeptides described above, trypsin digestion of the isolated glycoprotein generates a water-insoluble peptide which does not contain carbohydrate, but has instead a high content of non-polar amino acids. Amino acid sequence analysis of this peptide (described below) indicates that it is derived from the middle third of the polypeptide chain and represents a part of the "hydrophobic domain" of the glycoprotein.

Five unique fragments are produced when isolated glycophorin is subjected to cleavage by cyanogen bromide (6). Three of these fragments contain sugar residues and these appear to overlap the tryptic glycopeptides as illustrated in fig. 1. Only one of the fragments (C-2) lacks homoserine, the expected product of methionyl bond cleavage by cyanogen bromide, indicating that this fragment is the C-terminal peptide of the original molecule.

The purified C-2 fragment and the insoluble tryptic peptide described above have also been partially sequenced with a Beckman sequenator (7). Since these fragments overlap a continuous sequence of 50 residues has been determined, which represents approximately one-fourth of the polypeptide chain of the glycophorin molecule. Within this sequence we have found a stretch of 34 non-polar amino acids more than half of which are hydrophobic (ie. leu, ile, val, phe). We speculate that this segment might be considered the "hydrophobic domain" of the polypeptide chain since it appears to be a unique segment which is ideally suited to interact via hydrophobic association with either membrane lipids or hydrophobic regions of other proteins. This hydrophobic segment of the molecule seems to be located in the vicinity of residues 100 through 135 and is thus considerably removed from the C-terminal end of the molecule which is located at ~205. This has some bearing on how the molecule might be oriented in the intact membrane as described below.

Orientation of Glycophorin in the Membrane

It is possible to conceive of at least five different ways in which a glycoprotein having the structure shown in fig. 1 might be oriented at the cell surface. These are illustrated in fig. 2. Each model depicts the sugar-containing segment of the molecule as being exposed to the external environment of the cell which is consistent with the known accessibility of the antigens and receptors to the external medium. Experiments with neuraminidase also indicate that all of the glycoprotein-bound sialic acid is external to the lipid barrier (15). Results

reported at this symposium by Steck further confirm that
all of the carbohydrate residues are bound to the external
surface of the membrane.

The mode of attachment of glycophorin to the membrane
shown in fig. 2-A is unlikely for several reasons. The
polypeptide chain could fold so that the most hydrophobic
residues are internalized as is the case with many globular
proteins, but this would make them unavailable for
associations with lipids of the membrane. Binding of the
glycoprotein to the membrane would have to be via electrostatic interactions with the polar groups of the phospholipids. If this were so one would predict that the
glycoprotein could be extracted from the membrane with high
ionic strength buffers or by manipulating the pH, and this
is not the case. In addition, the results of labeling
parts of the glycoprotein with radioactive reagents,
described below, are not consistent with this model.

The three remaining possibilities represent variations
of the idea, originally proposed by Morawiecki (16) and
Winzler (8), that this glycoprotein might be bound to the
lipids of the membrane via hydrophobic associations
between non-polar segments of the polypeptide chain and
the membrane. This mode of attachment would be more
stable under physiological conditions than that maintained
by electrostatic interactions and this would also explain
why this molecule is difficult to extract from membranes.
The model proposed by Morawiecki and Winzler is essentially that depicted in fig. 2-C. Each has postulated
that one end of the glycoprotein molecule was "lipophilic"
and was buried in the lipid regions of the membrane.
However, the sequence information described above indicates
that the most hydrophobic segment of the glycoprotein is
located not at the C-terminal end of the polypeptide
chain but some 60 amino acid residues away. In addition
the polypeptide chain between the hydrophobic segment and
the C-terminal contains a number of charged amino acids
and, on thermodynamic grounds we would predict that this
segment would most likely reside in an aqueous environment.
Thus we feel that models B and D are more consistent with
the known structural features of the molecule.

Recently Bretscher has proposed that the major

glycoprotein of the red cell membrane (the molecule we call glycophorin) extends completely across the membrane as illustrated in D. This interpretation is based on experiments in which a water-soluble radioactive reagent was used to label exposed segments of membrane proteins (17). Bretscher found that the "hydrophobic segments" of the glycoprotein were not labeled when the reagent was added to intact red cells but these regions of the molecule could be labeled if the reagent was added to the membrane ghosts produced by osmotic lysis. Bretscher concluded that the altered permeability of the ghost membrane allowed the reagent to penetrate inside the cell and thereby label the internal segment. However, this interpretation has been questioned by other investigators on the grounds that membrane proteins might rearrange during osmotic lysis and thereby become more accessible to the labeling reagent. For example, it is conceivable that the orientation of the glycoprotein shown in fig. 2-C might shift to the orientation shown in fig. 2-B during or as a result of the osmotic shock.

To resolve this question we have tried to determine which segments of membrane-bound glycophorin are exposed to the exterior by using the lactoperoxidase catalysed iodation procedure described by Phillips and Morrison (18). Although this procedure theoretically iodinates all accessible tyrosine residues on proteins, we were able to determine which portions of the glycoprotein are labeled by extracting the labeled molecule and analysing the activity associated with individual cyanogen bromide peptides by autoradiography (19). When intact red blood cells are exposed to the iodination reaction mixture only the N-terminal one-half of the molecule becomes labeled. When "leaky" ghost membranes are incubated under the same conditions, the label is distributed throughout the molecule including the C-terminal region. However, if the ghost membranes are "resealed" after lysis by the addition of salt and divalent cations before iodination, only the N-terminal segment becomes labeled. From these experiments we tentatively conclude that the glycoprotein molecule probably extends completely across the lipid region of the membrane.

Distribution of Glycophorin over the Cell Surface

Freeze-etching and electron microscopy have been used
to locate individual glycophorin molecules on the surface
membranes of intact cells. Ferritin conjugates of PHA or
WGA can both be used for this purpose since each of these
reagents binds to a specific receptor site known to reside
on the exposed segment of glycophorin. Both receptors are
distributed uniformly over the surface of the red cell
membrane in a pattern which closely approximates that of
the intramembranous particle population. The relationship
between glycoprotein molecules and the membrane particles
has been studied in some detail in our laboratory (20) and
by others (21), and the evidence suggests that the
particles may represent an interaction between hydrophobic
segments of the glycoprotein and membrane lipids or other
membrane proteins. However, the physical basis for these
connections and the functional role of this complex is
still completely unknown.

Summary

The major glycoprotein of the human red cell membrane
is oriented at the cell surface so that the N-terminal
one-half of the polypeptide chain, bearing all of the
covalently-bound carbohydrate, extends into the external
medium, while the C-terminal one-half of the same molecule
is anchored within the lipid regions of the membrane. The
exposed oligosaccharides form some of the blood group
antigens (AB, I, MN) and also have receptor activities
(PHA, WGA, etc.). The buried portion of the molecule has
a unique segment of non-polar amino acids which is well
suited to interact with lipids. Glycoprotein molecules
seem to be connected to the 85 Å globular particles which
are present within the lipid regions of cell membranes.
The hydrophobic segment of the glycoprotein and the intramembranous particle together represent a new structural
unit of the red cell membrane which may have important
functional properties.

References

1. V. T. Marchesi, T. W. Tillack, R. L. Jackson, J. P. Segrest, and R. E. Scott, Proc. Nat. Acad. Sci., In press (1972).

2. V. T. Marchesi and E. P. Andrews, Science 174, 1247 (1971).

3. R. H. Kathan, R. J. Winzler, and C. A. Johnson, J. Exp. Med. 113, 37 (1961).

4. O. O. Blumenfeld, Biochem. Biophys. Res. Comm. 30, 200 (1968).

5. J. P. Segrest, R. L. Jackson, E. P. Andrews, and V. T. Marchesi, Biochem. Biophys. Res. Comm. 44, 390 (1971).

6. J. P. Segrest, R. L. Jackson, and V. T. Marchesi, In preparation.

7. J. P. Segrest, R. L. Jackson, W. Terry, and V. T. Marchesi, Fed. Proc. 31, 736, Abs. (1972).

8. R. J. Winzler, In Red Cell Membrane Structure and Function (G. A. Jamieson and P. J. Greenwalt, eds., J. B. Lippincott, Philadelphia, Pa. (1969).

9. R. Kornfeld and S. Kornfeld, J. Biol. Chem. 245, 2536 (1970).

10. M. M. Burger and A. R. Goldberg, Proc. Nat. Acad. Sci. 57, 359 (1967).

11. G. F. Springer, Biochem. Biophys. Res. Comm. 28, 510 (1967).

12. R. L. Jackson, J. P. Segrest, and V. T. Marchesi, In preparation.

13. D. B. Thomas and R. J. Winzler, J. Biol. Chem. 244, 5943 (1969).

14. G. M. Cook, Biol. Rev. Cambridge Phil. Soc. 43, 363 (1968).

15. E. H. Eylar, M. A. Madoff, O. V. Brody, and J. L. Oncley, J. Biol. Chem. 237, 1992 (1962).

16. A. Morawiecki, Biochim. Biophys. Acta 83, 339 (1964).

17. M. S. Bretscher, J. Mol. Biol. 59, 351 (1971).

18. D. R. Phillips and M. Morrison, Biochem. Biophys. Res. Comm. 40, 284 (1970).

19. I. Kahane, J. P. Segrest, and V. T. Marchesi, In preparation.

20. T. W. Tillack, R. E. Scott, and V. T. Marchesi, J. Exp. Med., In press (1972).

21. P. Pinto da Silva, S. D. Douglas, and D. Branton, Nature 232, 194 (1971).

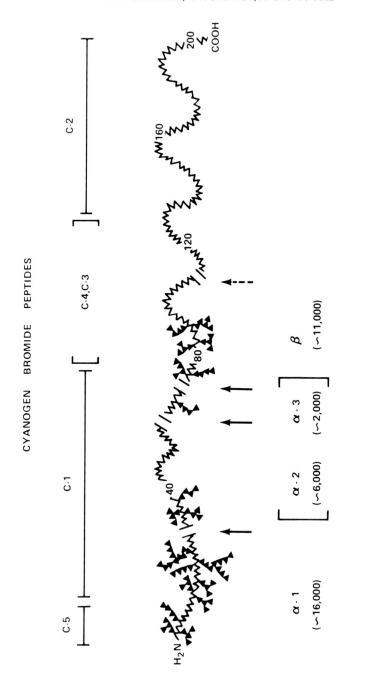

Figure 1

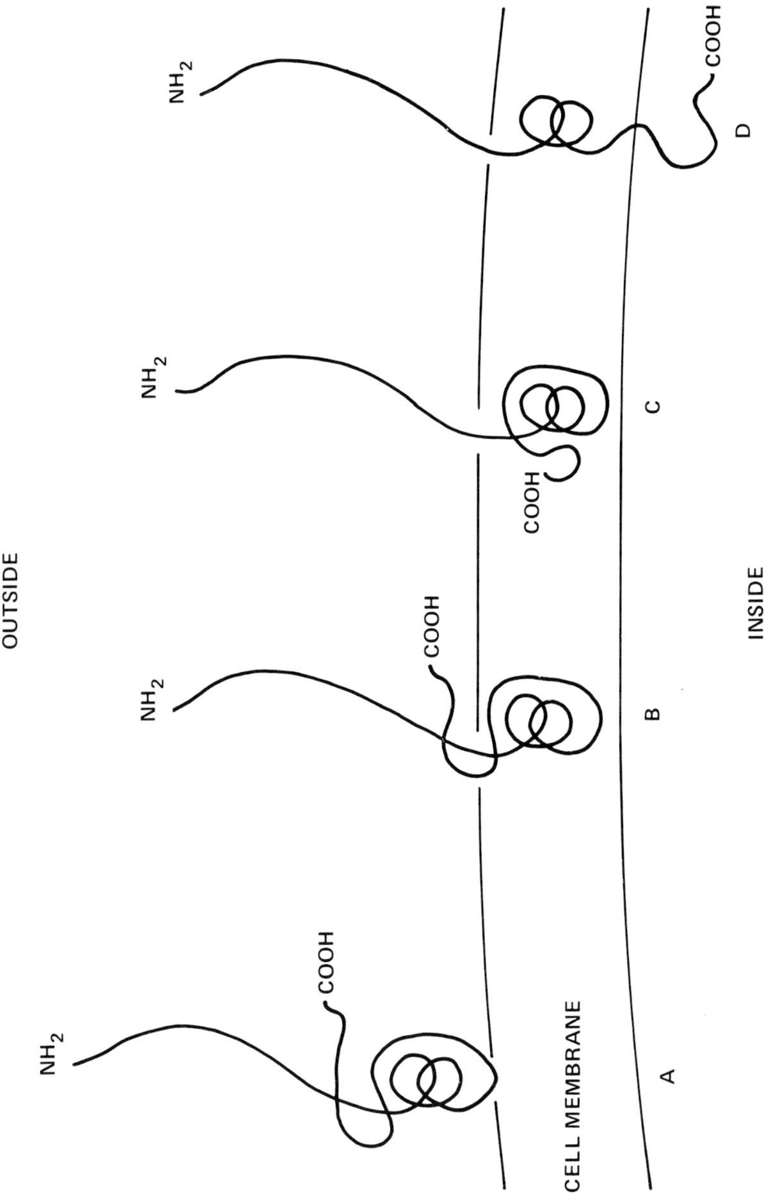

Figure 2

TOPOLOGICAL STUDIES ON THE STRUCTURE OF CELL MEMBRANES *

Garth L. Nicolson

The Salk Institute for Biological Studies
San Diego, California 92112

ABSTRACT

This article is concerned with the topological organization of plasma membranes and how these arrangements may determine certain functional aspects of membrane structure such as sensitivity to complement-mediated cell lysis, plant agglutinin cytotoxicity, cell agglutination and mitogenic stimulation. Topological distributions of membrane components are dependent on the state of the membrane; agents that change this state by perturbing membrane structure (e.g. proteolytic enzymes, lipases, etc.) also affect membrane topology. Changes in membrane surface distribution probably occur by lateral diffusion through the fluid membrane environment.

INTRODUCTION

To understand the molecular organization of biological membranes requires specific information on the location of components in the membrane. Most methodology used at present for elucidating membrane organization yields information on membrane cross-sectional structure (i.e. on which side of the membrane is the component located), without yielding data on membrane topological organization.

---

* Supported by a contract from the National Cancer Institute, USPHS, and grants from the Armand Hammer Fund for Cancer Research and the New York Cancer Research Institute.

In the latter category, electron microscopy has proved to be a useful tool in studying the topological organization of membranes using distinct electron-dense markers such as ferritin-conjugates (1), charged colloidal particles (2), small viruses (3), and histochemical precipitates (4-6). Each of these marker systems is limited in resolution to the size of the electron-dense marker and its specificity for the target site. For example, colloidal iron hydroxide (CIH) has a useful resolution of approximately 50-100Å (7) compared to hybrid antibody (anti-target/anti-virus)-virus labeling (8) which is limited to approximately the dimensions of the virus plus antibody (400-1,000 Å) (9). Obviously, these membrane marker systems are not suitable for determining organization at the level of a few Angstroms, but they are useful for certain molecules that are present in reasonable numbers in the membrane (10).

Three techniques have recently been developed to observe the topography of cell membranes. These are: serial thin-sectioning of single plastic-embedded cells followed by three-dimensional reconstruction of the surface (11), freeze-cleaving of membranes followed by etching to reveal the membrane surface (12-13), and mounting of lysed flattened membranes on thin support films (10, 14-18). We have used mainly the membrane mounting technique to study the surface distributions of membrane antigens using ferritin-conjugated antibodies (15-16), and oligosaccharides using ferritin-conjugated plant agglutinins (14, 16-17). I will discuss here some of the results obtained by these techniques and the relationship of membrane topological organization to some functional properties of cell membranes.

METHODS

Ferritin-conjugated antibodies and plant agglutinins were prepared and purified as previously described (13, 18). Colloidal iron hydroxide was made by the method of Gasic, et al. (21) and was used undiluted. Influenza virus was purified as described by Kien and Holland (22) and was used at a concentration of 10,000 hemagglutinating units.

Intact rabbit erythrocytes or ghosts were trypsinized for two min. at room temperature with 0.05% crystalline trypsin (Worthington) and washed several times with 100-fold excess of ovomucoid solution and stored at $0^{\circ}C$.

Human erythrocyte ghosts were prepared by the method of Dodge, et al. (23). The ghosts in 0.001 $\underline{M}$ tris-HCl-0.15 $\underline{M}$ NaCl (pH 7.4) buffer were incubated with protease-free (24) phospholipase C (Worthington) at a concentration of 2 mg/ml for 30 min. at $37^{\circ}C$. The phospholipase C-treated ghosts were washed in phosphate buffer and assayed for phosphorus by the method of Bartlett (25) and stored at $0^{\circ}C$. These ghosts had approximately 70% of the membrane phosphorus removed, but showed no change in the circular dichroism properties of the membranes (24).

The methods used for direct and indirect labeling of lysed mounted membranes are published elsewhere (14-18). Briefly, for direct labeling, cells (such as erythrocytes) are lysed on a low osmolarity buffer or water surface (14). Tissue culture cells are first strengthened by brief form-aldehyde fixation before lysis (18). Some of the lysed cell membranes float at the air-buffer interface (held flat by surface forces) and are picked up from above on coated electron microscopy grids. The membranes are directly labeled with the electron-dense marker stain against specific membrane targets and are washed to remove excess stain prior to examination by transmission electron microscopy. For indirect labeling the cells are incubated with, for example, anti-target γ-globulin antibodies to maximum uptake. After washing, the cells are strengthened and/or lysed and mounted for staining with marker stain directed against the membrane bound γ-globulin antibodies. In the experiments reported here the indirect marker stain is ferritin-conjugated anti-γ-globulins. The membrane target is visualized by a cluster of ferritin-antibody molecules, because the antibody bound to the target has several sites for the indirect ferritin-anti-γ-globulin marker.

For CIH staining of human erythrocyte ghosts at low pH, the ghosts were first fixed in 1.5% glutaraldehyde for 20 min. at room temperature and then washed in 0.15 $\underline{M}$ NaCl. The fixed ghosts were flattened at an air-water

interface and mounted in the usual way. After a brief
treatment with a 5% bovine serum albumin solution, the
mounted ghosts were stained with CIH (pH 1.8) for 1 min.
followed by a 1 min. wash on several fresh buffer (12%
acetic acid-acetate pH 2.0) surfaces and finally on a distilled water surface.

RESULTS

Using direct labeling of mounted membranes with influenza virus (Fig.1), colloidal iron hydroxide (CIH) (Fig.3)
and ferritin-conjugated plant agglutinins (Figs. 7, 10,
11); the surface distributions of myxovirus receptors,
negatively charged acidic residues, and certain terminal
saccharide residues are present in random distributions on
the membrane surface. Similarly, using indirect labeling
techniques, the distributions of human $Rh_0(D)$ antigens on
Rh^+ erythrocytes (Fig.5) and murine $H-2^d$ alloantigens on
Balb/c erythrocytes (Fig.6) are also random. The labeling
processes are specific; removal of membrane target components by enzymatic action (Figs.2 and 4), addition of competitive inhibitors to the staining solution (14-18, 20)
and substitution of marker stains of slightly different
specificities (such as ferritin-antibodies recognizing a
different allotype of histocompatibility antigens (16))
block labeling to the membrane.

Treatment with proteolytic enzymes and lipases affects
membrane topology, possibly by upsetting the normal equilibrium between components in the membrane plane. Treating
a variety of different types of cells with low concentrations of trypsin or other proteolytic enzymes causes topological rearrangements (clustering) of the surface CIH
sites, certain antigenic sites (such as the $Rh_0(D)$ antigenic sites of human erythrocytes) (unpublished observations) and ferritin-conjugated plant agglutinin sites
(Fig.8 and ref. 26). This same phenomenon is observed
after erythrocyte ghosts are treated with phospholipase C
(Fig.9).

Enzymatic treatments are not exclusive in modifying
membrane topology. When certain tissue culture cells are
transformed by oncogenic viruses, their surface properties
change concurrently with transformation. For example, the

56

plant agglutinin sites (e.g. concanavalin A sites) on normal murine Balb/c 3T3 fibroblasts are in a dispersed state (Fig.10) compared to the distribution of agglutinin sites after transformation with SV40 virus (Fig.11). This change in topological distribution is paralleled by an increase in specific lectin-mediated agglutinability of the SV40 transformed 3T3 cells (Table I).

TABLE I

Agglutination of Murine 3T3 Fibroblasts
by Concanavalin A After Transformation
by SV40 Virus or Trypsinization*

Cell	Trypsinization	Inhibitor Added	Concentration of Concanavalin A Required for Half Maximal Agglutination
3T3	-	-	>1,000 μg/ml
3T3	-	0.2$\underline{M}$ Sucrose	>1,000
3T3	0.001%-2 min	-	60-100
3T3	0.001%-2 min	0.2$\underline{M}$ Sucrose	>1,000
SV3T3	-	-	60-100
SV3T3	-	0.2$\underline{M}$ Sucrose	>1,000
SV3T3	0.001%-2 min	-	60-100

*Adapted from Nicolson (18, 26).

DISCUSSION

The topological distributions of membrane components appear to be in a random state, although associations may occur in the membrane plane to account for the "dispersed" versus "partially aggregated" status of some of these components. The status of association in the membrane is exemplified here in comparisons of the distributions of $Rh_0(D)$ antigens ("dispersed") (15) with the H-2 alloantigens ("partially aggregated") (16). Even in the partially aggregated H-2 antigen distributions it is apparent that

there is no fixed periodic topological distribution of
these cell surface antigenic sites or aggregates of antigenic sites. Thus, there is no long-range order to the
distribution of these membrane components which are un-
doubtedly proteins or glycoproteins (29-33). This is not
to say that all membrane components are distributed randomly in the membrane; there may exist special cases where
membrane structures are periodically arranged, such as the
structures involved in intercellular junctions (see section
by D. A. Goodenough) and viral envelope components.

Singer and I have suggested recently that the apparent
lack of topological order is due to the molecular structure
and fluid nature of biological membranes (17). In this
scheme of membrane structure, the membrane matrix is formed
by a discontinuous lipid bilayer interrupted by intercal-
ated amphipathic globular protein and glycoprotein components. Since the matrix is lipid, the distribution of the
amphipathic protein components is determined by free trans-
lational diffusion in the plane of the membrane. Of course
a requirement for this model is that the lipid matrix is
fluid under physiological conditions. This requirement is
supported by the studies of several investigators using
spin-label techniques (ref. 34-36 and section by H. M.
McConnell), differential calorimetry (37) and x-ray diffraction (ref. 38, 39 and section by J. K. Blasie). Also,
the observations of Frye and Edidin (ref. 40 and section by
Edidin) on fusion heterokaryons indicates that cell surface
antigens are able to laterally diffuse after fusion of unlike cells by Sendai virus. These experiments were performed by fusing human and mouse fibroblasts and localizing
human and mouse H-2 antigens with fluorescent antibody
techniques at various times after fusion. Shortly after
fusion, the heterokaryon cells had distinctly half of their
surface containing human antigens and half mouse H-2 antigens. Within 40 minutes the antigens had migrated, and
were completely intermixed across the entire cell hybrid
surface; this migration was independent of protein synthesis or energy sources generated by electron transport.

Evidence presented here also supports a fluid structure for cellular membranes; one in which the surface distribution of membrane components is determined by two-dimensional surface interactions (i.e. charge interactions,

etc.) between membrane components. When membranes are treated with proteolytic enzymes such as trypsin or lipases such as phospholipase C, the surface distributions of agglutinin sites, antigenic sites and CIH sites are radically changed; in the latter case without changing the circular dichroic properties of the membrane. Thus, a shift in the equilibrium state between membrane components may cause these drastic topological re-arrangements without causing dramatic changes in the average conformations of the membrane proteins.

An important problem of membrane structure and organization is the relationship of topological organization to certain functional requirements of cell surfaces. Phenomena such as changes in cell agglutinability, complement-mediated cell lysis, direct cytotoxicity by certain cell surface-specific proteins, stimulation of certain cells to proliferate, and differences between normal and oncogenic virus transformed cell surfaces may depend, in part, on the topological organization of cell membranes. A few examples of these functional aspects of membrane topology follow.

Cell agglutination occurs when agglutinating molecules are present and the balance of forces opposing agglutination (charge repulsion, etc.) are overcome by forces favoring agglutination (41). Among the forces favoring cell agglutination are the binding forces of the agglutinating molecules and the number of the molecules directly involved in the agglutination process. Without changing any parameters except the number of molecules involved in direct cross-bridging between adjacent cells, cell agglutinability can be increased several-fold (26). Modifying the topological distribution of the sites involved in agglutination to a clustered state results in an increase in intercellular cross-bridges (26). Conditions that change the distribution of membrane agglutination sites to a clustered state, such as proteolysis or viral transformation, also increase cell agglutinability.

The sensitivity of a cell to complement-mediated cell lysis is dependent on the topological distribution of antigens on the cell surface. Sanderson (42) found that the correct disposition of two adjacent γ-globulin molecules was necessary in order to bind complement. The membrane correlate of this observation is shown here (Fig.5 and 6). The complement-insensitive configuration of membrane-bound γ-globulin is exemplified by the topological distribution of $Rh_0(D)$ antigens (Fig.5). Even after sensitization to maximal uptake with anti-$Rh_0(D)$-γ-globulins, the Rh^+ erythrocytes are not lysed in the presence of excess complement. In contrast, the membrane distribution of anti-H-2 γ-globulins (Fig.6) on murine erythrocytes permits the binding of complement, resulting in cell lysis. This may be an example where evolutionary pressures have selected for effective barriers against incompatible H-2 allotypes, but against the most effective immunological barrier against mother and foetus Rh incompatibility.

Cell transformation by oncogenic viruses causes specific changes at the cell surface. Some of these changes, such as a difference in lectin agglutinability (43-46 and Table I), alterations in sugar transport (47), differences in the activities of cell surface glycosylases (48), and differential cytotoxicity of agglutinins toward tumor cells may be attributable to differences in the topological distributions of surface components after transformation. Recent studies have shown that a difference in the topological distribution of Con A sites on transformed cells (18) is related to increased Con A agglutinability without an increase in the number of Con A sites (49). Normal 3T3 cells have their Con A sites dispersed compared to the clustered distributions of Con A sites on SV40-transformed cells (18). General topological changes such as a clustering of membrane components may also account for the interference of Con A with amino acid transport in transformed cells (47) and the difference in galactosyl transferase activity on malignant cell surfaces (48). In the latter case the transfer of galactose from UDP-galactose to malignant cell surface acceptors occurs at all cell densities (both to the same membrane [cis-glycosylation], or to another cell [trans-glycosylation]). In normal cells the transfer of galactose occurs only when cells reach confluency (trans-glycosylation). Normal cells have the capacity to transfer galactose to an external low molecular weight acceptor, but unlike transformed cells cannot transfer galactose to membrane acceptors on the same membrane (cis-glycosylation).

Con A and other agglutinins are toxic to tumor cells (such as SV 3T3) at concentrations where normal cells are unaffected. The toxicity may also be related to the topological distribution of Con A sites, since there is no difference in the number of Con A sites on these cells (49). An explanation for the toxic properties of polyvalent agglutinins could be that they can kill cells by effectively cross-linking large areas of the cell surface, preventing normal plasma membrane turnover and interfering with transport processes that require lateral diffusion or translation of membrane components. In certain special cases this membrane cross-linking phenomenon might stimulate membrane turnover, as in the stimulation of lymphocytes to divide (50). Indeed, Taylor et al. have shown that γ-globulin antibodies stimulate lymphocyte processes that lead to mitosis, whereas monovalent Fab (non-cross-linking) antibodies do not (51). A fine line may exist in these systems where parameters such as the number and distribution of agglutinin sites and mechanisms of membrane turnover could determine whether a cell is killed or stimulated. Used at low concentrations phytohemagglutinin stimulates mitosis, but when higher concentrations are used the agglutinin is toxic (52).

The examples used here in relating membrane topology to a few functional properties of cell membranes, obviously only scratch the surface. But these examples indicate that membrane topological organization may be critically important in certain cell functions.

REFERENCES

1. S. J. Singer, Nature, 183, 1523 (1959).
2. R. W. Mowry, Lab. Investigation, 7, 566 (1958).
3. U. Hämmerling, T. Aoki, H. A. Wood, L. J. Old, E. A. Boyse and E. de Harven, Nature, 223, 1158 (1969).
4. L. A. Sternberger, J. Histochem. Cytochem., 15, 139 (1967).
5. P. K. Nakane and G. B. Pierce, J. Cell Biol., 33, 307 (1967).
6. S. Avrameas and G. Lespinats, C. R. Acad. Sci.,(Paris), 265, 1149 (1967).
7. R. C. Curren, A. E. Clark and D. Lovell, J. Anat., 99, 427 (1965).

8. U. Hämmerling, T. Aoki, E. de Harven, E. A. Boyse and L. J. Old, J. Exp. Med., 128, 1461 (1968).
9. T. Aoki, H. A. Wood, L. J. Old, E. A. Boyse, E. de Harven, M. P. Lardis and C. W. Stackpole, Virology, 45, 858 (1971).
10. G. L. Nicolson and S. J. Singer, Ann. New York Acad. Sci., In press.
11. C. W. Stackpole, T. Aoki, E. A. Boyse, L. J. Old, J. Lumley-Frank and E. de Harven, Science, 172, 472 (1971).
12. P. Pinto da Silva, S. D. Douglas and D. Branton, Nature, 232, 194 (1971).
13. T. W. Tillack, R. E. Scott and V. T. Marchesi, J. Cell Biol., 47, 213a (1970); J. Exp. Med., In press.
14. G. L. Nicolson and S. J. Singer, Proc. Nat. Acad. Sci. U.S., 68, 942 (1971).
15. G. L. Nicolson, S. P. Masouredis and S. J. Singer, Proc. Nat. Acad. Sci. U.S., 68, 1412 (1971).
16. G. L. Nicolson, R. Hyman and S. J. Singer, J. Cell Biol., 50, 905 (1971).
17. S. J. Singer and G. L. Nicolson, Science, 175, 720 (1972).
18. G. L. Nicolson, Nature New Biology, 233, 244 (1971).
19. S. J. Singer and A. F. Schick, J. Biophys. Biochem. Cytol., 9, 519 (1961).
20. G. L. Nicolson, V. T. Marchesi and S. J. Singer, J. Cell Biol., 51, 265 (1971).
21. G. J. Gasic, L. Berwick and M. Sorrentino, Lab. Investigation, 18, 63 (1968).
22. J. J. Holland and E. D. Kiehn, Science, 167, 202 (1970).
23. J. T. Dodge, C. Mitchell and D. J. Hanahan, Arch. Biochem. Biophys., 100, 119 (1963).
24. M. Glaser, H. Simkins, S. J. Singer, M. Scheetz and S. I. Chan, Proc. Nat. Acad. Sci. U.S., 65, 721 (1970).
25. G. Bartlett, J. Biol. Chem., 234, 466 (1959).
26. G. L. Nicolson, Nature, submitted for publication.
27. M. Inbar and L. Sachs, Nature, 223, 710 (1969).
28. M. Inbar and L. Sachs, Proc. Nat. Acad. Sci. U.S., 63, 1418 (1969).
29. F. A. Green, Immunochemistry, 4, 247 (1967).
30. F. A. Green, J. Biol. Chem., 247, 881 (1972).

31. B. Kaufman and S. P. Masouredis, J. Immunol., 91, 233 (1963).
32. A. Shimada and S. G. Nathenson, Biochem., 8, 4048 (1969).
33. T. Muramatsu and S. G. Nathenson, Biochem., 9, 4875 (1970).
34. W. L. Hubbel and H. M. McConnell, Proc. Nat. Acad. Sci. U.S., 61, 12 (1968).
35. A. D. Kieth, A. S. Waggoner and O. H. Griffith, Proc. Nat. Acad. Sci. U.S., 61, 819 (1968).
36. R. K. Kornberg and H. M. McConnell, Biochemistry, 10, 1111 (1971).
37. J. M. Stein, M. E. Fourtellotte, J. C. Reinert, R. N. McElaney and R. L. Rader, Proc. Nat. Acad. Sci. U.S., 63, 104 (1969).
38. J. K. Blasie and C. R. Worthington, J. Mol. Biol., 39, 417 (1969).
39. M. M. Dewey, P. K. Davis, J. K. Blasie and L. Bar, J. Mol. Biol., 39, 395 (1969).
40. C. D. Frye and M. Edidin, J. Cell Sci., 7, 313 (1970).
41. W. Pollack, Ann. N. Y. Acad. Sci., 127, 892 (1965).
42. A. R. Sanderson, Immunol., 9, 287 (1965).
43. M. M. Burger, Proc. Nat. Acad. Sci. U.S., 62, 994 (1969).
44. M. Inbar and L. Sachs, Proc. Nat. Acad. Sci. U.S., 63, 1418 (1969).
45. B. Sela, H. Lis, N. Sharon and L. Sachs, J. Membrane Biol., 3, 267 (1970).
46. G. L. Nicolson and J. Blaustein, Biochim. Biophys. Acta, In press.
47. M. Inbar, H. Ben-Bassat and L. Sachs, J. Membrane Biol., 6, 195 (1971).
48. S. Roth and D. White, Proc. Nat. Acad. Sci. U.S., 69, 485 (1972).
49. B. Ozanne and J. Sambrook, Nature New Biology, 232, 156 (1971).
50. A. E. Powell and M. A. Leon, Exptl. Cell Res., 62, 315 (1970).
51. R. B. Taylor, W. P. H. Duffus, M. C. Raff and S. de Petris, Nature New Biology, 233, 225 (1971).
52. D. A. Rigas and V. V. Tisdale, Experientia, 25, 399 (1969).

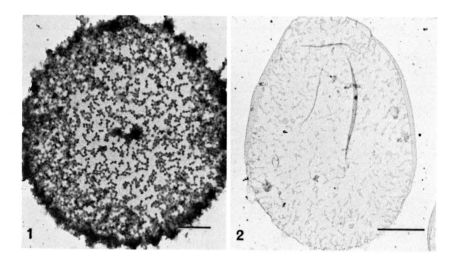

Fig. 1. A mounted rabbit erythrocyte was labeled with
influenza virus. After washing the specimen was
treated with 1% osmium tetroxide. Bar equals
1μm.

Fig. 2. A control experiment for Fig. 1. A neuraminidase
treated cell was incubated with influenza virus
in a parallel experiment. Bar equals 1μm.

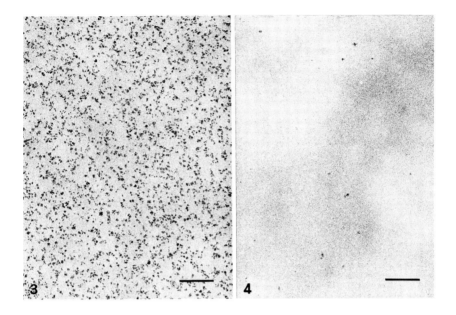

Fig. 3. The surface of a glutaraldehyde-fixed human erythrocyte stained with colloidal iron hydroxide at
pH 1.8. Bar equals 0.1μm.

Fig. 4. Control for Fig. 3. A glutaraldehyde-fixed neura-
minidase-treated human erythrocyte was stained
with colloidal iron hydroxide in a parallel experiment. Bar equals 0.1μm.

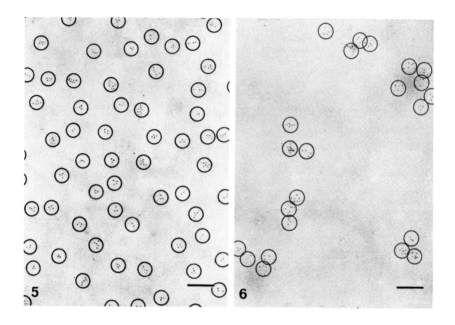

Fig. 5.   Human Rh$^+$ erythrocytes were maximally labeled with ^{125}I-human anti-Rh$_0$(D) γ-globulins. After mounting the anti-Rh$_0$(D) sensitized cells were stained with ferritin-conjugated goat anti-human γ-globulins. The number of ferritin clusters (in circles of 300Å radius) agrees with the number of membrane bound ^{125}I-anti-Rh$_0$(D) molecules. Bar equals 0.1µm.
[From Nicolson et al., 1971 (ref.14); Reproduced by permission from the Proceedings of the National Academy of Sciences]

Fig. 6.   Mouse erythrocytes (H-2^d) were maximally labeled with anti-H-2^d γ-globulin alloantibodies. After mounting the cells were stained with ferritin-conjugated anti-mouse 7s γ-globulins. The antigenic sites (in circles of 300 Å radius) are present in random clusters. Bar equals 0.1µm.
[From Nicolson et al., 1971 (ref.16); Reproduced by permission from The Journal of Cell Biology]

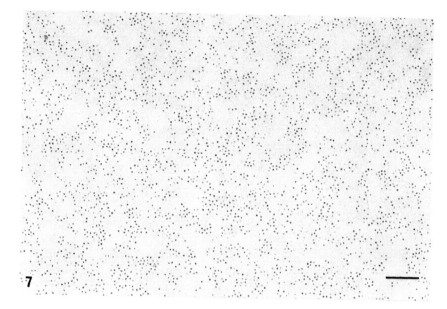

Fig. 7. A mounted rabbit erythrocyte was labeled with
ferritin-conjugated <u>Ricinus</u> <u>communis</u> agglutinin.
Bar equals 0.1μm.

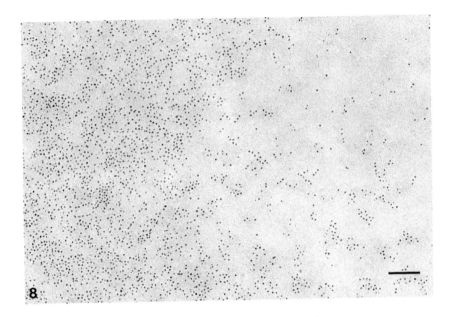

Fig. 8. Rabbit erythrocytes were treated with a dilute
trypsin solution (see METHODS) before mounting
and labeling with ferritin-conjugated <u>Ricinus</u>
<u>communis</u> agglutinin in a parallel experiment to
<u>Fig. 7.</u> Note regions of higher and lower density
of ferritin-agglutinin molecules compared to
Fig. 7, indicating a surface re-arrangement. Bar
equals 0.1μm.

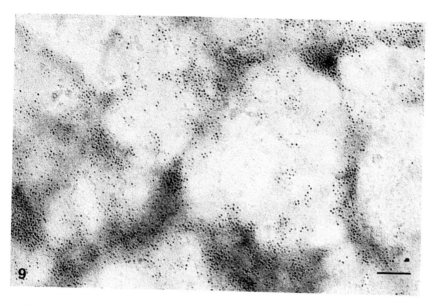

Fig. 9.   Human erythrocytes were treated with a phospholipase C solution (see METHODS) before mounting
and labeling with ferritin-conjugated concanavalin A.   Bar equals 0.1μm.

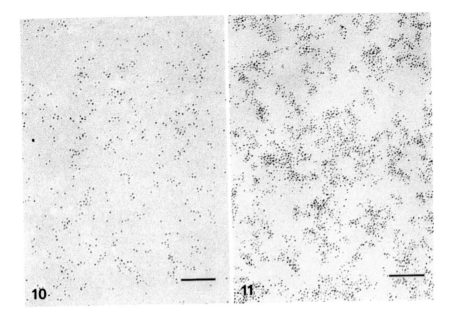

Fig.10. Murine 3T3 fibroblasts were briefly fixed with
0.1% formaldehyde, mounted and stained with ferritin-conjugated concanavalin A. The ferritin-
concanavalin A molecules are present in a dispersed state. Bar equals 0.1μm.
[From Nicolson, 1971 (ref.18); Reproduced by
permission from Nature New Biology]

Fig.11. Murine SV40 transformed 3T3 fibroblasts were
fixed and stained as in Fig.10. The ferritin-
concanavalin A molecules are in a more clustered
state on certain areas (shown here) of the transformed cell surface. Bar equals 0.1μm.
[From Nicolson, 1971 (ref.18); Reproduced by
permission from Nature New Biology]

THE ORGANIZATION OF PROTEINS IN HUMAN
ERYTHROCYTE MEMBRANES

Theodore L. Steck

Department of Medicine
University of Chicago
Chicago, Illinois 60637

ABSTRACT. The two surfaces of red blood cell membranes
were selectively probed by treating intact cells, resealed
ghosts, and inside-out vesicles with non-penetrating reagents. Sialic acid and acetylcholinesterase were found
only on the external face while a diaphorase and glyceraldehyde 3-phosphate dehydrogenase were confined to the internal surface. No rearrangement was detected. Some of
the polypeptides were labeled and digested at the outer
surface; these appeared to be glycoproteins. Other components, bearing no detectable carbohydrate, were accessible exclusively at the inner surface. Two glycoproteins
were digested at both membrane surfaces but showed different patterns of proteolysis at each face. All of these
components thus manifested an asymmetrical disposition
across the plane of the membrane; some polypeptides appear
to span the membrane thickness as well.

INTRODUCTION

Two tacit arguments underlie the particular approach
which we (and others) have pursued in studying membrane
organization. The first is that the functional asymmetry
existing across a membrane should be reflected in the
macromolecules comprising its two surfaces. Secondly,
membranes are of paucimolecular thickness, so that elucidating the large molecules at their two surfaces should go
far toward describing the structure as a whole.

We have isolated human erythrocyte membranes in the

form of resealed right-side-out (RO)* ghosts and inside-out (IO) vesicles. Each membrane face was selectively reacted with probes which do not cross the permeability barrier. In general, RO membranes resembled intact erythrocyte membranes while IO vesicles showed a distinctly different pattern of reactivity.

We have recently found that a small modification (1) of the original method for preparing IO vesicles (2) generates sealed RO vesicles instead (3). This undetected reversal probably led to serious misinterpretations of the earlier sidedness data (1) which the present investigation has tried to rectify.

METHODS

Unsealed ghosts were prepared from human erythrocytes (obtained from fresh or 3-4 week old bank blood) by hemolysis and washing in 5 mM Na phosphate, pH 8.0 (4). Resealed ghosts were prepared by two methods (3). The first was simply to include 1 mM $MgSO_4$ in the 5 mM Na phosphate hemolysis and wash solutions. In the second procedure, cells were lysed in 5 mM Na phosphate, pH 8.0. The ghosts were pelleted, resuspended in this buffer containing 0.15 M NaCl, and incubated at 37° for 40 minutes. The saline-resealed ghosts were washed twice more and stored in the same buffered saline solution.

Inside-out vesicles were prepared by a slight modification (3) of the original method (2). Briefly, unsealed ghosts were incubated for 1-2 hours in 0.5 mM phosphate buffer, pH 8.5-8.7, pelleted, homogenized, and fractionated on a Dextran-110 (Pharmacia) density gradient. The vesicles equilibrating at a density less than 1.03 g/ml were pooled, washed once and used directly. Sealed right-side-out vesicles were prepared by a slight modification of this method formerly presumed to generate IO vesicles (1). 0.1 mM $MgSO_4$ was introduced into the 0.5 mM Na phosphate vesiculation medium after the 1-2 hour incubation but before pelleting and homogenizing. All subsequent solutions were made 0.1 mM in $MgSO_4$. These preparations have an isopycnic

*The abbrevations used are: RO, right-side-out; IO, inside-out; PAS, periodic acid-Schiff; TD, tracking dye; H, hemoglobin chains; TNBS, trinitrobenzene sulfonate.

density in Dextran-110 gradients similar to that of the IO vesicles.

Proteolysis of membranes was performed essentially as described (1). After isolated membranes were digested, they were incubated with the appropriate protease inhibitor (1, 11), dissolved and electrophoresed directly. Following digestion, intact cells were washed 3 times with saline, incubated with inhibitor, then lysed and washed to prepare unsealed ghosts for electrophoresis.

Electrophoresis was performed in the presence of 0.2% sodium dodecyl sulfate on 5.0% polyacrylamide gels by a modification of our original protocol (4).

RESULTS

Sidedness and sealing of membrane preparations. The orientation and permeability of membrane preparations were assessed by measuring the accessibility of membrane-bound enzymes to their substrates or of sialic acid to sialidase (2, 3). Table 1 summarizes such an experiment. Because the acetylcholinesterase and sialic acid were entirely accessible in the right-side-out preparations (and intact cells) but only 10% available in the IO preparations, we judge the latter to be 90% pure sealed inside-out vesicles. The opposite pattern was seen with NADH:cytochrome c oxidoreductase and glyceraldehyde 3-phosphate dehydrogenase. Since these two enzymes were not accessible to their substrates in intact cells or sealed ghosts, they were taken to be confined to the inner (cytoplasmic) surface of the membrane. Sealed RO vesicles showed a pattern indistinguishable from resealed ghosts. Unsealed ghosts offered no barrier to any of the test substrates. They also appeared permeable to enzymes and to proteins as large as ferritin (5), yet seemingly not permeable to Dextran-110 (2). We conclude 1) that the isolated resealed ghosts, RO and IO vesicles are impermeable to molecules as small as simple substrates while unsealed ghosts are not; and 2) that no scrambling of the orientation of these proteins occurred during membrane isolation.

TNBS labeling of membranes. In recent investigations, intact cells and unsealed ghosts have been reacted with

73

with radioactive sulfanilic acid diazonium salt (6,7,8), formylmethionyl sulfone methyl phosphate (9), and iodine (using lactoperoxidase and H_2O_2) (10). These covalent labels appeared to be bound to only one or two major protein components in the intact cell membrane while the isolated ghost showed reactivity in many polypeptide components.

We have applied this approach using trinitrobenzene sulfonate (TNBS) as a covalent label. Bonsall and Hunt (11) recently offered evidence that the intact red cell membrane was rather impermeable to this reagent and was labeled to a limited degree, while the isolated membrane was more reactive. Figure 1 shows electrophorograms of membrane proteins labeled by TNBS in unsealed ghosts and in the intact cell. The patterns resemble those seen with the other labels (6-10): while all molecular weight classes were labeled in unsealed ghosts, only the region of about 90-100,000 daltons was conspicuously reactive in the intact cells. As predicted from Bretscher's studies (9,12), this broad zone could be resolved into two distinct TNBS labeled components-corresponding to bands 3 and PAS-1 (see below)-when electrophoresis was performed on 8.5% instead of 5.0% acrylamide gels. As in the other labeling studies, lipids were reactive as well as proteins, as shown in Figure 1 by the peak of label travelling just behind the tracking dye. Because TNBS appears to permeate resealed ghosts, its usefulness may be circumscribed.

Proteolytic digestion of intact erythrocyte and unsealed ghost membranes. A distinct difference in the susceptibility of intact and isolated membranes to proteolytic attack has been reported (1,7,8). As seen in Figure 2 (A vs. B), intact membranes were remarkably resistant to proteolysis under conditions where virtually every major component (except band 6) was digested in unsealed ghosts. Band 6 appears to have intrinsic conformational resistance, since it was not digested even when solubilized, unless denatured (1). Prolonged exposure of intact membranes to chymotrypsin or pronase (but not trypsin) brings about the digestion of band 3 and the reciprocal appearance of a major ~65,000 dalton product (migrating between bands 4.2 and 5) and a minor ~40,000 dalton product (running between bands 5 and 6, Fig. 2C and ref. 7,8,12,13).

There are membrane glycoproteins which do not take up appreciable Coomassie blue but are detected by PAS staining (4). These are quite sensitive to mild proteolytic digestion of the intact erythrocyte membrane (Fig. 3 and refs. 1,7,8). The sialic acid attached to the major PAS bands is also entirely exposed at the external face of the membrane (Table 1).

The unreactivity of many of the polypeptide species in intact membranes could be attributable to their localization at the inner membrane surface. In that case, resealed ghosts and RO vesicles should behave like intact erythrocytes while sealed IO vesicles would be extensively digested. If, on the other hand, the limited reactivity of the native erythrocyte membrane is a reflection of a special architectural packing arrangement or a protective factor which is lost during hemolysis, then resealed ghosts might also be digestible. Experiments were performed to investigate these possibilities.

Digestion of resealed ghosts. The response of saline-resealed and Mg-resealed ghosts to proteolysis resembled that of the membrane in the intact cell (Fig. 4A and B). While most components are spared, band 3 is slowly digested by chymotrypsin but not by trypsin. The 65,000 dalton proteolysis product was again conspicuous. Staining with PAS (not shown) showed the glycoproteins to be as susceptible to both enzymes as in the intact cell membrane (Fig. 3).

The possibility that the resealed ghosts became permeable during digestion was checked by the NADH:cytochrome c oxidoreductase assay. It was consistently found that the limited accessibility of this enzyme to its substrates did not increase during proteolysis.

Digestion of right-side-out vesicles. Electrophorograms of RO and IO vesicles (e.g., Fig. 4, gel 7 and Fig. 5, gel 1) demonstrate the partial or complete elution of bands 1, 2, 4.1, and 5 which accompanies vesiculation (1,4). The remaining components of the sealed RO vesicles were at least as resistant to proteolysis as in the resealed ghosts, as seen by comparing panel C with A and B in Fig. 4. In a previous study (1), RO vesicles were prepared by this method but were mistakenly taken to be inside-out; they showed a resistance to proteolysis similar to that seen in Fig. 4C, except that band 3 was more readily digested to the 65,000 dalton proteolysis product. The glycoproteins of RO vesicles show a response to proteolysis similar to that seen in intact cells.

Digestion of inside-out vesicles. Incubation of IO vesicle preparations with proteases resulted in a loss of bands 3, 4.1, 4.2, 7, and several minor components such as these running between bands 2 and 3 (Fig. 5). The proteolysis pattern varied among the enzymes but never evinced the 65,000 dalton band characteristically found following digestion of sealed RO species. The trace of bands 1 and 2 remained undigested. These components may represent a fraction of the molecules solubilized during vesiculation which became sealed inside the IO vesicles adventitiously.

The digestion of the glycoproteins in IO vesicles (Fig. 6) differed considerably from the extensive clearing observed in all of the right-side-out and unsealed membrane preparations. Trypsin and chymotrypsin had no effect on the major sialoglycoprotein bands (even when used at ten-times their usual concentration, 100 mcg/ml). Papain digestion led to the loss of the PAS-1 band and the reciprocal appearance of a zone of stainable material traveling at a slightly faster mobility.

Sidedness of band 6. Band 6 resists proteolysis (1) and has not been localized in digestion studies. However, we (14) have confirmed Gray and Tanner's (15) observation that this band represents the polypeptide chains of glyceraldehyde 3-phosphate dehydrogenase. This enzyme seems to be confined to the inner membrane surface (Table 1). It (and band 6) can be eluted with .15 M NaCl from unsealed but not from resealed ghosts. Furthermore, there seem to be high affinity attachment sites for this enzyme localized to the cytoplasmic membrane surface (14).

Labeling of carbohydrates. Oligosaccharides containing galactose, N-acetylgalactosamine, and related sugars can be oxidized by galactose oxidase and labeled by subsequent reduction with ^{3}H-NaBH$_4$ (16). Intact erythrocytes, resealed and unsealed ghosts were all labeled extensively

76

in this manner. Inside-out vesicles, however, were distinctly unreactive. Their incorporation of label varied with the fraction of acetylcholinesterase accessible to its substrate, suggesting that it was the trace of RO contaminants which were being tagged.

Figure 7 illustrates both the dependence of ghost labeling on the presence of galactose oxidase and the unreactivity of IO vesicles. The polypeptide labeling pattern in unsealed ghosts (as in intact erythrocytes) is different from the Coomassie blue and PAS profiles and reveals an unexpected multiplicity of glycoprotein species. Some peaks fall in the region of bands 3 and PAS-1 while the bulk of the tritium is distributed between bands 4.2 and 5. This region contains several poorly resolved Coomassie blue-stained bands and a broad zone of PAS-positive material (see below). Band 3, which has been found by direct analysis to contain carbohydrate (17), also can be shown to be labeled (18). Part of the label in the band 3 region may also be contributed by other components, such as PAS-1 and the minor bands which lie under band 3 (see Fig. 2, gel 9). Furthermore, the heterogeneity in both the staining and labeling profiles of band 3 may reflect multiple molecular forms of this component which might derive from variation in glycosylation.

A peak of tritium is also observed just behind the tracking dye (Fig. 7), suggesting that lipid is also labeled. Direct analysis of the lipids showed limited labeling of neutral lipids and phospholipids, which was poorly stimulated by galactose oxidase, and a greatly stimulated specific labeling of the tri- and tetrahexosyl-ceramide glycolipids (18). The glycolipids in IO membranes were just as unresponsive to galactose oxidase as were the proteins. Other experiments have indicated that the unreactivity was not due to loss of receptor molecules or to inhibitors but was overcome when the IO membrane permeability barrier was disrupted.

NaOH extraction. The distinctive labeling pattern of the ghost glycoproteins relates to the selective extraction behavior of certain membrane polypeptides. Ghosts treated with 6 M guanidine-HCl release most polypeptide species but retain certain proteins, along with the lipids, in a vesicular membrane residue. The unreleased species are principally 3, 7, a cluster of indistinct bands between 4.2 and 5, and the PAS-positive components (19). Brief exposure to 0.1 N NaOH effected the same selective extraction pattern, with almost quantitative separation of the two classes of polypeptides (Figures 8 & 9). Little or no phospholipid phosphorus, neutral sugar and sialic acid were released from the membrane (which retained the microscopic appearance of ghosts), while about 50% of the protein was solubilized. The tritiated glycoproteins are all recovered in the membranous NaOH residue and seem to correspond in mobility to most of the unextracted polypeptides (compare Fig. 7 with 8 & 9).

DISCUSSION

The complex dependence of vesicle formation on ionic milieu suggests that specific molecular transitions rather a random fragmentation of the membrane underlie this process. To generate sealed IO vesicles, unsealed ghosts are incubated and homogenized in 0.5 mM phosphate buffer at $pH \geq 8.0$. However, adding 0.1 mM $MgSO_4$ just before homogenization results in sealed RO vesicle formation. If the $MgSO_4$ is present from the outset, unsealed RO vesicles result. Sealing of the vesicles does not arise by a simple casting aside of the regions of the ghost bearing the "holes", since direct homogenization of ghosts (in the presence or absence of Mg) does not generate impermeable vesicles. It is clear that each vesicle preparation must be assessed with care; our earlier sidedness studies suffered from this deficiency (1).

All of the constituents studied here appeared oriented, neither randomly placed nor mobile, with respect to the two surfaces of the membrane. Most of our assignments agree with those made by others. However, much of the previous evidence is indirect or "unilateral". For example, enzymic digestion of intact red cells liberates the sialic acid (20) and inactivates the acetylcholinesterase (cf. 7). But the proof that these functions are confined to the external face of the membrane requires corollary experiments performed at the cytoplasmic surface. It cannot be assumed in advance that a protein is not mobile through the plane of the membrane (since this is part of what is to be proved) or that the probes do not penetrate or that because the acetylcholinesterase is inactivated its active site must be digested, etc. Inside-out vesicles offer a dimension of experimental flexibility and rigor to such sidedness studies.

Similarly, the assignment of a protein to the cytoplasmic surface of the membrane because it is unreactive in intact cells but reactive in ghosts remains an indirect inference until the inner surface is selectively probed. The ghost may differ from the intact cell membrane in many respects, permeability to probes being only one (21). For example, native packing, architecture and/or conformation could be perturbed upon membrane isolation. This type of interpretation was invoked by Roelofsen et al. (22) who found that phospholipids were not digestible in the intact red cell membrane but became fully susceptible to lipases following membrane isolation or sub-lytic detergent treatment. The present data at least seem to rule out the possibility that the resistance of the intact cell membrane against proteases derives from an external coat or cytoplasmic factor which is lost upon hemolysis. In addition, the fact that all the markers examined thus far (e.g., Table 1) conserve their sidedness indicates that a disordering or scrambling across the plane of the membrane does not occur during membrane isolation.

The present studies concur with recent reports that of the major polypeptides only band 3 and the glycoproteins are readily available to impermeable agents attacking the intact erythrocyte membrane (6-10, 12,13). Two lines of evidence suggest that the other components, which become accessible in unsealed ghosts, reside at the cytoplasmic face. The first is that their resistance to proteolysis is restored when ghosts or RO vesicles are resealed.* Secondly, the resistant bands are generally

---

* Several of these gel components (namely 4.1, 4.2, 5, 7 and minor components running between band 2 and 3) are, however, sporadically digested in resealed ghost preparations. While we tend to attribute this to artefact, it is also possible that these polypeptides may play a role at the external surface of this membrane yet elude our probes in most experiments.

reactive at the surface of IO vesicles (i.e., the cytoplasmic face of the membrane). Because bands 1, 2, and 5 are
released during IO vesicle formation (1,4) we cannot directly demonstrate them on the cytoplasmic face. However,
other findings (6-10), as well as the data presented above,
are in agreement with a recent study showing that ferritin-
conjugated antibodies against "spectrin" (presumably the
equivalent of our bands 1 and 2) are specifically bound to
the membrane's inner surface (23).

The present data are in disagreement with our previous sidedness study (1); however, that analysis suffered
in that the following points were not then appreciated:
1) Unsealed ghosts are permeable to proteins (and proteases) but apparently not to Dextran-110 (2,5). 2)
Tightly sealed RO and IO vesicles can arise from unsealed
ghosts (24). 3) Sealed RO and IO vesicles have a very
similar equilibrium buoyant density range (24). 4) The
timing of the addition of 10^{-4} M MgSO$_4$ adopted in that
study (1) results in nearly total replacement of the IO
fraction by sealed RO vesicles. Thus, RO vesicles were
consistently mistaken for IO and the flow of inferences
was reversed. By rejecting those interpretations, our
previous and present data are seen to be consonant and in
general agreement with other recent sidedness studies.

Our findings suggest that the polypeptide molecules
of bands 3 and PAS-1 are all exposed at both membrane
faces. The susceptibility to proteolysis of band 3 differs
distinctly at the two surfaces. Trypsin readily digests
this component only at the inner surface. The other proteases digest band 3 at both faces but produce the ~65,000
dalton band only in RO species. Band 3 thus appears to
span the membrane asymmetrically. This hypothesis is
strengthened by the observation that band 3 carbohydrate is
accessible to galactose oxidase only at the outer surface
(Fig. 7). Furthermore, Bretscher (13) has concluded that
the additional regions of band 3 which are labeled in isolated ghosts compared to intact cells reside at the cytoplasmic membrane surface.

The sialoglycoprotein PAS-1 is extensively digested
at the outer surface, but has a highly restricted susceptibility at the cytoplasmic face (see Fig. 3 vs. 6). The

fact that the major papain digestion products of PAS-1 in
IO vesicles retain all of the PAS stain and are only
slightly enhanced in mobility indicates that a rather
restricted portion of the non-sugar bearing (presumably
carboxy-terminal (25,26)) region of the polypeptide is
exposed at the cytoplasmic surface. Bretscher (12) has
compared the reactivity of the PAS-1 sialoglycoprotein in
intact cells and isolated ghosts and has attributed the
labeling of two extra peptide fragments in the ghosts to
their presence at the interior surface of the membrane.
Marchesi (26) has similarly found that this glycoprotein
can be iodinated at sites near its carboxyl terminus in
unsealed ghosts but not in erythrocytes. The evidence
thus favors the asymmetrical penetration of this component
across the thickness of the membrane, the carbohydrate
moieties being confined to the external surface. Our previous conclusion (1) that bands 3 and PAS-1 were asymmetrically penetrating proteins was prophetic but based on
incorrectly interpreted data.

Much or all of the galactose oxidase-sensitive carbohydrate of both glycoproteins and glycolipids is located
exclusively at the external surface. The distribution of
red cell sialic acid (2) and lectin binding sites (27) and
a wide range of indirect evidence (28) also support the
hypothesis that all of the membrane sugar is located at
the external surface. Thus far, the various species of
glycoproteins are the only major polypeptides identified
at the external surface; furthermore, they comprise that
half of the protein which is preferentially retained in
membranes exposed to 6M guanidine-HCl (19) and 0.1 N NaOH.
Several other polypeptides lack sugar, seem confined to the
internal surface, and are elutable by the aforementioned
denaturants. The single exception is band 7 which is
neither labeled with tritium nor reactive on the external
membrane surface, yet remains in the membrane upon exposure
to the denaturants.

Our working hypothesis based on these data is summarized in Figure 10. The outer membrane surface is
populated by several glycoproteins. They are oriented so
that their carbohydrate moieties are externally exposed.
They are firmly anchored, presumably hydrophobically, by
penetration into or, in at least two cases, through the

81

lipid stratum, as proposed for the major sialoglycoprotein (25,26). Such strong binding might serve to forestall release into the plasma or to stabilize their orientation. In contrast, the non-glycosylated proteins are accessible only at the inner surface. It is conceivable that their weaker binding relates to intramembrane mobility or to equilibration or interaction with the cytoplasm.

ACKNOWLEDGMENTS

The excellent technical assistance of Benita Ramos is gratefully acknowledged. This work was aided by grant no. P-578 from the American Cancer Society and a Fellowship from the Schweppe Foundation.

REFERENCES

1. Steck, T.L., Fairbanks, G. and Wallach, D.F.H., Bio-chemistry, 10, 2617 (1971).
2. Steck, T.L., Weinstein, R.S., Straus, J.H., and Wallach, D.F.H., Science, 168, 255 (1970).
3. Steck, T.L., and Kant, J.A., Methods Enzymol., in press.
4. Fairbanks, G., Steck, T.L., and Wallach, D.F.H., Bio-chemistry, 10, 2606 (1971).
5. Weinstein, R.S. and Steck, T.L., unpublished data.
6. Berg, H.C., Biochim. Biophys. Acta, 183, 65 (1971).
7. Bender, W.W., Garan, H. and Berg, H.C., J. Mol. Biol., 58, 783 (1971).
8. Carraway, K.L., Kobylka, D. and Triplett, R.B., Biochim. Biophys. Acta, 241, 934 (1971).
9. Bretscher, M.S., J. Mol. Biol., 58, 775 (1971).
10. Phillips, D.R. and Morrison, M., Biochemistry, 10, 1766 (1971).
11. Bonsall, R.W. and Hunt, S., Biochim. Biophys. Acta, 249, 281 (1971).
12. Bretscher, M.S., Nature New Biology, 231, 229 (1971).
13. Bretscher, M.S., J. Mol. Biol., 59, 351 (1971).
14. Steck, T.L., Yu, J. and Kant, J.A., unpublished data.
15. Tanner, M.J.A. and Gray, W.R., Biochem. J., 125, 1109 (1971).
16. Morell, A.G., Van Den Hamer, C.J.A., Scheinberg, I.H., and Ashwell, G., J. Biol. Chem., 241, 3745 (1966).
17. Guidotti, G., personal communication.
18. Steck, T.L., and Dawson, G., unpublished data.

19. Steck, T.L., Biochim. Biophys. Acta, 255, 553 (1972).
20. Eylar, E.H., Madoff, M.A., Brady, O.V., Oncley, J.L., J. Biol. Chem., 237, 1992 (1962).
21. Steck, T.L., in C.F. Fox and A.D. Keith (Editors), Molecular Biology of Membranes, Sinauer Associates, Stamford, Conn., 1972, in press.
22. Roelofsen, B., Zwaal, R.F.A., Comfurius, P., Woodward, C.B., and Van Deenen, L.L.M., Biochim. Biophys. Acta, 241, 925 (1971).
23. Nicolson, G.L., Marchesi, V.T., and Singer, S.J., J. Cell Biol., 51, 265 (1971).
24. Kant, J.A., and Steck, T.L., unpublished data.
25. Winzler, R.J., in G.A. Jamieson and T.J. Greenwalt (Editors), Red Cell Membrane Structure and Function, J.B. Lippincott Co., Philadelphia, 1969, p. 157.
26. Marchesi, V.T., this volume.
27. Nicolson, G.L., Proc. Nat. Acad. Sci. U.S.A., 68, 942 (1971).
28. Winzler, R.J., Intern. Rev. Cytol., 29, 77 (1970).

TABLE 1

ACCESSIBILITY OF MARKERS IN GHOSTS AND IO VESICLES

Marker	Unsealed Ghosts	Mg-Sealed Ghosts	Inside-Out Vesicles
		% accessible	
Sialic acid	99	99	10
Acetylcholine esterase	96	105	9
NADH diaphorase	77	4	139
Glyceraldehyde 3-P dehydrogenase	93	10	78

Accessibility of sialic acid to sialidase and of each membrane enzyme to its substrates was calculated from the extent of reaction in the absence and presence of Triton X-100 or saponin, assuming that the surfactant rendered the membrane 100% permeable to the probe. Adapted from reference 3, with permission.

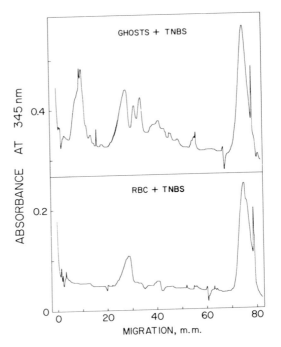

Fig. 1. TNBS labeling of unsealed ghosts and intact red cell membranes. Intact washed erythrocytes or unsealed ghosts were incubated at room temperature for 1 hour with .010 M TNBS in 0.10 M Na phosphate (pH 10.0). The reaction was stopped by washing with 50 volumes of 0.10 M Na phosphate (pH 6.0) - 0.05 M NaCl. The red cells were then hemolyzed and the membranes washed free of hemoglobin. About 80 mcg. protein of each membrane sample were electrophoresed. The gels were fixed overnight without stain in 25% isopropanol, soaked in water for 1-2 hours and scanned at 345 nm.

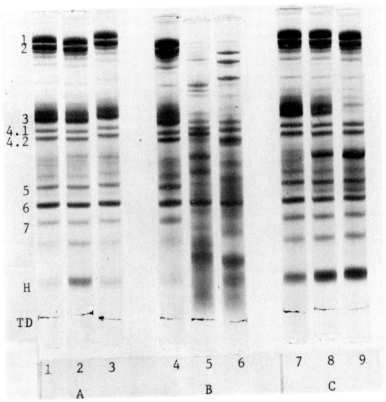

Fig. 2. Proteolysis of intact erythrocyte and unsealed ghost membranes. Washed red cells (A) and unsealed ghosts (B) (~ 1 mg. of membrane protein/ml) were incubated at room temperature for 1 hour with no enzyme (gels 1 and 4), 10 mcg/ml trypsin (gels 2 and 5) or 10 mcg/ml chymotrypsin (gels 3 and 6). In (C), red cells were incubated 18 hours with no enzyme (gel 7), chymotrypsin (gel 8) or pronase (gel 9) at 10 mcg/ml. Coomassie blue stain.

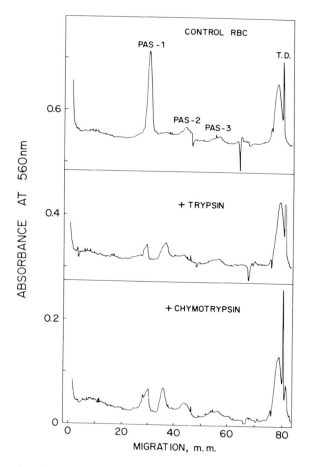

Fig. 3. Proteolysis of the glycoproteins in intact erythrocyte membranes. Washed red cells were treated as in Figure 2A. The gels were stained for carbohydrate with PAS and scanned at 560 nm. The major glycoproteins are designated PAS 1-3; the peak just behind the tracking dye corresponds to the lipid.

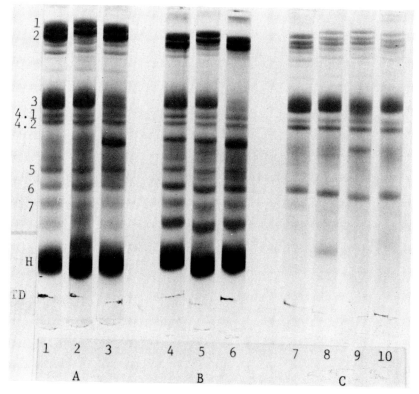

Fig. 4. Proteolysis of resealed ghosts and right-side-out vesicles. Saline-resealed (A) and Mg-resealed ghosts (B) (~ 1 mg membrane protein/ml) were incubated at room temperature for 1 hour with no enzyme (gels 1 and 4), 10 mcg/ml trypsin (gels 2 and 5) or 10 mcg/ml chymotrypsin (gels 3 and 6). In (C), sealed RO vesicles were incubated at room temperature for 1 hour with no enzyme (gel 7), trypsin (gel 8), chymotrypsin (gel 9) or papain (gel 10) at 10 mcg/ml. Coomassie blue stain. (The two extra bands seen in the Mg-sealed ghosts - e.g., gel 4 - between bands 4.2 and 5 and between bands 7 and H are of cytoplasmic origin.)

87

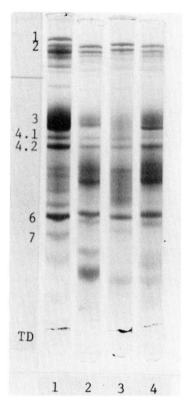

Fig. 5. Proteolysis of inside-out vesicles. IO vesicles (∼1 mg protein/ml) were incubated at room temperature for 1 hour with no enzyme (gel 1), trypsin (gel 2),
chymotrypsin (gel 3), and papain (gel 4) at 10 mcg/ml.
Coomassie blue stain.

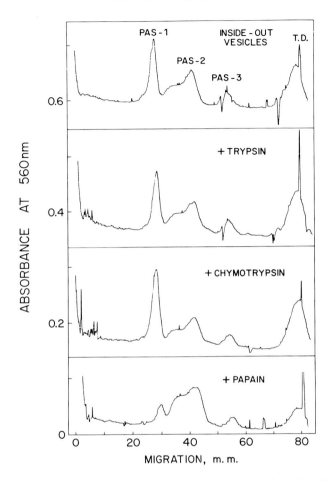

Fig. 6. Proteolysis of the glycoproteins in inside-out vesicles. Gels prepared as in Fig. 5 were stained with PAS and scanned (see Fig. 3 legend).

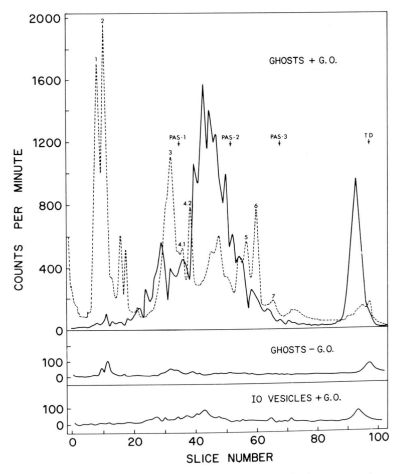

Fig. 7. Labeling of sugars. Unsealed ghosts and inside-out vesicles (~1 mg protein/ml) were incubated at room temperature for 1 hour in 0.050 M Na phosphate (pH 8.5) containing or lacking 2000 units/ml galactose oxidase (G.O.) (Kabi, Stockholm). One-fiftieth volume of 0.01 N NaOH containing fresh ^{3}H-NaBH$_4$ (nominally 304 mCi/m mole, 0.15 mM) was added. After 15 minutes, the membranes were washed three times in 0.005 M Na phosphate (pH 8.0) and electrophoresed. Gels were stained with Coomassie blue to verify the recovery of proteins, then sliced into 0.85 mm sections for counting. Solid lines, counts per minute; dashed line, densitometric scan of Coomassie blue-stained gel; arrows, PAS 1-3 peaks.

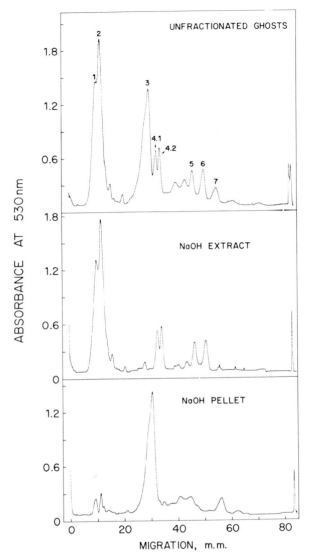

Fig. 8. NaOH extraction: Coomassie blue stain. 1 volume of packed unsealed ghosts (∿4 mg protein/ml) was incubated for 5 minutes on ice with 7 volumes of 0.1 NaOH-0.01 M 2-mercaptoethanol. The membrane pellet and the supernatant extract were collected following centrifugation and an aliquot equivalent to 40 mcg of whole ghosts was electrophoresed.

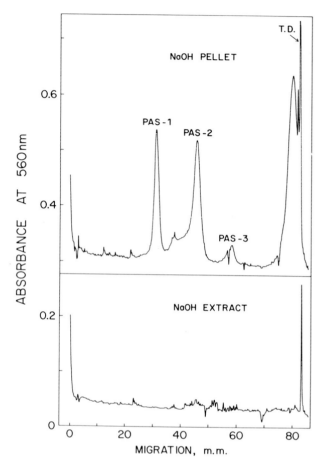

Fig. 9. NaOH extraction: PAS stain. Gels prepared as in Fig. 8 were stained with PAS and scanned (see Fig. 3 legend).

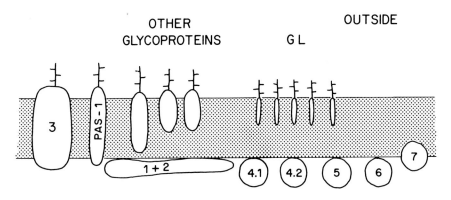

Fig. 10. A possible arrangement for the major membrane polypeptides. The shaded zone is the hydrophobic
permeability barrier. The glycosylated proteins and glycolipids (GL) face outwards and are deeply anchored in the
membrane; the remaining proteins are more loosely bound and
are accessible only at the cytoplasmic face.

III

MEMBRANE ASSEMBLY AND TURNOVER

RECONSTITUTION OF OXIDATIVE PHOSPHORYLATION AND VESICLES WITH RESPIRATORY CONTROL

E. Racker

Section of Biochemistry & Molecular Biology
Cornell University, Ithaca, New York 14850

ABSTRACT. The third site of oxidative phosphorylation was reconstituted by combining phospholipids, cytochrome oxidase, cytochrome c and preparations of the oligomycin-sensitive ATPase which was solubilized with cholate. After removal of cholate by dialysis, vesicles were formed from which external cytochrome c was removed by centrifiguration. After addition of coupling factors the reconstituted vesicles catalyzed oxidative phosphorylation with ascorbate-phenazine methosulfate as substrate.

Vesicles reconstituted without the oligomycin-sensitive ATPase oxidized ascorbate in the presence of cytochrome c at a slow rate which was accelerated 3 to 6 fold on addition of uncouplers and ionophorous agents such as nigericin and valinomycin. This respiratory control was abolished if the oligomycin-sensitive ATPase preparation had been incorporated during reconstitution. In these vesicles respiration was partially inhibited by oligomycin. In line with Mitchell's chemiosmotic hypothesis we propose that the reversible ATPase serves as a proton channel. Direct evidence in favor of this formulation is presented.

Two types of investigators have been working on membrane problems for the past 20 years. The first type knew that they were working on membranes, the second didn't. The first type were dedicated membranologists primarily interested in the function and structure of membranes. The second type were biochemists who happened to work on enzyme systems associated with membranes. In fact, they would have gladly disposed of the membrane if the pathways could have been solubilized. Since I belong to the second type of investigator, I doubt that I will ever qualify as a card-carrying membranologist.

The oxidation chain located in the inner mitochondrial membrane has been thoroughly studied for over 40 years. David Keilins' pioneering studies on the characteristic absorption spectra of the cytochromes opened the way to the isolation of these catalysts. Where there is an assay, there is a will and there is a way. Numerous investigators started to purify electron transport components with characteristic oxidation reduction spectra. It took many years, however, before biochemists realized that an enzyme derived from a membrane is not an ordinary protein. It is a multi-faced protein with several obligations. In addition to its catalytic activity, it must be able to attach to the membrane in such a manner that it can communicate with the biochemical pathway in which it participates. King, working in Keilin's laboratory, isolated a reconstitutively active succinate dehydrogenase by including succinate during isolation. He observed (1) that succinate dehydrogenase prepared in the absence of succinate, though enzymatically active with artificial electron acceptors, did not interact with the membrane. We have made a similar observation with cytochrome b which lost reconstitutive activity when fractionated in the absence of succinate (2). The reconstitutively active preparation of cytochrome b was rather impure compared to the highly purified preparations described in the literature. However, the latter were completely inactive in reconstitutions. I have urged my students in the past not to waste clean thinking on dirty enzymes, but in recent years I had to compromise. Now I tell them that I would rather have my enzyme dirty than impotent, incapable of coupling with his sisters under the membrane.

About five years ago, we initiated our work on the

separation of individual components of the respiratory chain using only one assay: the reconstitution of the oxidation chain (2). By relatively gentle procedures we separated individual components of this chain and reconstituted an active succinoxidase. One divident from these tedious labors was the discovery of a new oxidation factor which is required for the oxidation of succinate (3). But our main objective was to reconstitute oxidative phosphorylation. The succinoxidase complex wasted all energy of oxidation in heat. It showed no signs of energy conservation even when supplied with coupling factors and oligomycin-sensitive ATPase preparations.

In view of the well known complexities of the cytochrome b to cytochrome c_1 segment we attempted to reconstitute the third site of oxidative phosphorylation. We were encouraged in these efforts by the recent finding (4) that an insoluble protein fraction of the inner mitochondrial membrane which contained CF_o-F_1, the oligomycin-sensitive ATPase, could be reconstituted with phospholipids and coupling factors to yield a vesicular preparation capable of catalyzing an oligomycin-and uncoupler-sensitive $^{32}P_i$-ATP exchange. The key to this success was the observation that the cholate, which had been used to solubilize the hydrophobic proteins of CF_o, had to be removed very slowly to allow the formation of vesicles catalyzing the exchange.

The first experiment, in which cytochrome oxidase and cytochrome c were included during reconstitution of these vesicles, was a failure. Only after removal of external cytochrome c by centrifugation through a sucrose layer in the presence of bovine serum albumin, did we obtain a phosphorylating preparation. A recent experiment is shown in Table I. The P:O ratio with ascorbate-phenazine methosulfate (PMS) is 0.25, somewhat lower than we usually observe with submitochondrial particles, but considerably higher than we reported in our first communication (5). Phosphorylation was completely dependent on coupling factors (which were added to the vesicles 20 minutes prior to assay) and was abolished by uncouplers. In the presence of KCN, which inhibited respiration, there was a small incorporation of ^{32}P into ATP due to exchange, which was corrected for. This could be minimized by addition of further excess of hexokinase.

The phospholipids as well as the hydrophobic proteins used in these experiments were crude mixtures. We first explored the phospholipid requirement of the $^{32}P_i$-ATP exchange (6) and found that phosphatidyl ethanolamine and phosphatidyl choline together substituted for the crude mixture of soybean phospholipids. In fact, synthetic preparations of phospholipids were active provided they contained side chains with double bonds. Again, both phosphatidyl choline and phosphatidyl ethanolamine were required. Vesicles catalyzing oxidative phosphorylation appear to be more demanding since with crude phospholipids higher P:O ratios were obtained than with purified preparations. Here too, both phosphatidyl choline and phosphatidyl ethanolamine were required for oxidative phosphorylation.

The purification of the hydrophobic proteins of CF_o appears to be a formidable task. Many conventional methods of protein fractionation are unsuitable and problems of instability are plaguing us. However, we are making slow progress and we know now that more than one fraction is required for reconstitution.

Meanwhile, we made some rather startling observations with vesicles that were reconstituted with cytochrome oxidase and phospholipids but without hydrophobic protein and cytochrome c (7). As shown in Table II, these vesicles oxidized reduced cytochrome c added from the outside at a very rapid rate provided unconplers of oxidative phosphorylation or ionophorous agents were added. In fact, the rate of oxidation with these vesicles is equal to or better than any of the activation procedures with detergents used for cytochrome oxidase assay. In the absence of an uncoupler the rate of oxidation was diminished five to six fold. Thus these vesicles exhibit a form of respiratory control which is responsive to the same reagents that release respiratory control in mitochondria. In fact, the respiratory control of these vesicles oxidizing ascorbate is much better than that of mitochondria. The vesicles resemble mitochondria in one other respect. During oxidation of naphtoquinone sulfonate or ascorbate in the presence of cytochrome c, K^+ and valinomycin, the medium became acidic and K^+ was taken up. The $O:H^+:K^+$ ratios were 1:2:2 and all ion movements were abolished by uncouplers. The same ion movements were observed during oxidation of naphtoquinone sulfonate in mitochondria. On the other hand, with cytochrome c inside, the vesicles resembled submitochondrial particles in that the proton moved from the outside to the inside (8).

While there is general agreement that cytochrome c is located on the outer surface of mitochondria and on the inner surface of the membrane in submitochondrial particles, the location of cytochrome a and a_3 has been controversial. By exposing mitochondria and submitochondrial particles to ^{35}S-labeled diazobenzene sulfonate, we have recently obtained evidence indicating that cytochrome oxidase is transmembranous with only a small fraction exposed on each side of the membrane (9). Since cytochrome a interacts with cytochrome c it follows that it must be localized at the same surface, i.e., on the outside of mitochondria and on the inside in submitochondrial particles. Accordingly, cytochrome a_3 should be on the matrix side in mitochondria and on the outer surface in submitochondrial particles. All data seem consistent with this formulation except for one complicating feature: Reduced cytochrome c, added from the outside to submitochondrial particles, is rapidly oxidized. Since there is good evidence (10) that cytochrome c cannot penetrate from one side of the membrane to the other, this observation must mean that in submitochondrial particles cytochrome a is either "scrambled", i.e., present on both sides of the membrane, or that cytochrome a_3 can also interact with cytochrome c. In view of the observation that the K_m for external cytochrome c is different in mitochondria and submitochondrial particles (9) we favor the latter possibility. It is important to note that in submitochondrial particles the oxidation of added reduced cytochrome c is not associated with phosphorylation while the pathway via internal cytochrome c gives rise to ATP generation (see Table I). To reach internal cytochrome c a lipid soluble carrier (e.g., PMS) must be added. It therefore appears that a vectorial organization of cytochrome oxidase is required for functional oxidative phosphorylation.

It is obvious that a vectorial organization of cytochrome oxidase in one direction only could not be achieved by our crude method of reconstitution. We had to impart an asymmetry to the membrane by placing cytochrome c either

101

on the inside (by including it during reconstitution) or on the outside. Reconstituted vesicles with internal cytochrome c catalyzed oxidative phosphorylation provided we added required coupling factors from the outside. With external cytochrome c coupling factors would be required on the inside, and translocators for phosphate and adenine nucleotides, which are present in the mitochondrial membrane, would have to be incorporated as well. Although we hope to attack this problem eventually, it is not likely to be a simple task.

Although cytochrome oxidase vesicles with external cytochrome c are incapable of phosphorylation, they have served as a very valuable tool because of their high degree of respiratory control. Mitchell (11) proposed that during oxidation protons are conducted to one side of the membrane, electrons to the other, thereby building up a membrane potential and a proton gradient which inhibit further oxidation. Valinomycin (which collapses the membrane potential by conducting K^+ through the membrane) and uncouplers (which make the membrane permeable to protons) release respiratory control. Mitchell furthermore proposed that the transmembranous portion of the reversible ATPase plays the role of a proton channel which is blocked by energy transfer inhibitors such as oligomycin (11). This interpretation accounts for the dual role of oligomycin which inhibits the generation of ATP, yet permits the conservation of energy in "loosely coupled" submitochondrial particles. In such particles the reversible ATPase catalyzes in the presence of phosphate and ADP the formation of ATP but serves as a proton leak if the coupling mechanism is not operating.

This formulation suggested a simple experiment. The incorporation of the transmembranous portion of the reversible ATPase into cytochrome oxidase vesicles should eliminate respiratory control. This was found to be the case. As shown in Table III inclusion of increasing amounts of the hydrophobic protein fraction (CF_0F_1) during reconstitution resulted in a marked acceleration of the rate of oxidation, approaching the rate observed after addition of an uncoupler. As a control we included a crude mitochondrial extract which contained a large number of soluble proteins. There was little or no effect on respiration.

102

However, when the extract was heated and proteins thus rendered hydrophobic, they increased the rate of oxidation if included during reconstitution of the cytochrome oxidase vesicles. To avoid such non-specific effects we had to design an assay which differentiated between artifactual stimulation of respiration and the specific action of a proton channel. As shown in experiment 2 of Table III, the respiration in vesicles reconstituted in the presence of the coupling device was partially sensitive to rutamycin whereas that of control vesicles was insensitive. Mild heating of the CF_oF_1 preparation (5 min at 37°) completely eliminated the rutamycin inhibition, while a stimulation of respiration was still observed. If we correct for this non-specific stimulation of respiration, the inhibition of respiration by rutamycin is over 60%.

Table IV shows an analogous experiment with a hydrophobic protein fraction from chloroplasts which was isolated by the same procedure used for the preparation of CF_oF_1 from mitochondria. It can be seen that the chloroplasts fraction stimulated respiration although somewhat less effectively than the mitochondrial preparation. As expected, in view of the known resistance of chloroplasts to rutamycin, the respiration was not inhibited by rutamycin. However, DCCD which is an effective energy transfer inhibitor in chloroplasts (13) as well as in mitochondria (14) inhibited the respiration of cytochrome oxidase vesicles reconstituted with hydrophobic proteins from either mitochondria or chloroplasts. In both cases, the inhibitory effect of DCCD was lost when the hydrophobic protein was exposed to 37° for 5 minutes prior to reconstitution. The % inhibitions of respiration were small (24%) if calculated without correction of the respiration with heated hydrophobic proteins, but were appreciable (70%) after this correction.

We had to establish whether our preparation of CF_o in fact serves as a proton channel. To test this, the indicator phenol red (10 mM) was included during reconstitution of the cytochrome oxidase vesicles. After prolonged dialysis the proton conductivity of the red vesicles were tested in a dual wavelength Aminco spectrophotometer. On addition of acid, an initial drop of absorption was seen at 560-590 caused by a small amount of residual phenol red on

103

the outside of the vesicles. A subsequent very sluggish decrease in absorption in the absence or presence of valinomycin was indicative of the low permeability of these vesicles to protons. On addition of an uncoupler the red color disappeared instantaniously. If reconstitution of the cytochrome oxidase vesicles was performed in the presence of hydrophobic proteins there was a greater permeability to protons as indicated by the more rapid decrease in absorption on addition of a proton pulse as shown in Table V. Moreover, there was a distinct inhibition of proton translocation on addition of rutamycin.

I would like to discuss briefly the significance of these findings in the light of current hypotheses of oxidative phosphorylation. You are probably aware of the fact that Mitchell's chemiosmotic hypothesis is not very popular with the establishment in the field of oxidative phosphorylation. I have remained neutral, following my own advice, that it is more fun to work with 2 hypotheses than with one. However, our recent observations are much more readily interpreted in terms of Mitchell's formulations.

As shown in Fig. 1, the reconstituted vesicles with cytochrome c inside represent a model for submitochondrial particles. Electrons move from the inside surface of the membrane to the outside surface, where cytochrome a_3 interacts with oxygen. Protons are released inside from reduced PMS and are taken up from the medium to form water during the oxidation of reduced cytochrome a_3. Protons formed inside move via the proton channel of CF_0-F_1 leading to the formation of ATP.

With cytochrome c on the outside, the reconstituted vesicles represent a model for mitochondria. Ascorbate or hydroquinone (which Dr. P. Hinkle uses instead of ascorbate because it is a 2 hydrogen donor) are oxidized slowly because a membrane potential and pH gradient (OH^- on the inside) is building up. A collapse of the potential and of the gradient can be achieved in several ways. A combination of valinomycin-K^+ (which collapses the potential) and nigericin or FCCP (which collapses the gradient) is optimal but either FCCP or nigericin alone are quite effective. CF_0 which is proposed to be the natural proton channel also eliminates respiratory control.

How do we explain respiratory control and ion movements in the reconstituted cytochrome oxidase vesicles by the chemical hypothesis? If the ion pumps invoked by the chemical hypothesis exist, they must be present in the highly purified preparations of cytochrome oxidase. $A \sim X$ is the first high-energy intermediate of the chemical hypothesis formed between a member of the respiratory chain and a member (X) of the coupling device. To explain respiratory control in the reconstituted vesicles X must therefore be present in the cytochrome oxidase preparation. We have shown, however, (4) that a coupling device capable of $^{32}P_i$-ATP exchange, which according to the chemical hypothesis requires the formation of $X \sim Y$, can be reconstituted without cytochrome oxidase. Is X then an accidential contaminant of the cytochrome oxidase preparation? We have tested several cytochrome oxidase preparations isolated by different procedures and found that they all yielded vesicles with respiratory control.

If the proton and K^+ movements in mitochondria are caused by separate pumps as postulated by the chemical hypothesis why are the $O:H^+:K^+$ ratios in the reconstituted vesicles 1:2:2? Why do we require a vectorial organization of the respiratory chain for oxidative phosphorylation?

In Mitchell's formulation on the other hand the respiratory chain and the reversible ATPase are separate transmembranous entities, thus consistent with the findings that ion movements take place on addition of ATP to a coupling device which is depleted of respiratory activity (4) or during respiration of a preparation free of coupling factors (7).

In view of these considerations we have adopted for the time being those formulations of the chemiosmotic hypothesis which allow us to ask some more specific questions about the nature of the coupling mechanism. If the reversible ATPase is a proton channel and responsible for the formation of ATP what is the specific function of the coupling factors F_1 and CF_1? Ryrie and Jagendorf (15) have reported the incorporation of tritium into the chloroplast coupling factor CF_1 on exposure of chloroplasts to light. Kagawa (16) has recently observed tritium incorporation into F_1 in submitochondrial particles during succinate

105

oxidation. Perhaps these coupling factors serve as a
trapping device for the protons emerging from the channel
and utilize the protonmotive force for the generation of
ATP. In line with this proposition are the findings that
F_1 is required for respiratory control (17) and for proton
translocation in submitochondrial particles (8).

I know that there will be electron microscopists in
the audience who will want to know what the reconstituted
particles look like. Fig. 2 shows the cytochrome oxidase
vesicles with hydrophobic proteins (A) and without (B).
The latter are more homogenous in size and show less
structural details.

Finally, I should like to say that there is still a
great deal of work ahead of us. It was shown by Schatz
et al (18) that the heme of cytochrome oxidase is associated with small molecular subunits. The function of the
larger subunits is unknown. An elucidation of their function is essential to our understanding of how cytochrome
oxidase operates in reconstituted vesicles. Fractionation
of the hydrophobic proteins is high on our priority list.
We hope to obtain better inhibitions of respiration and
proton translocation by rutamycin or DCCD when we achieve
purification of the specific proton channel. Reconstitution of the second and first site will be attempted, when
resolution of these segments into individual components
has advanced further.

Although it is clear that we have a long way to go I
believe that the reconstituted vesicles catalyzing the
third site of oxidative phosphorylation and exhibiting
respiratory control will give us the badly needed acceleration to make more rapid progress.

REFERENCES

1. King. T. E., J. Biol. Chem., 238, 4037 (1963).
2. Yamashita, S., and Racker, E., J. Biol. Chem., 244,
   1220 (1969).
3. Nishibayashi-Yamashita, H., Cunningham, C., and
   Racker, E., J. Biol. Chem., 247, 698 (1972).
4. Kagawa, Y., and Racker, E., J. Biol. Chem., 246, 5477
   (1971).

5. Racker, E., and Kandrach, A., J. Biol. Chem., 246, 7069 (1971).
6. Kagawa, Y., Kandrach, A. and Racker, E., in preparation.
7. Hinkle, P. C., Kim, Jung Ja, and Racker, E., J. Biol. Chem., 247, 1338 (1972).
8. Hinkle, P. C., and Horstman, L. L., J. Biol. Chem., 246, 6024 (1971).
9. Schneider, D. L., Kagawa, Y., and Racker, E., J. Biol. Chem., in press.
10. Racker, E., Burstein, C., Loyter, A., and Christiansen, R. O. in Electron Transport and Energy Conservation (J.M. Tager, S. Papa, E. Quagliariello, E. C. Slater, eds.) Adriatica Editrice, Bari, p. 235, 1970.
11. Mitchell, P., Biol. Rev., 41, 445 (1966).
12. Mitchell, P., Fed. Proc. 26, 1370 (1967).
13. McCarty, R. E., and Racker, E., J. Biol. Chem., 242, 3435 (1967).
14. Racker, E., and Horstman, L. L., J. Biol. Chem., 242, 2547 (1967).
15. Ryrie, I. J., and Jagendorf, A. T., J. Biol. Chem., 246, 3771 (1971).
16. Kagawa, Y., unpublished observations.
17. Cockrell. R. S., and Racker, E., Biochem. Biophys. Res. Commun., 35, 414 (1969).
18. Schatz, G., Groot, G. S. P., Mason, T., Rouslin, W., Wharton, D. C. and Saltzgaber, J., Fed. Proc., Vol. 31, 1 (1972).

ACKNOWLEDGMENTS

This work was supported by United States Public Health Service Grant CA-08964 from the National Cancer Institute. We wish to thank Dr. C. Carmeli for the preparation of hydrophobic proteins from spinach chloroplasts and Dr. J. Telford for taking the electron micrographs.

TABLE I

Reconstitution of Oxidative Phosphorylation
at the Third Site

Experimental conditions were as described previously (5)
except that one half the amount of cytochrome oxidase was
used and 150 mM NaCl was included during the dialysis.

Additions	P:O ratio[*]
Complete system	0.25
" " minus coupling factors	0.014
" " plus FCCP	0.01
SMP	0.39

[*]Phosphorylation rates were corrected for the small rate
of $^{32}P_i$-ATP exchange observed in the presence of KCN.

TABLE II

Effect of Uncoupling and Ionophorous Agents
on the Oxidation Rate of Reconstituted Cytochrome Oxidase
Vesicles

Cytochrome oxidase vesicles were reconstituted as described
(7) except dialysis was carried out against 200 volumes of
50 mM KCl - 5 mM Tricine, pH 8.0. Assay conditions were
as described (7) except that 50 mM K phosphate buffer pH
6.7, 1.5 mg of cytochrome c and 50 mM K ascorbate were
used instead of the described lower concentrations, which
were sufficient at the more alkaline pH used in the earlier
experiments

Additions	n atoms oxygen/min
Cytochrome oxidase vesicles	39
+ valinomycin (2 μg)	72
+ FCCP[*] (2 μM)	169
+ nigericin (0.4 μg)	110
+ valinomycin + FCCP	255
+ valinomycin + nigericin	235

[*]FCCP (trifluoromethoxy carbonyl cyanide phenylhydra-
zone) was a gift from Dr. P. Heytler.

TABLE III

Effect of Hydrophobic Protein Fraction on Respiration
of Reconstituted Cytochrome Oxidase Vesicles

A suspension of 2.7% soybean phospholipid was sonicated in
the presence of 2% cholate as described previously (5).
To 0.3 ml of the clarified preparation 600 µg of cytochrome
oxidase and the indicated amounts of hydrophobic proteins
(4) expressed in µg per mg phospholipids were added as indicated. The final volume was adjusted to 0.5 ml with
distilled water and the mixture dialyzed as described (5)
except that dialysis was performed against 2 changes of
500 volumes of 5 mM Tricine pH 8.0 containing 50 mM NaCl.
Oxygen uptake was measured in an oxygraph in a final volume
of 1.1 ml in the presence of 50 mM K phosphate, pH 7.5,
50 mM Tris-ascorbate, 600 µg of cytochrome c, and 10 µl of
the reconstituted vesicles. Corrections were made for the
low rate of ferrocytochrome c autooxidation (about 14 n
moles/min). One µl of an alcoholic 0.4% solution of rutamycin and 2 µl of an alcoholic 0.02 M solution of 1799
were added where indicated.

Experiment 1	Oxygen uptake					
Additions	0	60	100	200	300	500µg HP*
	n atoms oxygen per min					
None	32	42	54	83	100	116
1799**	123	115	120	123	127	123
Experiment 2	0	500 µg HP		500 µg heated HP*		
None	40	182		93		
Rutamycin	38	126		93		
1799	182	176		142		

*HP = hydrophobic protein from mitochondria (5). The
amounts are expressed in terms of µg protein mg of
phospholipid.

**1799 (a 2:1 adduct of hexafluoroacetone and acetone) was
a gift from Dr. P. Heytler.

TABLE IV

Effect of DCCD* and Rutamycin on Cytochrome Oxidase
Vesicles Reconstituted with Hydrophobic Proteins
from Chloroplasts or Mitochondria

Experimental conditions were as described in Table III except that 100 μg of mitochondrial hydrophobic protein and
115 μg of chloroplast hydrophobic protein per mg of phospholipid was used.

Additions	Preparation of cytochrome oxidase vesicles		
	without CHP**	with CHP	with CHP (5'at 37°)
Exp. I			
None	24	56	27
DCCD*	24	36	27
1799	110	103	96
Exp. II			
None	31	54	
Rutamycin	31	54	
1799	115	104	
	without HP***	with HP	with HP (5'at 37°)
Exp. I			
None	25	81	49
Rutamycin	25	61	49
1799	116	122	127
Exp. II			
None	24	74	
DC CD	24	56	
1799	110	108	

*DCCD = N,N'-dicyclohexylcarbodiimide

**CHP = hydrophobic protein fraction from chloroplasts.

***HP = hydrophobic protein fraction from mitochondria.

TABLE V

Proton Translocation in Reconstituted Vesicles

The vesicles were reconstituted with phospholipids and
cytochrome oxidase as described previously (7) except that
10 mM of phenol red was included. Dialysis was carried out
against large volumes of 5 mM Tricine (pH 8.0) - 50 mM KCl
until no phenol red appeared in the outside fluid. A
sample corresponding to 120 μg of cytochrome oxidase was
placed in a dual wavelength spectrophotometer and recorded
at 560-590 nm in a final volume of 3 ml containing 5 mM
Tricine pH 8.0 - 50 mM KCl. Addition of 6 μl of 0.5 N HCl
was followed by valinomycin (2 μl of 0.1 μg per ml), rutamycin (2 μl of 4 μg per ml) and 1799 (1 μl of 0.01 M solution). The changes are expressed in arbitrary units as
cm per minute.

| Additions | Reconstituted vesicles | |
	−HP	+HP
HCl	< 0.1	0.2
Valinomycin	0.8	2.6
Rutamycin	0.8	2.0
1799	> 30	> 30

RECONSTITUTED VESICLES

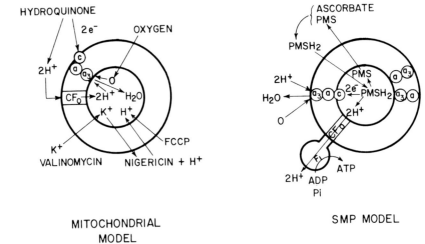

Fig. 1. Oxidative phosphorylation and respiratory control in reconstituted vesicles.

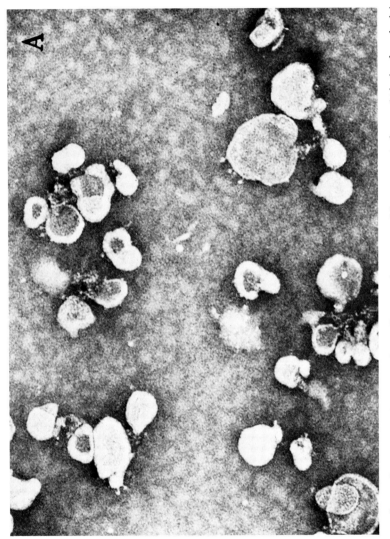

Fig. 2. Electron micrographs of cytochrome oxidase vesicles with hydrophobic proteins (A) and without hydrophobic proteins.

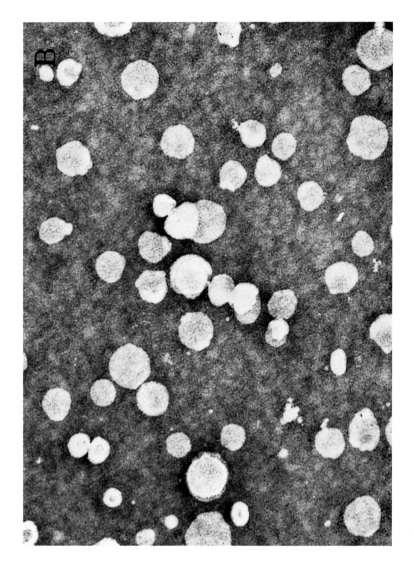

Fig. 2 (B). Negative stains with phosphotungstate. Final magnification X 120,000.

TURNOVER OF MEMBRANE PROTEINS OF ANIMAL CELLS

Robert T. Schimke and Peter J. Dehlinger*

Department of Pharmacology
Stanford University School of Medicine
Stanford, California 94305

ABSTRACT. Both the lipid and protein constituents of endoplasmic reticulum are turning over rapidly. Such a rapid
flux raises fundamental questions concerning how membranes
are synthesized (or assembled), how they are degraded, and
how the constituents of membranes change in response to the
administration of pharmacological agents.
    The model of membrane eventually proposed is one in
which protein constituents continually associate and dissociate from the membrane, and only in the dissociated state
are they subject to degradation by normal intracellular degradative processes. In addition the rates at which individual proteins are synthesized and assembled into the membrane can be modulated by drugs. Thus membranes, including
plasma membrane, are neither synthesized as a unit, nor degraded as a unit, but represent a dynamic mosaic of macromolecular constituents.

INTRODUCTION

Various investigations have demonstrated amply that
turnover of cytoplasmic proteins in animal tissues is continual and extensive (1). Thus a given protein content,
i.e. enzyme level, results from a continually changing balance between its synthesis and its degradation, either of
which can be altered by hormonal, nutritional, developmental, or genetic variables (see review of Schimke and Doyle
(1)). More recent has been the realization that the compo-

---

* Present address: Institute of Molecular Biology,
University of Oregon, Eugene, Oregon 97403.

nents of internal membranes of rat liver, i.e. the endoplasmic reticulum, including both phospholipids and proteins are also being continually synthesized and degraded, and hence the apparent structured continuity of membranes must be viewed in the light of this continual flux. Such rapid flux raises fundamental questions concerning how membranes are synthesized (or assembled), how they are degraded, and how the membrane constituents change in response to various exogenous variables, including the administration of pharmacological agents.

In this paper we shall compare and contrast the properties of turnover of cytoplasmic and membrane-associated proteins and propose a model in which the protein constituents continually associate and dissociate from the membrane, and only in the dissociated state are they subject to degradation by normal intracellular degradative processes. Thus membranes, including plasma membrane, are neither synthesized as a unit nor degraded as a unit, but represent a dynamic mosaic of macromolecular constituents.

PROPERTIES OF TURNOVER OF CYTOPLASMIC AND MEMBRANE ASSOCIATED PROTEINS

Studies from a number of laboratories have established the following general properties of degradation of cytoplasmic proteins of rat liver.

1. *Turnover is extensive*. Studies of Swick (2), Buchanan (3), and Schimke (4) have indicated that essentially all proteins of rat liver take part in the continual replacement process. For instance, Buchanan (3) estimated that approximately 70% of rat liver protein was replaced every 4 to 5 days from a dietary source.

2. *Turnover is largely intracellular*. The life span of hepatic cells is of the order of 160 to 140 days (3,5,6), and hence cell replacement is precluded as the explanation for the extensive protein turnover that occurs in 4 to 5 days.

3. *There is a marked heterogeneity of rates of replacement of different proteins*. Table 1 gives a representative listing of rates of degradation of various cytoplasmic proteins (enzymes). More extensive listings of various proteins are given by Schimke and Doyle (1) and Rechcigl (7).

Remarkable is the wide diversity of half-lives, ranging from 11 minutes for ornithine decarboxylase to 60 days for LDH$_5$. There is no necessary relationship between half-lives and metabolic functions of the enzymes. Thus glycokinase and LDH$_5$, both involved in carbohydrate metabolism, vary in half-lives by 30 hours and 16 days. Likewise tyrosine amino-transferase and arginase, both involved in amino acid catabolism, have half-lives of 1.5 hours and 5 days.

4. *There is a correlation between the* in vivo *rate of degradation and susceptibility to proteolytic attack.*
This is depicted in Figure 1, in which cytoplasmic proteins have been double labeled, using single intraperitoneal administration of ^{3}H-leucine and ^{14}C-leucine separated in time by 5 days, where the animals were killed 6 hours after the administration of the ^{3}H-leucine. In this technique proteins that turn over rapidly have high ^{3}H/^{14}C ratios. (See Arias *et al.* (21) for discussion of this technique.) The mean ratio of the double labeled proteins was initially 4.8. Digestion of such double labeled cytoplasmic proteins by pronase preferentially released to a trichloroacetic soluble form amino acids with a high ^{3}H/^{14}C ratio. As digestion proceeds towards 100% solubilization, the ratios decrease progressively from initial values of 7.0 such that when 60% of the protein has been solubilized the ratio has decreased to 4.6, and the remaining undigested protein has a ratio of 4.4. That the specificity for attack resides with the structure of the cytoplasmic proteins is indicated by the control experiment in which the proteins were first denatured (8 M urea, and blocking of sulfhydryl groups). Proteolysis was more rapid, in keeping with the concept that denatured proteins are better substrates for proteases (23). More importantly, the radioactivity released with time showed no discrimination of ^{3}H/^{14}C ratios, i.e. the ratio remained 4.5 at all times. These studies suggest that the susceptibility of the protein molecule to proteolytic attack is one parameter contributing to the mechanism of the heterogeneity of turnover of cytoplasmic proteins.

5. *There is a correlation between the size of a protein and its rate of degradation* in vivo. Figure 2 depicts this general phenomenon for the soluble fraction from rat liver as displayed on Sephadex G-200 columns. This correlation was found whether the proteins were fractionated as multi-

meric proteins (in the absence of SDS, Figure 2A, or in its presence, Figure 2B). More recently Dice and Schimke have found this same correlation for proteins of rat liver ribosomes (25), as well as with cytoplasmic proteins from kidney, brain, and testis (unpublished observation, Dice and Schimke). Such studies have led these workers to propose that the correlation of size and rate of degradation is based on the overall greater chance of a larger protein being "hit" by a protase, producing an initial rate-limiting peptide bond cleavage. Since the relative rate of degradation is of the same range of magnitude for the dissociated subunits as for the multimeric proteins (i.e. $^3H/^{14}C$ ratios are similar in Figures 2A and 2B), it was further suggested that proteins were degraded in a dissociated state, whether the proteins are "cytoplasmic" or associated with ribosomes (24).

There are many similarities as well as certain dissimilarities between degradation of membrane-associated and cytoplasmic proteins:

1. *There is a marked heterogeneity of turnover of membrane-associated proteins.* Table 1 also lists the calculated half-lives of a number of proteins (enzymes) associated with the endoplasmic reticulum. The half-lives of these proteins vary from 2 hours for hydroxymethylglutaryl CoA reductase to 16 days for NAD glycohydrolase. The differences between the half-lives of NADPH cytochrome c reductase (60 hours) and cytochrome b_5 (80 hours) has also been found by Arias *et al.* (21) using the double isotope technique described above.

2. *There is a correlation between the size of the proteins of the membrane and the relative rate of degradation* in vivo. Dehlinger and Schimke have further analyzed the question of the heterogeneity of turnover rates of membrane proteins using the double isotope technique (26). By this technique, if all proteins are turning over at the same rate, the $^3H/^{14}C$ ratios of the proteins should be similar, whereas if the rates differ, the $^3H/^{14}C$ ratios will vary. Figure 3 displays that fraction of the smooth endoplasmic reticulum fraction that is insoluble in 1% Triton X-100 and which has been subsequently solubilized in SDS and subjected to electrophoresis on SDS-acrylamide gels. Details are pre-

sented in the figure legend and in reference (26). In this fraction there are a large number of protein species with a major band(s) migrating with an apparent molecular weight of 50,000 daltons. Figure 4 indicates a control experiment in which the ^{3}H and ^{14}C isotopic amino acids were administered 4 hours prior to killing of the animals and subsequent preparation of the smooth endoplasmic reticulum fraction. This experiment gives an indication of the counts CPM determined, and the limits of the ^{3}H/^{14}C ratios in such a control. By the end of 4 hours the specific radioactivity of the smooth and rough endoplasmic reticulum fractions has approached the same ralue and has reached a plateau. Therefore, we conclude that the radioactivity present in this fraction does not constitute plasma proteins trapped in the membrane. In addition there is essentially no adsorption of cytoplasmic proteins during preparation, since the addition of unlabeled membrane fractions to labeled cytoplasm results in at most 1% contamination of the final washed membrane fraction (Dehlinger, unpublished observations). Figure 5 gives the results from the experiment in which the two isotopes were separated in time by 4 days. Two points are of significance. Firstly, as indicated by the variation in ^{3}H/^{14}C ratios, there is a marked heterogeneity of rates of degradation of the proteins. Perhaps more significant is the finding that there is a general correlation between the size of the protein (subunit) and the relative rate of degradation. This finding was also made for the Triton X-100 soluble fraction (26). In addition, as shown in Figure 6, this general correlation also exists with proteins of the plasma membrane. Thus, the general correlation between subunit size and relative rate of degradation exists with a number of proteins, both cytoplasmic and those associated with organelles, including membranes and ribosomes.

In contrast to cytoplasmic proteins (Figure 1), the digestion of endoplasmic reticulum by pronase does not show a correlation between rate of turnover *in vivo* and susceptibility to proteolysis. As shown in Figure 7, using conditions similar to those employed for cytoplasmic proteins (Figure 1), there is less susceptibility to degradation, and if anything, there is an inverse correlation, since the first radioactivity has a low ^{3}H/^{14}C ratio in contrast to the finding of Figure 1. In addition there is no indication of endogenous proteolysis as indicated by absence of solubilization of radioactivity in the absence of added

R. T. SCHIMKE AND P. J. DEHLINGER

protease. Furthermore, there is no alteration in the electrophoretic pattern of the SDS solubilized proteins following incubation of the membrane fraction alone.

SYNTHESIS OF MEMBRANE ASSOCIATED PROTEINS

The studies from other laboratories referred to above
as well as our own experimental results demonstrate that
membrane proteins are turning over at different rates.
There is now ample evidence that there are marked variations
in the rates of synthesis of various membrane proteins.
Most recent are the findings of Higgins and Rudney who have
shown that the rate of synthesis of hydroxymethylglutaryl
CoA reductase varies markedly during a 24 hour period,
thereby accounting for the alterations in activity of this
membrane-bound enzyme activity (20). Other evidence comes
from studies on the effect of phenobarbital on the so-
called mixed function oxidase activity of endoplasmic reticulum. Studies by Arias *et al.* (21), and Omura and Kuriyama
(27), and Jick and Shuster (28) have shown that phenobarbital administration increases the rate of synthesis of NADH
cytochrome c reductase some 2 to 4-fold without altering
the rate of synthesis of cytochrome b_5. More recently
Dehlinger and Schimke have analyzed this question more
generally using the fractionation and solubilization procedures employed in the studies of Figure 5 (29). In the
result shown in Figure 8, the smooth endoplasmic fraction
from control animals labeled *in vivo* with ^{14}C-leucine was
mixed with a comparable fraction from an animal treated
16 hours with phenobarbital and labeled with ^{3}H-leucine,
and subsequently separated into Triton X-100 soluble and
insoluble fractions. If phenobarbital increased the rate
of synthesis of all membrane-associated proteins to the
same extent, then the ^{3}H/^{14}C ratios should be similar in
all proteins. It is evident that there is preferential synthesis (or insertion) of certain proteins, and most strikingly those of the 50,000 dalton protein band(s). We have
tentatively identified this band, or a portion of it, as
P450 on the basis of the finding that P420, the inactive
purification product of P450 (30), likewise demonstrates
the same preferential increased rate of synthesis and also
migrates on SDS acrylamide gels in the 50,00 molecular
weight region.
Such studies indicate, then, that just as the rate of

turnover of membrane proteins varies markedly, the rates of synthesis of various membrane proteins are not constant but can be altered by various factors, including drug administration (NADPH cytochrome c reductase and P450) or time of day (hydroxymethylglutaryl CoA reductase).

A MODEL OF MEMBRANE BIOGENESIS AND TURNOVER

The general properties of membrane synthesis and turnover will now be discussed in terms of several alternative models:

1. *Membrane is synthesized (assembled) at discrete points on a membrane and are degraded at (other) discrete points.* This model suggests that the membrane somehow moves or flows within the cell from the point of genesis to the point of its destruction. If the number of growing points was small, then the kinetics of decay of isotope from pulse-labeled membrane proteins should approximate those of a life span, i.e. the isotope would be incorporated into the membrane fraction, remain a finite time, and then rapidly be lost from the membrane fraction. However, in all cases studied, the decay of labeled and membrane proteins follows (pseudo) first-order kinetics (18,21,27). Such a result implies that the process of destruction of the membrane is a random process. That membrane proteins are not inserted at a few discrete growing points is supported by the studies of Leskes *et al.* (31), showing that during the accumulation of glucose 6-phosphatase activity in rat liver endoplasmic reticulum following birth, activity detected by electron microscope histochemistry appears randomly throughout the endoplasmic reticulum, suggesting that enzyme is inserted into the membrane at a large number of places, perhaps randomly. However, it should be cautioned that the random process need not be the insertion of proteins into the membrane or subsequent random degradation of the entire segment of the membrane, if lateral movement or diffusion of proteins within the plane of the membrane is a rapid event, as proposed by Singer (32) and by Kornberg and McConnell (34). Thus, lateral diffusion could be sufficiently rapid to randomize proteins within the membrane and yet have discrete points of insertion and destruction.

2. *Membrane proteins are selectively inserted into a membrane and selectively degraded.* This model is consistent with the fact that the synthesis of individual membrane proteins, i.e. hydroxymethylglutarly CoA reductase, NADPH cytochrome C reductase, and P450 are not constant but vary with time of day (20) and administration of various drugs (21,27) (Figure 8).

Perhaps more important is the fact that membrane proteins are not degraded as a unit but rather have remarkedly different half-lives, varying from 2 hours to 16 days (Table 1). Thus the membrane is neither synthesized as a unit nor degraded as a unit.

Several mechanisms can be conceived for the heterogeneity of degradation rates for membrane proteins. One possibility is that the more "peripheral" proteins in the membrane, being more exposed, are subject to attack by a proteolytic process, whether cytoplasmic or lysosomal. The results of Figure 6 argue against such a simplistic explanation, since there is no correlation between rate of turnover *in vivo* and susceptibility to attack by exogenous proteases. Another possibility is that there is an endogenous protease activity in the membrane which results in selective destruction of membrane proteins, depending on the geometric relationship of the protease to other membrane proteins. This suggestion is based on the existence of protease activity in the red cell membrane (34). This possibility does not seem likely, since we have been unable to demonstrate protease activity in the washed endoplasmic reticulum fraction (Figure 6). Other explanations for the heterogeneity might involve specific modification of certain of the proteins, such as formylation, phosphorylation, adenylation, deamidation (35). Another possibility is that the specific protein is damaged by lipid peroxidation. Dehlinger and Schmike have suggested that such a mechanism may underlie the apparent rapid turnover of P450, i.e. the 50,000 dalton band of Figure 5. In addition these workers have also suggested that the effect of heme to inhibit the synthesis (insertion) of P450, may result from heme (or iron) catalyzed peroxidative damage to the membrane proteins, either resulting in rapid turnover or failure of normal membrane insertion (26).

Perhaps the most striking finding is the general correlation between the size of a protein and its relative rate of degradation *in vivo* (Figures 2, 5, 6). That this observation has been found for proteins associated with the membrane as well as ribosomal and cytoplasmic proteins, suggests to us that membrane proteins are subject to the same mechanism of turnover as cytoplasmic proteins. In the case of ribosomal proteins, Dice and Schimke have provided evidence for a significant pool of ribosomal proteins in the cytoplasm of rat liver (25). Warner has made a similar observation for ribosomal proteins from yeast (36). This finding plus a lack of correlation between rates of turn-over *in vivo* and susceptibility to proteolytic attack of membrane associated proteins (Figure 7), as opposed to the existence of this correlation with cytoplasmic proteins (Figure 1), leads us to the proposal that membrane-associ-ated proteins are subject to the same controls concerning degradation as are cytoplasmic proteins, and that they are degraded not in association with the membrane but rather in a dissociated state. This general concept is summarized in Figure 9.

We, therefore, propose that a membrane is in a dynamic flux of association and dissociation, and that the protein components are degraded only in the dissociated state. This concept is in keeping with the known rapid exchange of membrane phospholipids (37), association-dissociation phenomenon of complex organelles (ribosomes) (25), as well as more simple systems involving only proteins. Such a model implies a finite cytoplasmic pool of membrane pro-teins. It also raises the question of whether membrane-associated proteins are synthesized on ribosomes intimately bound to the membrane or synthesized on so-called free polysomes, with subsequent insertion of protein into the membrane dependent on the properties of the protein fol-lowing release from the ribosome, or perhaps as it is under-going folding during peptide chain elongation.

ACKNOWLEDGMENTS

This work was supported by Grant GM 14931 from the National Institutes of Health. P.J.D. is the recipient of National Institutes of Health Training Grant GM 0712.

REFERENCES

1. R.T. Schimke and D. Doyle, *Ann. Rev. Biochem. 39*, 929 (1970).
2. R.W. Swick, *J. Biol. Chem. 231*, 751 (1958).

3. D.L. Buchanan, *Arch. Biochem. Biophys. 94*, 500 (1961).
4. R.T. Schimke, *J. Biol. Chem. 239*, 3808 (1964).
5. R.W. Swick, A.L. Koch and D.T. Handa, *Arch. Biochem. Biophys. 63*, 226 (1956).
6. R.A. MacDonald, *Arch. Intern. Med. 107*, 335 (1961).
7. M. Rechcigl, Jr., *Enzyme Synthesis and Degradation in Mammalian Systems*, M. Rechcigl, Jr. (Ed.), University Park Press, Baltimore, 1971, p. 236.
8. D. Russell and S.H. Snyder, *Proc. Nat. Acad. Sci. U.S. 60*, 1420 (1968).
9. H.S. Marver, A. Collins, D.P. Tschudy and M. Rechcigl, Jr., *J. Biol. Chem. 241*, 4323 (1966).
10. R.W. Swick, A.K. Rexroth and J.L. Stange, *J. Biol. Chem. 243*, 3581 (1966).
11. V.E. Price, W.R. Sterling, V.A. Tarantola, R.W. Harley, Jr. and M. Rechcigl, Jr., *J. Biol. Chem. 237*, 3468 (1962).
12. F.T. Kenney, *Science 156*, 525 (1967).
13. R.T. Schimke, E.W. Sweeney and C.M. Berlin, *Biochem. Biophys. Res. Commun. 15*, 214 (1964).
14. H. Niemeyer, *Natl. Cancer Inst. Monograph 27*, 29 (1966).
15. H.L. Segal and Y.S. Kim, *Proc. Nat. Acad. Sci. U.S. 50*, 912 (1963).
16. P.J. Fritz and E.S. Vesell, *Proc. Nat. Acad. Sci. U.S. 62*, 558 (1969).
17. P.W. Majerus and E. Kilburn, *J. Biol. Chem. 244*, 6254 (1969).
18. T. Omura, P. Siekevitz and G.E. Palade, *J. Biol. Chem. 242*, 2389 (1967).
19. K.W. Bock, P. Siekevitz and G.E. Palade, *J. Biol. Chem. 246*, 188 (1971).
20. M. Higgins, T. Kawachi and H. Rudney, *Biochem. Biophys. Res. Commun. 45*, 138 (1971).
21. I.M. Arias, D. Doyle and Schimke, R.T., *J. Biol. Chem. 244*, 3303 (1969).
22. D. Cole, *Methods in Enzymology XI*, S.P. Colowick and N.O. Kaplan (Eds.), Academic Press, New York, 1967, p. 315.
23. K. Linderstrøm-Lang, *Cold Spring Harbor Symp. Quant. Biol. 14*, 117 (1950).
24. P.J. Dehlinger and R.T. Schimke, *Biochem. Biophys. Res. Commun. 40*, 1473 (1970).
25. J.F. Dice and R.T. Schimke, *J. Biol. Chem. 247*, 98 (1972).

26. P.J. Dehlinger and R.T. Schimke, *J. Biol. Chem. 246*, 2574 (1971).
27. T. Omura and Y. Kuriyama, *J. Biochem. (Tokyo) 69*, 651 (1971).
28. H. Jick and L. Shuster, *J. Biol. Chem. 241*, 5366 (1966).
29. P.J. Dehlinger and R.T. Schimke, *J. Biol. Chem. 247*, 1257 (1972).
30. T. Omura and R. Sato, *J. Biol. Chem. 239*, 2370 (1964).
31. A. Leskes, P. Siekevitz and G.E. Palade, *J. Cell. Biol. 49*, 264 (1971).
32. S.J. Singer and G.L. Nicolson, *Science 175*, 720 (1972).
33. R.K. Kornberg and H.M. McConnell, *Biochemistry 10*, 1111 (1971).
34. W.L. Morrison and H. Neurath, *J. Biol. Chem. 200*, 39 (1953).
35. A.B. Robinson, J.H. McKerrow and P. Carry, *Proc. Nat. Acad. Sci. U.S. 66*, 753 (1970).
36. J.R. Warner, *J. Biol. Chem. 246*, 447 (1971).
37. K.W.A. Wirtz and D.B. Zilversmit, *J. Biol. Chem. 243*, 3596 (1968).

Table I

HALF-LIVES OF SPECIFIC ENZYMES AND SUBCELLULAR FRACTIONS OF RAT LIVER

Enzymes	Half-life	Reference	Method of measure
Ornithine decarboxylase (soluble)	11 min	8	Loss of activity after puromycin
δ-aminolevulinate synthetase (mitochondria)	70 min	9	Loss of activity after puromycin
Alanine-aminotransferase (soluble)	0.7 - 1.0 d	10	Time course of enzyme change
Catalase (peroxisomal)	1.4 d	11	Recovery of activity after irreversible inhibition of activity
Tyrosine aminotransferase (soluble)	1.5 hr	12	Isotope decay*
Tryptophan oxygenase (soluble)	2 hrs	13	Isotope decay*
Glucokinase (soluble)	1.25 d	14	Change of enzyme activity
Arginase (soluble)	4-5 d	4	Isotope uptake and decay*
Glutamic-alanine transaminase	2-3 d	15	Time course of enzyme change
Lactate dehydrogenase isozyme-5	16 d	16	Isotope uptake*
Acetyl CoA carboxylase (soluble)	2 d	17	Isotope decay*
Cytochrome c reductase (endoplasmic reticulum)	60-80 hrs	18	Isotope decay
Cytochrome b_5 (endoplasmic reticulum)	100-120 hrs	18	Isotope decay
NAD glucohydrolase (endoplasmic reticulum)	16 d	19	Isotope decay
Hydroxymethylglutaryl CoA reductase	2-3 hrs	20	Activity decay after cycloheximide and isotope decay

* Denotes use of immunoprecipitation techniques.

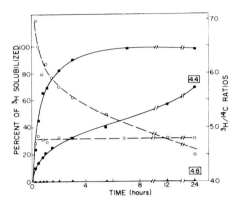

Fig. 1.   Susceptibility of double labeled supernatant
proteins of rat liver to proteolysis by pronase.   A rat
weighing 120 g was given 200 µC of ^{14}C-leucine (uniformly
labeled, specific activity 300 mC/mM).   The animal was
killed 6 hours later, and following homogenization in 0.25
M sucrose (2.5:1, sucrose to wet weight liver), and initial
centrifugation at 1,000 and 10,000 x g for 30 minutes each,
a 100,000 x g supernatant fraction was obtained.   This fraction was freed of amino acids by passage through a column
of Sephadex G-25 equilibrated with 0.05 M potassium phosphate, pH 7.5.   To a fraction of the supernatant, urea was
added to a final molarity of 8.0, and pH adjusted to a pH
of 9.5 with 3.0 M Tris·OH.   After standing at room temperature for 60 minutes, the sulfhydryl groups were blocked by
amino-ethylation as described by Cole (22).   The denatured
protein was dialyzed over night against a large excess of
2.0 M urea in 0.05 M potassium phosphate, pH 7.5.   The concentration of both the native and denatured proteins was
adjusted to 15 mg/ml, and pronase was added to a final concentration of 80 µg/ml.   2 ml samples were incubated at 37°
with and without pronase, and at the times indicated 100 µl
aliquots were removed and added to 0.5 ml of 10% trichloroacetic acid.   The precipitates were allowed to sediment at
4° over night, centrifuged, and 0.25 ml samples removed
and extracted 3 times with 2 ml of ethylether.   0.20 ml
samples were counted in a standard dioxane scintillation
mixture in the presence of 0.5 ml of NCS solubilizer (21).
Samples of protein were solubilized in 0.5 ml of NCS solubilizer and counted in the same manner.   The initial ^{3}H/^{14}C
ratio of the cytoplasmic proteins was 4.8.   The ratio of
the undigested protein was 4.4 (indicated by bracketed numbers).   △——△ no added pronase, ●——●CPM of "native"
proteins, ■——■ CPM of "denatured" proteins, ○---○^{3}H/^{14}C
ratios of native proteins, ☐---☐ ^{3}H/^{14}C ratios of "denatured" proteins.

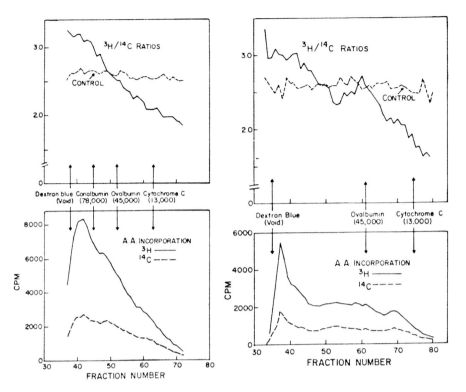

Fig. 2. Relative rate of degradation of "soluble" proteins of rat liver as a function of molecular size. Relative rates of degradation were estimated by the double isotope method of Arias *et al.* (21) in which ^{14}C-leucine is administered to rats four days prior to ^{3}H-leucine administration, with death of animals four hours later. The "control" indicates rats receiving both isotope forms of leucine at the same time. High ^{3}H/^{14}C ratio indicates relatively high rates of degradation. Proteins in absence (a) and presence (b) of SDS to disrupt multimeric proteins, were chromatographed on Sephadex G-200 columns. Details are given in Dehlinger and Schimke (24).

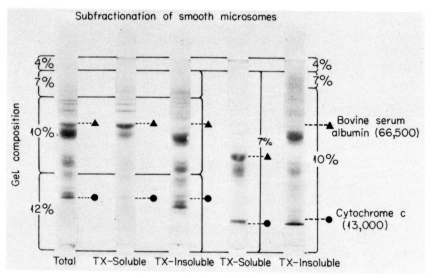

Fig. 3. Electrophoretic patterns of smooth endoplasmic reticulum and subfractions after fractional solubilization in Triton X-100. Smooth endoplasmic reticulum was solubilized (approximately 50%) by 1% Triton X-100. The remaining fraction was then solubilized in 0.5% sodium dodecyl sulfate, and all samples were electrophoresed in the presence of 0.1% sodium dodecyl sulfate. The percent composition of acrylamide was varied as indicated. Approximately 225 μg of protein was applied to each gel. See reference (24) for detail.

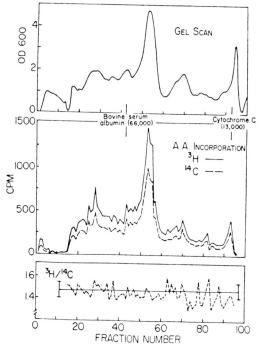

Fig. 4. Electrophoretic pattern of radioactivity and
optical density of Triton X-100 insoluble proteins of smooth
endoplasmic reticulum. An animal received a simultaneous
injection of 50 µCi of ^{14}C-L-leucine and 250 µCi of ^{3}H-
leucine, and was sacrificed 4 hours later. In all experiments food was withheld from the animals for 18 hours prior
to isotope injection. In those experiments lasting for 4
days, food was replaced 4 hours after isotope administration.
The TX-soluble protein fraction was prepared and subjected
to electrophoresis as outlined in Figure 3, for determination of radioactivity for the optical scan. 5 mg of protein was applied to a gel 19 x 75 mm to provide sufficient
material for counting accurately. The peaks of radioactivity were matched to the optical scan. This control gives
an indication of any systematic counting error produced by
using gels of varying concentrations of acrylamide (see
Figure 5), and also gives an estimate of variability to be
expected in the counting procedure. Each gel was removed
from the glass tube, frozen in Dry Ice, and sliced by hand
with a razor blade. Each gel slice was dissolved in H_2O_2
and hyamine and counted in a liquid scintillation counter.
^{3}H efficiency was 15% and ^{14}C efficiency was 55%. A 35%
spillover of ^{14}C counts into the ^{3}H window was subtracted
before calculation of ratio. Ratios were not indicated in
those fractions in which the counting error exceeded 5%.
(See reference (26) for detail.)

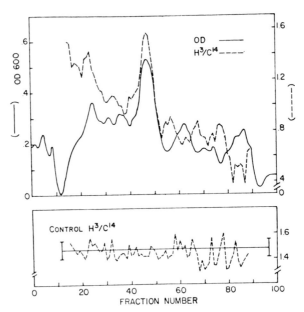

Fig. 5. Relative turnover of Triton X-100 insoluble proteins of rat liver smooth endoplasmic reticulum. A single male, albino rat weighing 150 g was given 250 μCi of ^{3}H-L-leucine 4 days after administration of 100 μCi of ^{14}C-L-leucine. The rat was killed 4 hours after the second isotope administration and a smooth endoplasmic reticulum fraction was separated into Triton X-100 soluble and insoluble fractions. The lower box shows the results of the control experiment of Figure 4. See reference (26) for detail.

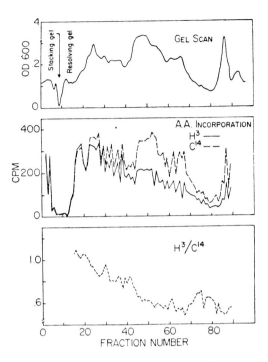

Fig. 6. Electrophoretic patterns of double labeled plasma membrane of rat liver. The labeling schedules were similar to those described in Figure 5. The upper box shows the scan of an analytical gel to which 225 μg of protein were applied. The middle box indicates actual counts obtained. The lower box indicates the calculated $^3H/^{14}C$ ratios obtained. 5 mg of plasma membrane protein were applied to a gel, 19 × 75 mm. See reference (26) for detail.

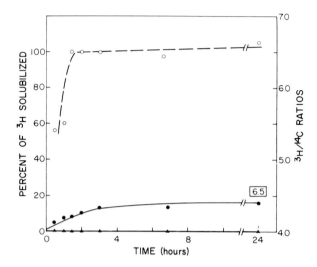

Fig. 7. Susceptibility of double labeled endoplasmic reticulum proteins to proteolysis by pronase. Details of isotope administration and digestion and counting by pronase are the same as described in the legend to Figure 1. The membrane fraction was a smooth endoplasmic fraction prepared as described by Dehlinger and Schimke (26). The $^3H/^{14}C$ ratio of the proteins of the membrane was 6.5. ▲—▲ no added pronase; ●—● ^3H CPM released. ◯---◯ $^3H/^{14}C$ ratios of released radioactivity.

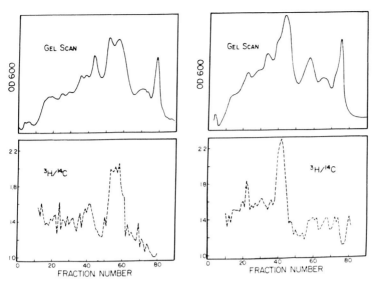

Fig. 8.   Relative amino acid incorporation into TX-in-soluble (right) SER proteins in response to phenobarbital administration.   The experimental animal was injected intra-peritoneally with 100 mg per kg of phenobarbital dissolved in water, and a control animal received a corresponding volume of water 16 hours prior to killing.   Four hours before killing the experimental animal received 250 µCi of ^{3}H-leucine and the control received 75 µCi of ^{14}C-leucine. The SER fractions from each animal were combined and elec-trophoretically treated as described in reference (29). Approximately 225 µg of protein were applied to the analy-tical gels and 5 mg of protein to the preparative size gels. See reference (29) for details.

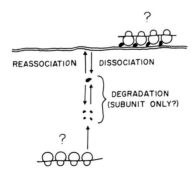

Fig. 9.   Schematic model of membrane genesis and de-gradation.

133

THE ROLE OF PHOSPHOLIPASE A IN *E. COLI*

Shoshichi Nojima, Osamu Doi,
Nakako Okamoto and Mihoko Abe

Department of Chemistry
National Institute of Health
Shinagawa-ku, Tokyo, Japan

ABSTRACT. Two kinds of phospholipases A, detergent-resistant (dr) phospholipase A and detergent-sensitive (ds) phospholipase A, were found in *E. coli* K12. Three kinds of mutants on phospholipases A were isolated; dr⁻, ds⁻ and dr⁻ds⁻. It was difficult to find marked differences between the properties of the parent and the three mutants with respect to growth, phospholipid and fatty acid compositions, turnover of phosphatidylgylcerol in the logarithmic phase of growth, and phage-induced lysis. Fatty acid releasing activity, after incubation of the cell extracts at 37°, was still observed even in the case of the dr⁻ds⁻ mutant.

In biochemical studies of mutants defective in phospholipase A, we have recently found evidence for two kinds of phospholipiase A, detergent-resistant phospholipase A (dr) which seems to be similar to the membrane-bound phospholipase A1 recently purified to near-homogeneity by Kornberg and Scandella (1) and found in the outer membrane by Osborn; and degergent-sensitive phospholipase A (ds) (2). We have tentatively named the former *dr*-enzyme and the latter *ds*-enzyme. The properties of the *ds*-enzyme are quite different from those of the *dr*-enzyme. The principal differences between them are summarized in Table 1. *dr*-Phospholipase A is activated more than 100-fold by methanol. This enzyme is stable to or activated by heat treatment. Both phosphatidylethanolamine and phosphatidylglycerol are hydrolyzed by the *dr*-enzyme. On the other hand, *ds*-enzyme activity can not be detected in the presence of organic solvents or detergents, and is labile to heat treatment. So far as the crude extract is concerned, the

ds-enzyme can hydrolyze only phosphatidylglycerol. Phosphatidylethanolamine is not degraded by ds-enzyme and inhibits the hydrolysis of phosphatidylglycerol. After disruption of the cells by sonication, dr-enzyme activity was found mainly in the membrane fraction, whereas ds-enzyme activity was in the supernatant fraction (Table 2). The ds-enzyme, however, could be loosely bound to the membrane.

Three classes of phospholipase A mutants were isolated in our laboratory: dr^- (3), ds^- and dr^-ds^- (Fig. 1). These were isolated from the parental K12 strain by nitrosoguanidine mutagenization or by conjugation. The principle of the method of selection of the mutant is as follows: Endogenous phospholipids in the mutagenized cells are hydrolyzed by autolysis to form free fatty acids, probably after the combined action of phospholipase A and lysophospholipase. The release of fatty acids was detected with an unsaturated fatty acid auxotroph of $E.\ coli$, isolated by the method of Silbert and Vagelos (4), as the indicator strain. Thus, cells from colonies or a few drops of the culture of the mutagenized or recombinant cells were autolyzed on an agar plate, and the plate was overlaid with soft agar containing the unsaturated fatty acid auxotroph.

The detectable phospholipase A activity in each of the mutants dr^-ds^+, dr^-ds^- was less than 1% of that of the parent as confirmed by mixing experiment with the corresponding parent strains. Each of the three mutants grows well at 30° or 42° in λ-broth as well as in Penassay medium. The comparison of the doubling time in the logarithmic phase of growth as measured by viable cell count showed no differences between the parent and mutant strains (Table 3). Analysis of phospholipid content and composition, and the fatty acid composition in the logarithmic phase of growth as well as in stationary phase yielded identical results for the parent and the three mutant strains.

In the logarithmic phase of growth, the acyl portion of phosphatidylglycerol does turn over (5). The possibility that phospholipase A plays a role in the mechanism of the $in\ vivo$ degredation of phosphatidylglycerol was therefore tested. Cells in logarithmic phase were labeled with ^{14}C-acetate and the amount of loss of the label in the acyl moieties in phosphatidylglycerol after a chase with

^{12}C-acetate was determined for the parent and the three mutant strains. Approximately 10% loss of the label in phosphatidylglycerol in one doubling time was observed for each of the three mutant strains as well as for the parent strain (Table 3).

The contribution of the detergent-resistant and detergent-sensitive enzymes to the release of fatty acids in the lysis of the bacterial cells was studied. The following two properties were tested: 1) the time of and burst size after phage-induced lysis and 2) autolysis of endogenous phospholipids after incubation of the extracts at 37°.

The phage used for the lysis of cells was $\lambda imm^{434}cl$, which grows in λ-lysogenic cells. The results (6) showed that both mutants (dr$^-$ and dr$^-$ds$^-$) support the normal growth of the phage. The time of cell lysis and the burst size were similar to those of the infected parental cells. In contrast to the parental cells, however, no significant release of fatty acids from ^{14}C-acetate prelabeled phospholipids was observed at the time of lysis of the mutant cells (dr$^-$ds$^+$, dr$^-$ds$^-$). The release of fatty acids in the parent was presumed to be caused by dr-phospholipase A. It was very difficult to find a difference between the dr$^-$ds$^+$ and dr$^-$ds$^-$ mutants in the phage-lysis experiment.

When we tested the growth of the unsaturated fatty acid auxotroph on the agar plate, some growth of the auxotroph was noted even where the dr$^-$ds$^-$ was spotted. This indicates that autolyzed cells of the mutant still release free fatty acid although the total phospholipids of $E.$ $coli$ were not hydrolyzed at all to form free fatty acids, when they were added to the extract as exogenous substrates. Each of the extracts of ^{14}C-acetate labeled cells of the parent and the three mutants were incubated for 2 hr at 37° and the amount of free fatty acid released was compared. At the same time, the effect of various compounds including detergents, organic solvents, EDTA etc., on the release of fatty acids was examined. The nature of the hydrolysis of endogenous substrates was, in general, similar to the results obtained from exogenous substrates already mentioned. For instance, 20% methanol enhanced the free fatty acid formation in the case of dr$^+$ds$^+$ or dr$^+$ds$^-$ mutants, while it inhibited the fatty acid formation in the case of dr$^-$ds$^+$ or dr$^-$ds$^-$ mutants.

After 20 min incubation, the extent of the fatty acid

released from endogenous substrate in the cases of dr^-ds^+ and dr^-ds^- mutants was about 10% of the extent in the case of dr^+ds^+. After 2 hr, it was about 30% (Fig. 2). From Fig. 2, the difference between dr^+ds^+ and (dr^+ds^-, dr^-ds^+, dr^-ds^-) or between dr^+ds^- and (dr^-ds^+, dr^-ds^-) is clear. However, it was difficult to find differences between dr^-ds^+ and dr^-ds^- as observed in phage-induced lysis. Our explanation of these experimental results is as follows: Detergent-sensitive enzyme may be more latent than the detergent-resistant one. The former activity may be inhibited by a large amount of phosphatidylethanolamine (a very slight difference between dr^-ds^+ and dr^-ds^-). The activity of detergent-sensitive enzyme can be observed in the endogenous substrate experiments only when detergent-sensitive enzyme hydrolyzes the phosphatidylglycerol "released" from the membrane after the initial attack of detergent-resistant enzyme on membrane phospholipids (difference between dr^+ds^+ and dr^+ds^-). At present, the nature of the fatty acid releasing activity observed in the dr^-ds^- mutant is unknown.

REFERENCES

1. C.J. Scandella and A. Kornberg, *Biochemistry 10*, 4447 (1971).
2. O. Doi, M. Ohki and S. Nojima, *Biochim. Biophys. Acta 260*, 244 (1972).
3. M. Ohki, O. Doi and S. Nojima, *J. Bacteriol.*, in press.
4. D.F. Silbert and P.R. Vagelos, *Proc. Nat. Acad. Sci. U.S.A. 58*, 1579 (1967).
5. D.C. White and A.N. Tucker, *J. Lipid Res. 10*, 220 (1969).
6. Y. Sakakibara, O. Doi and S. Nojima, *Biochem. Biophys. Res. Commun. 46*, 1434 (1972).

TABLE 1

Comparison of the properties of the detergent-resistant and detergent-sensitive phospholipases A (2).

	Detergent-resistant phospholipase A	Detergent-sensitive phospholipase A
Detergents	Stable or activated	Inactivated
Organic solvents	Activated	Inactivated
Preincubation at 60°	Stable	Inactivated
Requirement for Ca^{2+}	+	±
Substrate specificity	Phosphatidyl-ethanolamine Phosphatidylglycerol	Phosphatidyglycerol
Localization	105,000 x g precipitate	105,000 x g supernatant

TABLE 2

Localization of detergent-resistant and detergent-sensitive phospholipase A and lysophospholipase in *E. coli* K12 (2).

Subcellular fraction	Protein mg (%)	Detergent-resistant phospholipase A		Detergent-sensitive phospholipase A		Lysophospholipase	
		Spec. act. (%)	Total act. (%)	Spec. act. (%)	Total act. (%)	Spec. act. (%)	Total act. (%)
18,000 sup.	114	0.300(100)	34.2	1.06(100)	122	39.6(100)	451
105,000 sup. (A)	83(67)	0.057(19)	4.7(16)	1.68(159)	139(78)	37.7(95)	313(62)
105,000 ppt. (B)	39(33)	0.628(209)	24.5(84)	1.02(96)	40(22)	48.9(123)	191(38)
A + B	122(100)		29.2(100)		179(100)		504(100)

E. coli K12 strain S15 was cultured in one liter of Penassay broth medium to the late logarithmic phase of growth (7.8 × 10^8 cells per ml). The cells, collected by centrifugation and washed with saline, were suspended in 16 ml of borate buffer (5 mM, pH 7.3) and sonicated for 2 min at 0°. The sonicate was centrifuged at 18,000 × g for 10 min at 0° and the resulting supernatant fraction was centrifuged in the same way. The supernatant fraction obtained (10 ml) was designated 18,000 sup. This fraction was further centrifuged at 105,000 × g for 2 hr at 4°. The supernatant fraction (9.5 ml) was designated 105,000 sup., and the pellet 105,000 ppt. (5 ml), after suspending in 4.5 ml of borate buffer. Each of the fractions, 18,000 sup., 105,000 sup. and 105,000 ppt., were assayed for enzyme activity. ^{14}C-Phosphatidylethanolamine, ^{14}C-phosphatidylglycerol and ^{14}C-lysophosphatidylethanolamine were used as substrates for the assay of detergent-resistant phospholipase A, detergent-sensitive phospholipase A and lysophospholipase, respectively. The reactions were performed at 37° for 1 hr, 30 min and 15 min, respectively. Spec. act. and total act. denote specific activity (nmoles of phosphatidylethanolamine, phosphatidylglycerol and lysophosphatidylethanolamine hydrolyzed per hr per mg of protein), and total activity (nmoles of phosphatidylethanolamine, phosphatidylglycerol and lysophosphatidylethanolamine hydrolyzed per hr), respectively.

TABLE 3

Comparison of doubling times and degradation percentages of phosphatidylglycerol in one doubling time of logarithmic phase of growth in mutants of *E. coli* K12 defective in phospholipase A.

E. coli K12 strains	Doubling time (min)	Degradation of phosphatidylglycerol (%)
dr^+ds^+ (parent)	23	11
dr^+ds^-	27	15
dr^-ds^+	27	11
dr^-ds^-	23	11

The cells of *E. coli* K12 strains S15 (dr^+ds^+, parent) (3), 21 (dr^+ds^-), 23 (dr^-ds^+) (3) and 1712 (dr^-ds^-) were cultured in Penassay broth medium at 37° overnight. Each of these cultures was added to 15 ml of the same medium as 0.5% (v/v) and an aqueous solution of $1-^{14}C$-acetate (Na) was added to the culture (2.5 µCi per ml of the culture). Each of these cultures were incubated at 37° for 3 hr and centrifuged at room temperature for 5 min at 4000 rpm. The pellet of cells from each sample was suspended in 5 ml of Penassay broth; the resulting suspension was filtered through a Millipore filter at room temperature; and the washing was repeated once more. The labeled cells on the filter were suspended in 40 ml of Penassay broth and cultured at 37° for 165 min. At appropriate time intervals, a sample of 0.05 ml was assayed for viable cell count on λ-agar plate and 2 samples of 1.0 ml were removed for analysis of phospholipids (2). The radioactivity in phosphatidylglycerol at the start of the chase was approximately 1000 cpm when counted with 10 ml of the solution described by Bray using a Beckman liquid-scintillation counter.

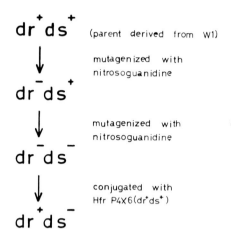

Fig. 1. Mutants of *E. coli* K12 defective in phospholipase A.

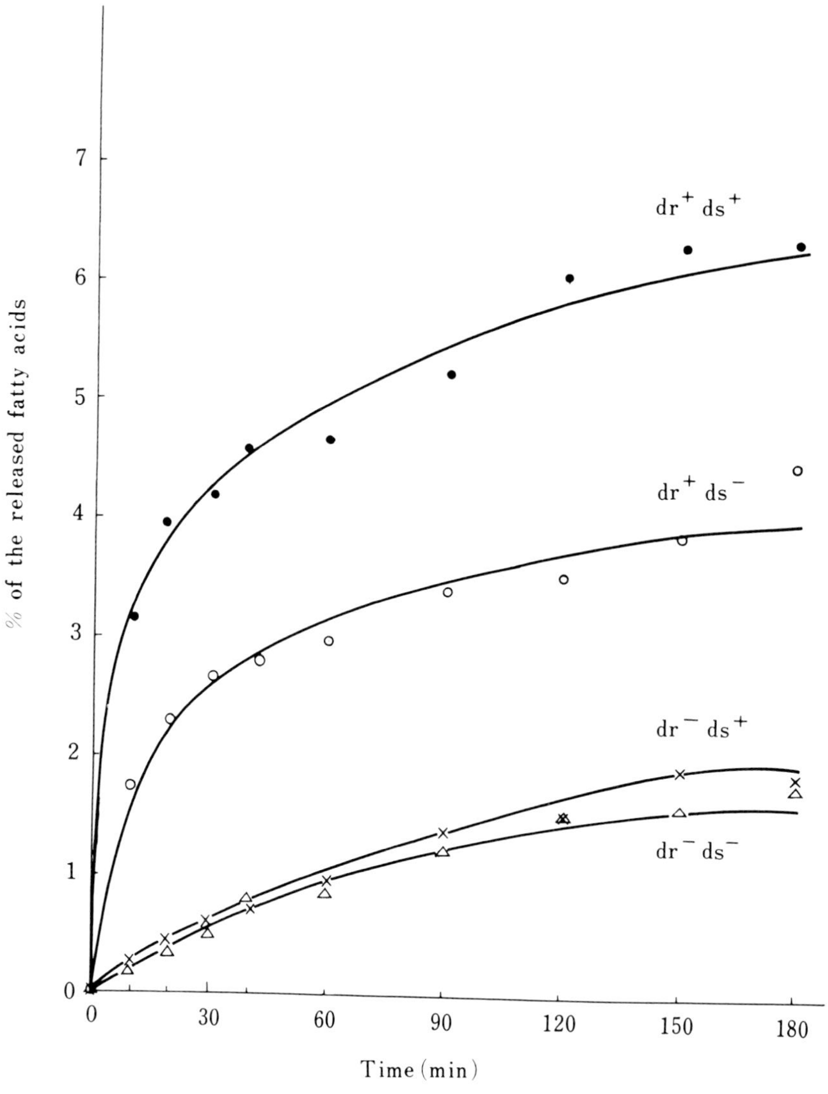

Fig. 2. Caption on following page.

Fig. 2. Hydrolysis of endogenous substrates of parent and the three mutant strains after incubation of their extracts at 37°.

The cells of *E. coli* K12 strains S15 (dr^+ds^+, parent) (3), 21 (dr^+ds^-), 23 (dr^-ds^+) (3) and 1712 (dr^-ds^-) were cultured in Penassay broth medium at 37° overnight. Each of these cultures was added to 50 ml of the same medium as 1.0% (v/v), and an aqueous solution of $1-^{14}C$-acetate (Na) was added to the culture (2.0 μCi per ml of the culture). Each of these cultures was incubated at 37° for 3.5 hr. Each culture was then chilled and centrifuged at 7,000 to 8,500 x g for 10 min at 0°. The cells were washed with cold saline. The washed cells were suspended in 1.5 ml of 5 mM Tris·HCl buffer (pH 7.0) and sonicated (Umeda Sonicator, 20 kcycles) for 2 min at 0°. The sonicate was centrifuged at 18,000 x g for 15 min at 0°. The supernatant solution was used as the labeled enzyme solution (13 to 10 mg protein per ml).

The incubation mixture contained 0.2 ml of the labeled enzyme solution, 0.1 ml of 10 mM $CaCl_2$ and 0.4 ml of 200 mM potassium phosphate buffer (pH 7.0), made up to a final volume of 1.0 ml with distilled water.

A series of the incubation mixture was incubated at 37° with shaking for 2 hr. At appropriate times, the incubation mixtures were chilled, and 9 ml of chloroform-methanol (2:1, v/v) was added. This mixture was shaken for 30 sec and centrifuged at 2,000 x g for 10 min for separation into two layers. To the upper layer 6.0 ml of chloroform was added, and the mixture was again shaken for 30 sec and centrifuged for separation. The combined chloroform extract was evaporated to dryness *in vacuo* at 30°, and the residue was dissolved in 0.3 ml of chloroform-methanol (2:1, v/v). Samples (50 μl) of the solution were spotted on a silica gel thin-layer plate and developed with chloroform-methanol (98.5:1.5, v/v). The spots corresponding to fatty acids and total phospholipids (the origin of the spot at 0 time incubation) were scraped and counted with 10 ml of the solution described by Bray using a Beckman liquid-scintillation counter. The ordinate represents the percentage of fatty acids released from the total phospholipids of the extracts of each strain.

THE INFLUENCE OF LIPID PHASE TRANSITIONS
ON MEMBRANE FUNCTION AND ASSEMBLY

C. Fred Fox and Norihiro Tsukagoshi

Department of Bacteriology
University of California
Los Angeles, Ca. 90024

SUMMARY. The properties of the lactose transport system, induced in an unsaturated fatty acid auxotroph of *Escherichia coli*, have been studied after shifting from growth in elaidic acid to oleic acid supplemented medium or vice versa. When induction proceeds at 37° after the fatty acid shift, Arrhenius plots describing transport rate as a function of temperature are biphasic in slope. The intersect of the slopes defines a transition which indicates that the newly formed transport system is influenced by a mixed bulk lipid phase. When induction proceeds at temperatures below 30°, the phase transition temperature for elaidate derived lipids, the Arrhenius plots are triphasic, and a substantial portion (approximately half) of the transport activity has a behavior reflecting the properties of the lipids synthesized after the fatty acid shift, even in cases where the membrane lipids synthesized after the shift account for less than 10% of the total membrane phospholipids. If the cells induced at 25° are subsequently incubated at 37° for 10 min before transport assay, the Arrhenius plot becomes biphasic, and the transition temperature defined by the intersect reflects the properties of a mixed bulk lipid phase. Thus the phospholipids in the newly formed domains of membrane can mix with the bulk of the membrane lipids above the elaidate lipid phase transition, but not below it. Furthermore, since a one min incubation at 37° does not suffice to alter the Arrhenius plots for transport induced at 25° (after the fatty acid shift) from triphasic to biphasic, the lateral diffusion of lipids in biological membranes may proceed more slowly than in artifical lipid bilayer membranes.

145

INTRODUCTION

We have published evidence consistent with the view that assembly of a fully functional lactose transport system in *E. coli* proceeds by the coordinated incorporation of newly synthesized phospholipids and transport proteins into membrane. Induction of the lactose operon in an unsaturated fatty acid auxotroph of *E. coli* during periods of starvation for an essential fatty acid leads to normal synthesis of β-galactosidase and thiogalactoside transacetylase. At the same time, however, the induction of a fully functional lactose transport system is blocked (1). Further studies of this type have been performed with a glycerol auxotroph of *E. coli*. During glycerol starvation, protein synthesis continues for approximately one generation of growth, whereas lipid biosynthesis ceases. Induction of the lactose operon during glycerol starvation results in the synthesis of normal quantities of β-galactosidase, thiogalactoside transacetylase, and a membrane associated lactose transport protein (M protein) capable of binding β-galactosides, but non-functional in transport (2).

Induction of lactose transport has also been studied in experiments where an unsaturated fatty acid auxotroph of *E. coli* was shifted from growth in oleic acid supplemented medium, to growth in linoleic acid supplemented medium at 37°. The ratio of β-galactoside transport rate measured at 28° and 10° is markedly different for cells grown with either oleic or linoleic acid supplements. When transport was induced before the shift from oleic to linoleic acid medium, the 28°/10° ratio in transport rate remained that of oleic acid grown cells even after one-fifth of a generation of growth in linoleic acid medium. When transport was induced after the fatty acid shift, however, the 28°/10° ratio in transport rate was identical to that observed with cells grown entirely in linoleic acid supplemented medium. These experiments were interpreted to indicate that the newly formed transport proteins associate and remain associated with the newly synthesized lipids, i.e., those derived from linoleic acid (3). A recent report by Overath and his colleagues (4) and more thorough investigations in this laboratory, however, show clearly that in the experiment described above, the transport system formed after the fatty acid shift is influenced by the properties of the bulk lipid phase rather than by the properties of the newly synthesized lipids. We describe here modifications of this experiment, which do provide data consistent with the view that the newly formed transport proteins associate and remain associated with newly synthesized lipids under defined conditions.

RESULTS

Transport rate as a function of assay temperature has been studied in unsaturated fatty acid auxotrophs of *E. coli*. The properties of transport systems are markedly influenced by the structure of the unsaturated fatty acid incorporated into the membrane lipids (5-7). Fig. 1 describes the influence of temperature on the rate of β-glucoside transport in an auxotroph grown at 37° in either elaidic acid or linoleic acid supplemented medium. The Arrhenius plots are biphasic intersecting at points termed transition temperatures. In the examples shown, the transition temperature for transport by elaidic acid grown cells is 30°, and that for linoleic acid grown cells is 7°. Table 1 describes the transition temperatures for β-glucoside and β-galactoside transport in cells grown with a variety of unsaturated fatty acid supplements. Though the two transport systems share neither common protein components nor common modes of energy coupling, the transition temperatures for transport are identical in each case. Overath and his colleagues have shown that the transport transition temperature correlates well with a physical transition detected at the air-water interface with lipids extracted from the cells used for transport assay (6). Taken together, these experiments show that the transport transition temperatures are determined by a physical property of the membrane lipid phase, rather than by a lipid-protein interaction unique for a given transport system.

The lipid phase transition which determines the transition in transport rate is also a determinant of the efficiency of transport system assembly as a function of the temperature at which the cells are induced for transport. We have observed that transport system assembly is abortive when induction of the transport system proceeds below the temperature defining the transition in transport rate (8,9). If cells grown in elaidic acid medium at 37° are transferred to oleic acid medium before commencing induction of lactose operon at 25°, however, the loss in efficiency of induction of the fully functional transport system is prevented. No

147

loss in the efficiency of induction of β-galactoside transport is observed when cells initially grown in oleic acid medium are subsequently shifted to incubation in elaidic acid medium at 25° for short periods of induction.

We have investigated further the properties of the transport system formed at 25° after shifting from growth in oleic to elaidic acid medium or vice versa. Fig. 2B describes the result of an experiment where cells growing in oleic acid medium were shifted to elaidic acid medium for induction of transport at 25°. In contrast to the Arrhenius plots shown in Fig. 1, this plot is triphasic in slope, exhibiting two transition temperatures, one similar to that of elaidic acid grown cells, the second similar to that of oleic acid grown cells. This experiment may be compared with that shown in Fig. 2A. Here, transport was induced at 37° after the same fatty acid shift, and the Arrhenius plot of transport rate as a function of temperature is biphasic, with an intersect (transition temperature) of approximately 18°. In the experiments shown in Fig. 2A and 2B, the extent of elaidic acid derived lipid formed after the oleic acid to elaidic acid shift was identical, as determined by the extent of [³H]elaidic acid incoporated into membrane phospholipids. The elaidic acid derived lipid comprised less than one-sixth of the total membrane lipids formed after the fatty acid shift in these experiments. Fig. 2C shows the result of incubating the cells used for the experiment described in Fig. 2B for 10 min at 37°. The transition temperatures of 26° and 14.6° are no longer observed, and the new transition temperature is identical within the limits of error to that obtained when induction proceeded at 37° after the fatty acid shift. A one min incubation at 37°, however, is not sufficient to change the Arrhenius plots from triphasic to biphasic. We have also done similar experiments where the order of growth with the fatty acid supplements was reversed. In other words, the cells were first grown in elaidic acid medium, and then shifted to oleic acid medium for a short period of transport system induction. The same qualitative type of result was obtained here as in the experiments described in Fig. 1.

We interpret these experiments in the following fashion. The transport system induced at 37° resides in a mixed, bulk lipid phase. As shown in Fig. 1A, induction of transport for less than one-fifth of a generation after shifting from growth in oleic to growth in elaidic acid medium yields

a transport transition temperature of approximately 18°.
This is intermediate between the transition temperatures of
oleate or elaidate grown cells, 13° and 30° respectively
(Table 1), and is much closer to the oleic acid transition
temperature as would be expected for a membrane consisting
largely of oleic acid derived lipids. Induction at 25°
after the fatty acid shift, on the other hand, leads to the
production of transport activity displaying two transition
temperatures. One of these (26°) is 8° higher than the
transition temperature expected for the mixed bulk lipid
phase, and closely resembles the transition temperature
characteristic of elaidic acid grown cells. Thus induction
at 25° after the oleic acid to elaidic acid shift (or vice
versa) leads to the assembly of two populations of transport
sites, the lipids of one of these consisting primarily of
newly synthesized lipids. (The production of the two populations of transport sites is certainly an assembly related
process. When cells from the experiment described in Fig.
2A were subsequently incubated at 25° for a period of time
equal to that used for transport induction in the experiment described in Fig. 2B, the shape of the Arrhenius plot
remained that described in Fig. 2A.) When the cells induced for transport at 25° are subsequently incubated for
10 min at 37°, the Arrhenius plots change from triphasic
to biphasic. This indicates that the membrane lipids which
were segregated into two different populations (Fig. 2B)
have mixed, producing a mixed bulk lipid phase identical
to that obtained during induction at 37°.

We have also determined the critical temperature range
which allows the segregation of newly formed transport proteins with newly synthesized lipids formed after the fatty
acid shift in experiments similar to that described in Fig.
2B. Biphasic Arrhenius plots were observed after induction
at temperatures above 31°, and triphasic plots after induction at temperatures below 29°. This indicates that the
critical temperature for segregation of newly formed transport proteins with newly formed lipids is approximately
30°, the phase transition temperature for elaidate derived
lipids (Table 1).

DISCUSSION

We have described the results of experiments where
density labeling and other techniques were used to test

models for "localized" membrane assembly (10-12). These
experiments were interpreted to indicate that during growth
above the temperature of the physiological and physical
(lipid monolayer) lipid phase transitions, the *E. coli* membrane is either not assembled at any distinct point or
points on the membrane surface, or that assembly does proceed by a localized process but is followed by a mixing of
newly synthesized and presynthesized membrane constituents.
The experiments reported here provide support for the latter of these interpretations. When the newly synthesized
lactose transport system is assembled after an essential
fatty acid shift at a temperature below the transport and
physical (monolayer) transition for one of the two lipid
compositions (newly and presynthesized), the newly synthesized transport system is affected by either the newly synthesized lipids, or by a lipid phase which resembles the
presynthesized lipids, rather than the lipids of the bulk
mixed lipid phase. We interpret this finding as an indication that the transport system is incorporated into membrane
with other newly synthesized constituents in units which
differ widely in size distribution. The lipids of the smaller of these units mix with the presynthesized lipid phase,
and those of the larger units remain intact, at least for
a time equal to that required for our experimental procedures. When the cells are incubated above a critical temperature, which is apparently identical to the transport
transition temperature, the presynthesized and newly synthesized lipids mix, perhaps by the lateral diffusion process described by McConnell and his associates (13-14).
The experiments described here allow the development of a
reasonable rationale by which to test for localized membrane
assembly, circumventing the difficulties imposed by lateral
diffusion of membrane components.

ACKNOWLEDGMENTS

This work was supported by U.S.P.H.S. research grants
GM-18233 and AI-10733, by grant 569 from the California
Division of the American Cancer Society, by grant BC 79
from the American Cancer Society and grant DRG-1153 from
the Damon Runyon Fund. C.F.F. is the recipient of U.S.P.H.S.
Research Career Development Award GM-42359. We thank
Harden McConnell for highly valuable discussions on numerous
occasions.

REFERENCES

1. C.F. Fox, *Proc. Nat. Acad. Sci. U.S. 63*, 850 (1969).
2. C.C. Hsu and C.F. Fox, *J. Bacteriol. 103*, 410 (1970).
3. G. Wilson and C.F. Fox, *J. Mol. Biol. 55*, 49 (1971).
4. P. Overath, F.F. Hill and I. Lamnek-Hirsch, *Nature New Biol. 234*, 264 (1971).
5. H.U. Schairer and P. Overath, *J. Mol. Biol. 44*, 209 (1969).
6. G. Wilson, S. Rose and C.F. Fox, *Biochem. Biophys. Res. Commun. 38*, 617 (1970).
7. M. Esfahani, A.R. Limbrick, S. Knutton, T. Oka and S. Wakil, *Proc. Nat. Acad. Sci. U.S. 68*, 3180 (1971).
8. C.F. Fox, *Fed. Proc. 30*, 1032 (1971).
9. C.F. Fox, *Membrane Assembly*, in *Membrane Molecular Biology*, C.F. Fox and A.D. Keith (Eds.), Sinauer Associates, Stamford, Conn., 1972, p. 345.
10. N. Tsukagoshi, P. Fielding and C.F. Fox, *Biochem. Biophys. Res. Commun. 44*, 497 (1971).
11. G. Wilson and C.F. Fox, *Biochem. Biophys. Res. Commun. 44*, 503 (1971).
12. N. Tsukagoshi and C.F. Fox, *Fed. Proc. 30*, 1120 (1971).
13. R.D. Kornberg and H.M. McConnell, *Proc. Nat. Acad. Sci. U.S. 68*, 2564 (1971).
14. H.M. McConnell, P. Devaux and C. Scandella, this volume.

TABLE 1

Transition temperatures for β-galactoside
and β-glucoside transport

Fatty acid supplement	Transport transition temperatures	
	β-Glucoside	β-Galactoside
Elaidate	30	30
9-(and 10-)Bromostearate	22	22
Oleate	13	13
Dihydrosterculate	11	11
cis-Vaccenate	10	10
Linoleate	7	7

Cells were grown and induced for transport at 37° in medium supplemented with the stated essential fatty acids. Transport transition temperatures were determined as described in Fig. 1. [After Wilson and Fox (3) and C.F. Fox, J.H. Law, N. Tsukagoshi and G. Wilson, $Proc.$ $Nat.$ $Acad.$ $Sci.$ $U.S.$ 67, 598 (1970).]

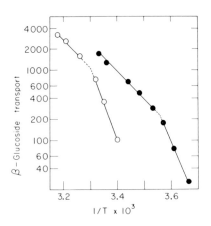

Fig. 1. Arrhenius plots describing β-glucoside transport in cells of an unsaturated fatty acid auxotroph grown at 37° in medium supplemented with elaidic (◯) or linoleic (●) acid [from Wilson and Fox (3)].

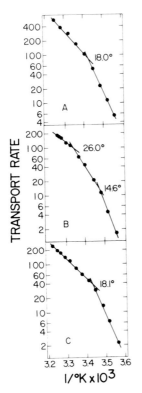

Fig. 2. Properties of the lactose transport system induced after shifting from growth in oleate supplemented medium at 37° to growth in elaidate supplemented medium. The
amount of elaidic acid incorporated into phospholipids was
determined using [^{3}H]elaidic acid and was the same in (A)
and (B). In (A) induction was at 37° and in (B), 25°. In
both cases, the cells were chilled to ice bath temperature
after induction, and held at that temperature during all
processing steps before transport assay at the indicated
temperatures. In (C) cells from (B) were incubated without
aeration for 10 min at 37° before commencing transport assays.

MODE OF INSERTION OF LIPOPOLYSACCHARIDE INTO THE OUTER MEMBRANE OF *ESCHERICHIA COLI*

Charles F. Kulpa, Jr., and Loretta Leive

Laboratory of Biochemical Pharmacology
National Institute of Arthritis and Metabolic Diseases
National Institutes of Health
Bethesda, Maryland 20014

ABSTRACT. The low carbohydrate content of the lipopolysaccharide (LPS) of a mutant of *E. coli* can be increased by growing the cells in the presence of galactose. This increase causes the outer membrane in which this LPS is located to increase in density. This phenomenon has been used to study the mode of insertion of LPS molecules into the outer membrane. When galactose was added to cells grown in its absence, and the density of fragments from outer membrane measured at various times thereafter, it was found that some dense fragments were formed after only 10% of a doubling time. These results suggest that LPS enters the outer membrane via discrete specialized regions, rather than by diffuse intercalation.

Recent studies of M. J. Osborn and her collaborators (1) have indicated that although the final location of lipopolysaccharide (LPS) in coliform bacteria is in the outer membrane, it is synthesized in the inner membrane. LPS must therefore be translocated from its site of synthesis to its final location, and in some manner, as yet unknown, inserted into the outer membrane.

To determine how the LPS is inserted, we used a mutant of *E. coli* isolated by Elbein and Heath (2) that lacks UDP-galactose-4-epimerase. Only when provided with galactose can this organism synthesize LPS with a full complement of sugar. Carbohydrate has a density much greater than either protein or lipid, which are the other components of the outer membrane. Therefore, we would predict that increasing the carbohydrate content of the outer membrane should increase its density.

To test this prediction, we measured the density of outer membranes from cells grown in the presence or absence of galactose, as shown in Fig. 1. Outer membrane from cells grown in the presence of galactose (dense outer membrane) is much denser than outer membrane from cells grown in the absence of galactose (light outer membrane). The density of dense outer membrane was 1.24, the density of light outer membrane was 1.22, and on these gradients their positions differed by 16 tubes.

We therefore attempted to determine the mode of entry of LPS into the outer membrane by growing cells in the absence of galactose, adding galactose, and at various times thereafter isolating their membranes. The membranes were sonicated extensively to obtain fragments of an appropriate size, and centrifuged to equilibrium in sucrose gradients. The outcome of such an experiment depends on the manner in which LPS is inserted into the outer membrane. The results predicted for two possible modes of insertion are illustrated in Fig. 2. The first hypothesis states that LPS enters the outer membrane by diffuse intercalation, that is, by being inserted randomly into existing membrane. The second hypothesis states that the LPS enters into the outer membrane via discrete specialized regions of the outer membrane. If the first hypothesis were true, we would predict that as new carbohydrate-rich LPS molecules were incorporated into the outer membrane, there would be a gradual and uniform shift in the density of the outer membrane fragments with increasing time of exposure to galactose as shown in Fig. 2A. At one generation, the membrane fragments would band at a density between that of light and dense outer membrane. The second hypothesis (Fig. 2B) predicts that if the fragments were of appropriate size, then at some early time, 1/10 of a generation for example, there would be some fragments of outer membrane containing a high concentration of new LPS. The rest of the fragments would contain relatively little new LPS relative to old, carbohydrate-poor, LPS. Therefore, some of the fragments of outer membrane would band at the density of dense outer membrane. Other fragments, even at early times, might have a density intermediate between dense and light outer membrane as new LPS diffused from the point of entry into existing membrane. The majority of membrane fragments would have a density similar to light outer membrane.

The experiment was performed as described in the legend to Fig. 3. At early times after galactose addition, such as 2 minutes, no material was found to band at the density of dense outer membrane (fraction 23). After 4 minutes (1/10 of a doubling time), a shoulder was observed at the density of dense outer membrane, and another shoulder was present at an intermediate density. The remainder of the material formed a peak at the density of light outer membrane (fraction 40). After 10 minutes (not shown), the profile was similar to that observed at 4 minutes, except that the material at the density of dense outer membrane formed a more discrete peak. To verify that these regions were truly membrane fragments of differing densities, regions corresponding to dense, intermediate, and light outer membrane were pooled from a gradient identical to the lower gradient in Fig. 3 (4 minutes) and recentrifuged. The material from the dense region was heterogeneous, but did contain some material banding at the density of dense outer membrane. The material isolated from the region of intermediate density gave a sharp band at an intermediate density. Material isolated from the region of light outer membrane gave a sharp band at the density of light outer membrane. Diffuse intercalation, in which LPS is inserted throughout the outer membrane, would be predicted to give rise to outer membrane fragments of intermediate density only after one doubling time. Fragments of intermediate or greater density could result after such a short (10% of a doubling time) exposure to galactose if new LPS molecules enter the outer membrane at discrete points and diffuse from these sites into the existing membrane as postulated in Fig. 2B. Final proof that these dense fragments constitute points of entry of LPS will require further studies on the size and composition of the isolated fragments.

In summary, our results, though of a preliminary nature, do not appear to be compatible with a random mode of entry for lipopolysaccharide into the outer membrane of *E. coli*. Instead, they suggest that lipopolysaccharide may enter the outer membrane of *E. coli* by means of discrete specialized regions, then diffusing from these regions into the existing outer membrane.

References

1. Osborn, M. J., Gander, J. E., Parisi, E., and Carson, J., *J. Biol. Chem.*, *247*, in press.

2. Elbein, A. D., and Heath, E. C., *J. Biol. Chem.*, *240*, 1919 (1965).

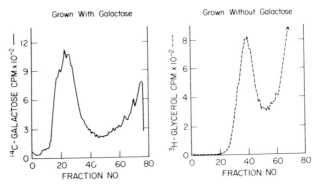

Fig. 1. Membranes from cells grown in the presence or absence of galactose. Cells were grown as described by Osborn *et al.* (1) with the following additions: *Grown with galactose*: [¹⁴C]galactose, 2.5 mM (0.1 μCi/μmole), [¹²C]glycerol, 10 mM. *Grown without galactose*: [³H]glycerol, 10 mM (0.2 μCi/μmole). Membranes were prepared as described by Osborn *et al.* (1). Membranes in 40% sucrose containing 5 mM EDTA, were layered on gradients of 42-55% sucrose, 5 mM EDTA. The gradients were centrifuged to equilibrium (18 hours) in a Spinco SW 41 rotor at 183,000 x g (av). At these concentrations of sucrose, inner membrane remains at the top of the gradient which is at the right of each graph.

How does lipopolysaccharide enter the outer membrane of *E. coli* ?

A. If by diffuse intercalation, we would predict:

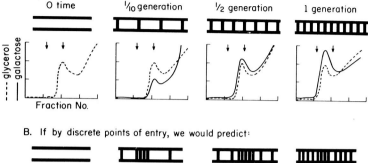

B. If by discrete points of entry, we would predict:

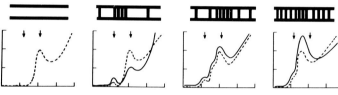

Fig. 2. Possible modes of entry of lipopolysaccharide into the outer membrane. The figures show the predicted patterns of membrane fragments centrifuged to equilibrium from cells harvested various times after addition of galactose. The membrane diagrams above each figure show new galactose-containing LPS as dark bars within the matrix of the membrane. Details in text.

See figure on following page
Fig. 3. Membranes of cells grown 2 minutes and 4 minutes in the presence of galactose. Cells were grown without galactose (but with [3H]glycerol) exactly as described for Fig. 1, except that fucose (2 mM) was added to induce the cells for the galactose metabolic pathway. The cells were centrifuged and resuspended in fresh medium containing the same concentration of [3H]glycerol, plus [14C]galactose, 40 μM (12.5 μCi/μmole). After the indicated times, cells were rapidly chilled and membranes prepared as for Fig. 1, except that sonication was for 6 minutes instead of 1 minute. Gradients were prepared and centrifuged as for Fig. 1. The positions of dense outer membrane (tube 23) and light outer membrane (tube 40) are indicated by arrows. The insets show the region of dense outer membrane on an expanded scale.

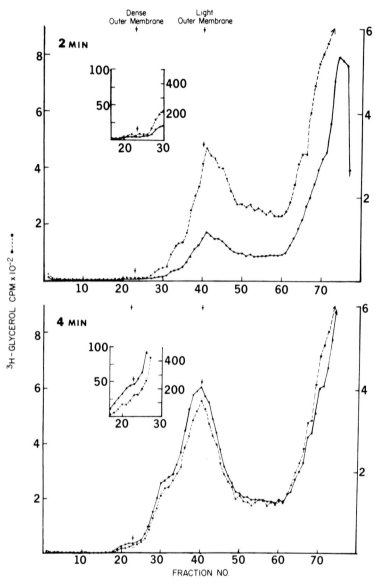

Fig. 3

IV

STRUCTURE AND ASSEMBLY OF
VIRAL MEMBRANES

STRUCTURE AND ASSEMBLY OF VIRAL MEMBRANES

Purnell W. Choppin, Richard W. Compans, Andreas Scheid,

James J. McSharry, and Sondra G. Lazarowitz

The Rockefeller University

New York, New York 10021

ABSTRACT. Influenza and parainfluenza virions consist of a helical nucleocapsid surrounded by a membrane covered with spike-like projections composed of glycoproteins. The glycoproteins possess hemagglutinating and neuraminidase activities and are external to the membrane lipids, which the available evidence suggests form a bilayer. Associated with the inner surface of the viral membrane is a non-glycosylated protein which is thought to play the major role in imparting stability to the membrane. The viruses are assembled at the plasma membrane of the host cell by first the incorporation of viral envelope proteins into discrete areas of membrane, followed by the alignment of the nucleocapsid beneath these areas of altered membrane, and then envelopment of the nucleocapsid by the membrane and release of the virion through a budding process. Although the viral and cellular membranes are morphologically similar and continuous during budding, the virion contains only virus-specific proteins. Methods have been developed for the isolation of these viral proteins in biologically active form. In contrast to the absence of cellular proteins in the virion, the lipids of the virion are, with a few exceptions, those of the plasma membrane of the host cell. There is a suggestive correlation between cell membrane lipid composition and susceptibility to virus-induced cell fusion and virus yield. In infected cells continuously releasing parainfluenza virus, there is inhibition of sphingomyelin synthesis and stimulation of glycosphingolipid synthesis, changes that occur

to a much lesser extent with time in uninfected confluent
cell monolayers. These studies illustrate the usefulness
of enveloped RNA viruses in the study of membrane
structure and biogenesis.

INTRODUCTION

The myxovirus group includes the various types of influenza virus which infect man and animals. The paramyxovirus group includes measles, mumps, Newcastle disease
virus, and several parainfluenza viruses. These viruses
are biologically important because of the variety of
acute and chronic diseases they cause, and this has stimulated work in our laboratory on structure, replication,
and mechanisms of cell damage, using influenza and the
parainfluenza virus SV5 as representative viruses. In
addition, we have studied these viruses because of their
potential value as model systems for investigation of
membrane structure and biogenesis, since they possess a
membrane which is acquired when the virion is assembled
at the plasma membrane of the host cell.

Studies on enveloped viruses (such as influenza and
parainfluenza) have revealed that they have several properties which are advantageous for the study of the
structure and assembly of plasma membranes of vertebrate
cells. These properties which will be discussed in this
paper, include the following: 1) The viral membrane contains only a few proteins which are virus-specific; thus
the protein composition of the viral membrane can be
varied by selecting strains of virus. 2) The lipid composition of the viral membrane reflects that of the plasma
membrane of the host cell, thus the lipids can be varied
by selecting the appropriate host cell. 3) Highly purified virus preparations can be obtained in a form suitable not only for biological and biochemical studies,
but also for biophysical studies such as X-ray diffraction
and electron spin resonance. 4) Methods have been recently developed for the isolation of each of the proteins
associated with the viral membrane in biologically active
form. 5) Some parainfluenza viruses, such as SV5, can
replicate continuously in certain cells without causing
cell death or obvious cell damage, thus providing a system
for studying the turnover of plasma membrane lipids by

utilizing the infected cells and released virus. This report will summarize work carried out in our laboratory over the past several years on SV5 and influenza viruses. Other viruses have also been employed at times to establish the generality of a finding among the major groups of enveloped RNA viruses. These studies have been described in detail elsewhere (1-20) and the earlier work has been reviewed previously (21-22).

METHODS

The W3 strain of the parainfluenza virus SV5, the WSN strain of influenza A_0, and the Indiana strain of vesicular stomatitis virus (VSV) were employed. The procedures for growth, assay, and purification of these viruses have been described (1, 2, 5, 18). Primary rhesus monkey kidney (MK) cells, the BHK-21F line of baby hamster kidney cells, the HaK line of adult hamster kidney cells and the MDBK line of bovine kidney cells were grown and infected as described previously (1, 3, 6).

The experimental and analytical procedures used in the studies discussed in this communication have been described in detail elsewhere. The procedures and the appropriate references are as follows: electron microscopy (4, 17); isotopic labeling of cells and viruses (12, 18, 19); polyacrylamide gel electrophoresis (12, 19); isolation of plasma membranes (6, 19); lipid analyses (6-8); carbohydrate analyses (8, 12); proteolytic enzyme treatment of virions (18, 13); electron spin resonance studies (20).

RESULTS

Structure of Influenza and Parainfluenza Viruses. The structure of these enveloped RNA viruses has recently been reviewed (21, 22). Those points pertinent to the topic of membranes will be briefly summarized here. Influenza virions are 800-1000 A and parainfluenza virions 1200-4500 A in diameter (Figs. 1 and 6). Most virions are roughly spherical in shape, however long filamentous forms up to several microns in length may be formed. The latter may contain multiple copies of the virus genome. Although influenza and parainfluenza viruses differ in several

structural features, most notably in their RNA content and
the structure of the helical nucleocapsids which are coiled
within the virions, the envelopes of these viruses share
important properties. The viral envelope consists of a
membrane which morphologically resembles the plasma membrane of the host cell (Fig. 2), and which is covered by
a layer of surface projections or spikes 80-100 A in
length. These spikes possess hemagglutinating and neuraminidase activities and the strain-specific antigenicity of
the virion. With influenza virus, evidence has been obtained that the neuraminidase and hemagglutinating activities reside on the virion in separate spikes which are
morphologically distinct (23). In appropriately stained
thin sections of influenza virions, an electron dense
layer is seen immediately beneath the structure which has
the triple-layered image typical of cell membranes (Fig.
8). This is thought to be a protein layer which will be
discussed below.

The Assembly of Virions. Before examining the chemical
composition and structure of the virions in more detail,
it will be useful to consider briefly the current knowledge
concerning the sequence of events in the assembly of enveloped RNA viruses. The assembly of influenza and parainfluenza virions has been described in detail (4, 17)
and recently reviewed (21, 22). The following summary is
based on work with both the parainfluenza virus SV5 and
influenza virus.

The first morphologically identifiable step in the
multistep process of virus assembly is the appearance of
the nucleocapsid in the cytoplasm of the infected cell.
The nucleocapsid then becomes aligned in a regular,
ordered array beneath areas of plasma membrane which
contain viral envelope proteins (Fig. 3). The conclusion that the nucleocapsid aligns specifically beneath those areas of membrane in which viral proteins
are already present is based on the ordered arrangement
of nucleocapsid beneath only certain areas of membrane,
suggesting that specific recognition is occurring, and
the demonstration of viral antigen and viral hemagglutinin
in discrete areas of membrane which do not yet have nucleocapsid beneath them (21, 22). However, the viral spikes
are seen as projections only on the surface of those areas
of membrane beneath which nucleocapsid is aligned,
suggesting that the appearance of the morphologically

identifiable spikes occurs after the nucleocapsid aligns
under those areas of membrane which contain the viral
proteins. The spikes may form by a co-operative phenomenon, such as assembly from subunits at the cell surface.
The next step in virus assembly is a budding process
in which the nucleocapsid is enveloped in the altered
membrane (Fig 4). Both roughly spherical and long filamentous particles are produced by this process (Fig 5).
During budding, the membrane of the emerging virion is
continuous with, and morphologically indistinguishable,
from the plasma membrane of the host cell. Not only does
the membrane of the virion as seen in thin section resemble that of the cell, but the membranes of virions
grown in different cells whose plasma membranes show distinctive staining properties can also be distinguished.
Another striking feature of the process of virus assembly
of parainfluenza virions is that the emerging virions
are covered with the viral spikes down to the base of the
budding particle, but they are not seen on the adjacent
areas of cell membrane. Similarly, ferritin-labeled
antiviral antibodies stain only the budding virions and
not the surrounding cell membrane (Fig. 5). This em-
phasizes the discrete, well-localized nature of the membrane alterations in parainfluenza virus infected cells
(9, 21, 22).

Lipids of the Viral and Plasma Membranes. Highly
purified SV5 virions consist of 0.9% RNA, 73% protein,
20% lipid and 6.1% carbohydrate (5). We have carried out
extensive comparative studies of the lipids of SV5 virions
grown in four different cell types and of the plasma
membranes of these cells. These studies were done not
only to determine the lipid composition of the virus and
of plasma membranes of cultured cells, but also to investigate the contribution of the plasma membrane to
the virion, and to correlate chemical composition of the
membrane with its biological response to the virus.
Monkey kidney (MK) and bovine kidney (MDBK) cells, which
produce a high yield of virus and are relatively resistant
to virus-induced cell fusion, and hamster kidney (BHK21-F
and HaK) cells, which produce a low yield and are very
sensitive to cell fusion, were used in these studies.
These results have been reported in detail (6-8, 21) and
are summarized below.
A comparison of the lipids of plasma membranes of the

whole cells with those of isolated plasma membranes revealed that the membranes have a higher sphingomyelin and cholesterol content, a higher molar ratio of cholesterol to phosopholipid, and more saturated fatty acids than the respective whole cells. A comparison of the distribution of the major lipid classes in virions grown in the four cell types with those of the respective plasma membranes showed, in each case, a close correlation, including the molar ratios of cholesterol to phospholipid. There are distinctive differences in the lipid patterns of the different cell membranes, particularly with regard to phosopholipids, e.g., MK cells have a very high phosphatidylethanolamine (PE) content, even higher than phosphatidylcholine (PC), relatively high phosphatidyleserine (PS), and little phosphatidylinositol (PI). However, the reverse is true in hamster cells which have relatively lower PE and PS, and higher PC and PI contents. These distinctive differences in the membrane lipid patterns of the various cells were clearly reflected in the lipids of the virions grown in these cells. The fatty acids of the phospholipids of the plasma membranes and of the virions were also examined, and with a few exceptions the fatty acids of the virions closely resembled those of the membrane of the cell in which they were grown. The fatty acid patterns of the membranes and the virions reflect that of the serum present in the medium in which the cells were grown. The fatty acid pattern of the virion could be significantly altered by addition of an essential fatty acid, linoleic acid, to the medium.

These results suggested that during the maturation of the virion, the lipids of the plasma membrane are incorporated essentially quantitatively into the virion. The host cell thus appears to be the chief determinant of the lipid composition of the viral envelope. However, some exceptions found in the MDBK cell system suggest that, within narrow limits, there may be some selective incorporation of available phospholipids, and that there may be an optimal membrane lipid pattern for virus production, characterized by a relatively high phosphatidylethanolamine and lower phosphatidylcholine content (7, 21).

Analysis of the glycosphingolipids in the different cells revealed distinctive qualitative and quantitative differences. MK and MDBK cells contain relatively high concentrations of neutral glycolipids and little neuraminic

acid-containing glycolipids (gangliosides), whereas BHK21-F
and HaK cells had relatively high concentrations of
gangliosides and very little neutral glycolipid. The
neutral glycolipids of SV5 virions resembled those of the
plasma membrane of the host cell, however the virions did
not contain the neuraminic acid-containing gangliosides.
The virions also did not contain neuraminic acid bound to
glycoproteins, and this was shown to be due to the loss
of neuraminic acid residues in those areas of plasma
membrane in which the viral enyzme neuraminidase had been
incorporated (8-10). The localized loss of these residues
represents a significant change in the cell membrane,
since they account for a major portion of the negative
change on the cell surface. Such a change might play a
role in subsequent steps in the incorporation of viral
proteins into the membrane or other aspects of viral
assembly.

The results of these lipid analyses have thus indicated
that the lipid composition of the viral membrane closely
resembles that of the plasma membrane of the host cell.
Thus virions can be produced which contain the same viral
proteins, but markedly different lipids, depending on the
cell in which the virus was grown.

The Proteins of Influenza and Parainfluenza Virions
and the Structure of Viral Membranes. In contrast to the
above described situation concerning the viral lipids,
evidence from many laboratories using different enveloped
RNA viruses suggests that there is not a significant
amount of host cell protein in the virions. This conclusion is based on experiments indicating that the same
pattern of proteins is found when a given virus is grown
in several different cell types, that different viruses
grown in the same cell contain completely different proteins, and that no cell protein prelabeled before infection is incorporated into the virions. Thus, in spite
of the mechanism of assembly of the virus at the plasma
membrane, those areas of membrane which become the viral
membrane contain only virus-coded proteins. These areas
thus represent localized domains which are in continuity
with the surrounding plasma membrane, but from which
cell membrane proteins are excluded.

The total number of proteins in the virions and the
number of proteins associated with the envelope varies
among the various enveloped RNA viruses. We have found

three major polypeptides associated with the envelope of parainfluenza virions, and five with that of influenza virions (11, 14, 18, 19). Glycoproteins have been found on all enveloped RNA viruses thus far studied; the number varies from one in vesicular stomatitis virions (VSV) to as many as four in influenza virions. By selective digestion of the spikes on the virus surface with proteolytic enzymes, it has been established with every RNA virus examined that these spikes are glycoproteins. This has been found with influenza (18), Sindbis, an arbovirus (16), parainfluenza (13), VSV (15) and Rous sarcoma viruses (24). Removal of the spikes by proteolytic enzyme treatment leaves a smooth-surfaced, membrane-enclosed particle (Fig. 7), which contains the viral lipids but has lost hemagglutinating and neuraminidase activities in the case of influenza and parainfluenza viruses, and infectivity, presumably due to the loss of the adsorptive mechanism. Although the glycoproteins can be completely removed, none of the other viral proteins are affected by proteolytic enzyme treatment unless the virus is first treated with a detergent, suggesting that these proteins are protected by the lipids of the membrane.

These studies have revealed several important features of the structure of enveloped viruses which can be summarized as follows. The viral spikes, but no other viral proteins, are glycoproteins. Since the spikes can be removed without disintegration or great distortion of the virion, they are not essential for maintaining the shape of the virus or the integrity of the viral membrane. In contrast to the other viral proteins which are protected, the glycoproteins are freely accessible to proteolytic enzymes, suggesting the spikes are largely, if not entirely, external to the lipid.

Further evidence on the organization of the lipid phase of the influenza viral membrane and the relationship of the glycoproteins to the lipids has been obtained in electron spin resonance studies done in collaboration with F. R. Landsberger and J. Lenard using three different spin-labels, two derivatives of stearic acid with a nitroxide ring at position C_5 or C_{16} and androstane (20). The data obtained suggest that the lipid phase of the viral membrane is slightly more rigid than that of erythrocyte ghosts, and that the viral membrane is in the form of a bilayer. Complete removal of the spikes with a proteolytic enzyme caused no alterations in the spectra of any of the three spin-labels, suggesting that the spikes are not involved in determining the organization of the lipid phase, and thus reenforcing the above conclusion that the spikes are superficial, and do not penetrate deeply into the bilayer.

If the glycoprotein spikes do not play a major role in imparting stability to the viral membrane is there another protein that does? In influenza, parainfluenza, and vesicular stomatitis virions there is a non-glycosylated protein associated with the viral envelope which appears to perform the function, although the evidence is still circumstantial. These proteins have been referred to as viral membrane proteins. In the case of influenza virus, there is evidence suggesting that this protein forms a layer on the inner surface of the viral membrane (17-19). In thin sections an electron-dense layer is seen beneath the triple-layered image of the viral membrane (Fig. 8), and there is only one protein present in sufficient quantity in the virion to form such a layer (18). Preliminary X-ray diffraction studies by Harrison and Compans are compatible with the presence of such a protein layer at that radius in the virion. Finally, with fixation and detergent treatment to remove the lipid from influenza virions, Schulze (25) and Skehel (26) have isolated shell-like structures of the appropriate size which consist of the above-described protein surrounding the nucleocapsid and another internal protein.

If as evidence suggests, the non-glycosylated envelope protein of an RNA virus with a helical nucleocapsid is associated with the inner surface of the viral membrane, one would expect that during viral assembly this protein would be the entity that the viral nucleocapsid recognizes and interacts with when it aligns beneath those areas of cell membrane which contain viral envelope proteins. Some evidence that this is the case has been obtained in phenotypic mixing experiments with cells doubly infected with the parainfluenza virus SV5 and VSV (15). It was found that there was no restriction on phenotypic mixing of the glycoprotein spikes, but that virions containing the VSV nucleocapsid contained only the non-glycosylated VSV membrane protein, not the corresponding SV5 protein. This suggests that in the assembly process a specific interaction between the VSV nucleocapsid and the VSV membrane protein must occur.

171

Fig. 9 shows polyacrylamide gel electrophoresis patterns of the proteins of representatives of four major groups of enveloped RNA viruses. The proteins were labeled with ^{14}C-amino acids and the glycoproteins identified by labeling with ^{3}H-glucosamine. The molecular weights of the glycoproteins vary from $\sim$ 30,000 to 80,000 daltons, the extremes being represented by the smallest and largest influenza virus proteins. The molecular weights of the non-glycosylated membrane proteins are $\sim$ 29,000 with VSV, 27,000 with influenza virus, and 41,000 with SV5. The nucleocapsid of each virus is composed of a single polypeptide species. Several points should be made about Sindbis virus. As illustrated in Fig. 9, only glycoprotein and a nucleocapsid protein are present. It was formerly thought (27, 28) that there was only a single glycoprotein, however it was recently reported that there are two glycoproteins present (29). For the purposes of this discussion, however the important point is that all the glycoprotein is superficially located in the viral spikes. As mentioned above Compans (16) found that all the glycoprotein could be removed from the virion by proteolytic enzyme treatment, leaving a smooth-surfaced, membrane-enclosed particle. The only protein remaining in the virion after removal of the spikes was the nucleocapside protein. Thus if any protein were responsible for the stability of the viral membrane, it would be the nucleocapsid protein. Important evidence on the structure of the Sindbis virus has been recently obtained in X-ray diffraction studies by Harrison and coworkers (30), which indicate that the lipid is in the form of a bilayer, and that the spikes do not extend through this bilayer.

Much remains to be learned about the biological functions, as well as the chemical and physical properties of viral envelope proteins. Although it is known with the influenza and parainfluenza viruses that the neuraminidase and hemagglutinating activities reside in the spikes, with only a few strains of influenza virus has it been possible to isolate these structures in biologically active form. There are thus important questions to be answered concerning which viral polypeptides possess the various biological activities, particularly in the case of the paramyxovirus group. For example, an important property associated with the envelope of these viruses is the ability to cause cell fusion and hemolysis of erythrocytes,

and the viral component responsible for this remains to be
established. In addition, little is known about the properties of the non-glycosylated membrane proteins of the
enveloped viruses.

We have recently succeeded in developing methods for
the isolation of both the glycoproteins (31) and the non-
glycosylated membrane protein (32) of SV5 virions by a
gentle procedure which preserves biological activity. The
procedure involves disruption of the virus by the non-ionic
detergent Triton X-100 in the presence of 0.5-1.0 M KCl,
separation of the two glycoproteins from each other by
high speed centrifugation in the presence of Triton X-100
and high salt, and separation of the non-glycosylated proteins from glycoproteins by precipitation of the former by
reducing the salt concentration. These procedures are
being perfected and scaled up so that sufficient quantities
can be obtained for analysis of the various chemical,
physical and biological properties of these proteins, including reconstitution experiments involving the viral proteins and lipids.

Association of Influenza Virus Proteins with the Plasma
Membrane in the Infected Cell. Although much remains to
be learned about the exact sequence of the incorporation
of the various envelope proteins into the viral membranes,
a few points will be made briefly here. In influenza
virus infected cells, all the viral structural proteins
can be found associated with the infected cell membranes
(19), however the two smallest glycoproteins are the last
to be detected at the plasma membrane. This has been
shown in pulse-chase experiments to be due to the cleavage
of the largest glycoproteins (mol. wt. ∿ 80,000) to two
smaller proteins (mol. wts. ∿ 50,000 and 30,000). The
available evidence suggest that this cleavage occurs at
the plasma membrane and is performed by a host cell enzyme.
The efficiency of this cleavage varies with the host cell,
and both cleaved and uncleaved molecules can be incorporated into virions. These polypeptides are associated with
the hemagglutinating activity of the virus, and this
activity is present regardless of whether the large uncleaved glycoprotein molecules or the two cleavage products are present on the virion.

Another significant finding in cells infected with
influenza virus is that the non-glycosylated membrane protein, which is the protein present in largest amount in

the virion, is present in relatively small amounts in the
infected cells (19). These results suggest that the
synthesis of this protein is tightly controlled, and that
this could be a rate-limiting step in viral replication
and assembly. Once the protein is synthesized it appears
to become associated very efficiently with the plasma
membrane.

Lipid Metabolism in Parainfluenza Virus Infected Cells.
SV5 causes a persistent infection in MK and MDBK cells
without extensive cell damage, and with continued release
of virus for many days (1, 3, 21). Since the lipids of
the virion are similar to those of the plasma membrane,
this provides a system for investigating the turnover of
plasma membrane lipids. In SV5 infected MDBK cells, it
was found that there was normal synthesis of glycerophos-
pholipids, but there was a decrease in the synthesis and
the total cellular content of sphingomyelin (33). There
was a reciprocal increase in the synthesis and total cell
content of the glycosphingolipid, globoside, which is the
major glycosphingolipid in MDBK cells. These results not
only reveal a virus-induced alteration in lipid metabolism,
but also suggest that there is an interdependence in the
metabolism of sphingolipids, lipids which are concentrated
in the plasma membrane. It was also found that a decrease
in sphingomyelin synthesis and an increase in globoside
synthesis occurred in ageing monolayers after they reached
confluence, but to a much lesser extent than was found in
infected cells. Robbins found (34) that in rapidly growing cells, whether transformed by a tumor virus or not,
there may be failure to synthesize the longer chain glycolipids that are normally synthesized after the cells reach
confluence. The results with SV5 suggest that infection,
which involves the continued release of plasma membrane
lipids in virions, greatly stimulates a process which
occurs normally in cells after they reach confluence, a
time when membrane turnover normally occurs as opposed to
the membrane synthesis associated with cell growth. A
final point regarding the decrease in sphingomyelin content
and the increase in globoside content in cells in SV5 infected cells is that the lipid composition of the virions
produced changes in a similar fashion with time, thus providing further evidence for the dependence of the viral
lipids on the composition of the host cell membrane.

DISCUSSION

Through the use by a number of workers of a variety of experimental and analytical techniques, a picture of the structure of the membranes of enveloped RNA viruses has begun to emerge. With no single virus is the knowledge complete, and one must draw on information obtained with several viruses. Recognizing that there may be important differences among the major groups of viruses, the following general model can be proposed which can serve as a basis for further investigation. The surface of the virion is covered with spikes which are composed of glycoproteins; the number of glycoproteins varies with the type of virus. These spikes do not penetrate through the lipid layer, but the possibility that they penetrate into it slightly cannot be excluded. The lipid of the viral membrane is organized as a bilayer. Associated with the inner surface of the lipid layer there is a non-glycosylated protein which plays the major role maintaining the integrity of the viral membrane. In those viruses with helical nucleocapsids, such as influenza, parainfluenza, and VSV, there is a non-glycosylated envelope protein which can perform this function. In the case of the arboviruses which possess an icosahedral nucleocapsid, the only protein that could serve is the protein subunit of the nucleocapsid.

The exact role of the carbohydrate on the glycoproteins is not yet clear. The finding that the spikes of all enveloped viruses are glycoproteins and the other proteins lack carbohydrate, suggests that the carbohydrate may be essential for the positioning of the spike on the surface of the viral membrane. A variety of lines of evidence obtained in several laboratories suggest that the composition of carbohydrate moieties of the glycoprotein is determined by the host cell. This evidence includes the finding that a host cell carbohydrate antigen is covalently linked to the hemagglutinin of influenza virus (35, 36), that Sindbis or vesicular stomatitis virions grown in different cells have a different carbohydrate composition (28, 37, 38), that the Sindbis virus glycoprotein can serve as an acceptor for host specified glycosyl transferases (37), and that influenza virus glycoproteins show slightly different mobilities in polyacrylamide gels, depending on the cell in which they were grown (18).

Furthermore, the genetic information available in most of these viruses is insufficient to code for a series glycosyl transfereases. There is also no evidence that the carbohydrate is responsible for any of the virus-specific immunological properties of the virion. These facts all suggest that the presence of carbohydrate determined by the host is important for the surface location of the viral spikes, but that a virus-specific sequence of sugars is not required for the various biological activities of the virus, although the possibility remains that the presence of host-specified carbohydrate chains on the viral polypeptides could permit the expression of some virus-specific activity such as adsorption.

As described above much has been learned about the assembly of enveloped viruses, but important details remained to be elucidated, such as in what order are the viral envelope proteins incorporated into the membrane, and how are the regions of membrane which become the viral enveloped formed? The possibilities for the latter include: 1) step-wise replacement of plasma membrane proteins with viral proteins; 2) the assembly of a new area of membrane about a focus beginning with the incorporation of the first viral protein; 3) the en bloc insertion in the plasma membrane, by fusion of a vesicle, of a patch of membrane containing viral proteins which was synthesized elsewhere in the cell, such as in the Golgi complex; and 4) the random insertion of viral proteins into the plasma membrane with their diffusion in the plane of the membrane to form patches of membrane containing only viral proteins. Although there is some suggestive evidence available in favor of mechanisms 2) or 3), none of the possibilities can be definitely established or excluded at the present. This question of viral membrane assembly obviously bears directly on the mechanisms of the assembly of plasma membranes in general, and it is being actively investigated in several laboratories using tools such as antisera specific for individual viral proteins and labeled with electron dense markers, and temperature sensitive mutants. One point that does seem clear is that once an area of membrane containing viral protein is assembled, it appears to represent a domain into which diffusion of host cell proteins does not occur. This could be due to a specific interaction among the viral membrane protein molecules, or between these proteins and the lipids of the membrane.

Studies on the lipid composition of plasma membranes of cultured mammalian cells have indicated that there are significant differences among different cell types. Comparison of our data on four different cells has suggested a correlation between plasma membrane lipid composition and the biological responses of the membrane to the virus including sensitivity to virus-induced cell fusion and yield of virus. Cells with a relatively high ganglioside content and low phosphatidylethanolamine content (BHK21-F and HaK) were sensitive to cell fusion and produced little virus, whereas cells with a low ganglioside and high phosphatidylethanolamine content (MK and MDBK) were relatively resistant to fusion and produced a high yield of virus (6-8, 21). Although these results do not definitely establish a direct relationship between these lipids and the behavior of the membranes, they are sufficiently suggestive to stimulate further investigation.

As described above, significant and interesting changes have been found in the metabolism of glycosphingolipids in cells infected with a parainfluenza virus. Further studies of the type should provide knowledge not only of virus-induced alterations in cell membranes, but also of the control of biosynthesis of plasma membrane lipids in general.

ACKNOWLEDGMENT

Research by the authors was supported by Research Grant AI-05600 from the National Institutes of Health and Contract No. AT(11-1)-3504 from the U. S. Atomic Energy Commission.

REFERENCES

1. Choppin, P. W., and Stoeckenius, W., Virology, 23, 195 (1964).
2. Choppin, P. W., Virology 23, 224 (1964).
3. Holmes, K. V., and Choppin, P. W., J. Exptl. Med. 124, 501 (1966).
4. Compans, R. W., Holmes, K. V., Dales, S., and Choppin, P. W., Virology, 30, 411 (1966).

5. Klenk, H.-D., and Choppin, P. W., Virology, 37, 155 (1969).
6. Klenk, H.-D., and Choppin, P. W., Virology, 38, 255 (1969).
7. Klenk, H.-D., and Choppin, P. W., Virology, 40, 939 (1970).
8. Klenk, H.-D., and Choppin, P. W., Proc. Natl. Acad. Sci. U.S., 66, 57 (1970).
9. Klenk, H.-D., Compans, R. W., and Choppin, P. W., Virology, 42, 1158 (1970).
10. Klenk, H.-D., and Choppin, P. W., J. Virol., 7, 416 (1971).
11. Caliguiri, L. A., Klenk, H.-D., and Choppin, P. W., Virology, 39, 460 (1969).
12. Klenk, H.-D., Caliguiri, L. A., and Choppin, P. W., Virology, 42, 473 (1970).
13. Chen, C., Compans, R. W., and Choppin, P. W., J. Gen. Virol., 11, 53 (1971).
14. Mountcaslte, W. E., Compans, R. W., and Choppin, P. W., J. Virol., 7, 47 (1971).
15. McSharry, J. J., Compans, R. W., and Choppin, P. W., J. Virol., 8, 722 (1971).
16. Compans, R. W., Nature, 229, 114 (1971).
17, Compans, R. W., and Dimmock, N. J., Virology, 39, 499 (1969).
18. Compans, R. W., Klenk, H.-D., Caliguiri, L. A. and Choppin, P. W., Virology 42, 880 (1970).
19. Lazarowitz, S. G., Compans, R. W., and Choppin, P. W., Virology, 46, 830 (1971).
20. Landsberger, F. R., Lenard, J., Paxton, J., and Compans, R. W., Proc. Nat. Acad. Sci. USA, 68, 2579 (1971).
21. Choppin, P. W., Klenk, H.-D., Compans, R. W. and Caliguiri, L. A., In "From Molecules to Man", Perspectives in Virology VII, M. Pollard, Ed., Academic Press, New York, p. 127 (1971).
22. Compans, R. W., and Choppin, P. W., In "Comparative Virology", K. Maramorosch and F. Kurstak, Ed., Academic Press, New York p. 407 (1971).
23. Laver, W. G., and Valentine, R. C., Virology, 38, 105 (1969).
24. Rifkin, D. B., and Compans, R. W., Virology, 46, 485 (1971).

25. Schulze, I. T., Virology, 47, 181 (1972).
26. Skehel, J. J., Virology, 44, 409 (1971).
27. Strauss, J. H., Jr., Burge, B. W., Pfefferkorn, E. R., and Darnell, J. E., Jr., Proc. Nat. Acad. Sci. U.S.A., 59, 533 (1968).
28. Strauss, J. H., Jr., Burge, B. W., and Darnell, J. E., J. Mol. Biol., 47, 437 (1970).
29. Schlesinger, M. J., Schlesinger, S., and Burge, B. W., Virology, 47, 539 (1972).
30. Harrison, S. C., David, A., Jumblatt, J., and Darnell, J. E., J. Mol. Biol., 60, 523 (1971).
31. Scheid, A., Caliguiri, L. A., and Choppin, P. W., Bacteriol. Proc. V 181 (1972).
32. McSharry, J. J., Compans, R. W., Lackland, H., and Choppin, P. W., Bacteriol. Proc., V 178 (1972).
33. Scheid, A., and Choppin, P. W., Bacteriol. Proc. V 205 (1971).
34. Robbins, P. W., and Macpherson, I., Nature, 229 569 (1971).
35. Howe, C., Lee, L. T., Harboe, A., and Haukenes, G., J. Immunol., 98, 543 (1967).
36. Laver, W. G., and Webster, R. G., Virology, 30, 104 (1966).
37. Burge, B. W., and Huang, A. S., J. Virol., 7, 412 (1970).
38. McSharry, J. J., and Wagner, R. R., J. Virol., 7, 412 (1971).
39. Grimes, W. J., and Burge, B. W., J. Virol., 7, 309 (1971).
40. Compans, R. W., and Choppin, P. W. In "Atlas of Virus Morphology", F. Haguenau and A. J. Dalton, eds., Academic Press, (1972) in press.

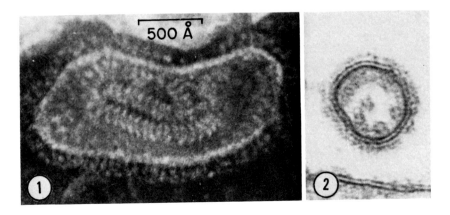

Fig. 1. SV5 virion negatively stained with sodium phosophotungstate. The helical nucleocapsid is coiled inside the envelope which is covered with a layer of projections or spikes ∿ 100 A in length. X 300,000. From Choppin and Stoeckenius, 1964 (1).

Fig. 2. A circular profile of a sectioned SV5 virion adjacent to the surface of a rhesus monkey kidney cell. The membrane in the viral envelope is resolved clearly, and is similar in appearance to the membrane of the host cell surface. The outer surface of the viral envelope is covered with a layer of projections which correspond to the spikes seen with negative staining; such projections are absent on normal plasma membrane. X 210,000. From Compans and Choppin, 1972 (40).

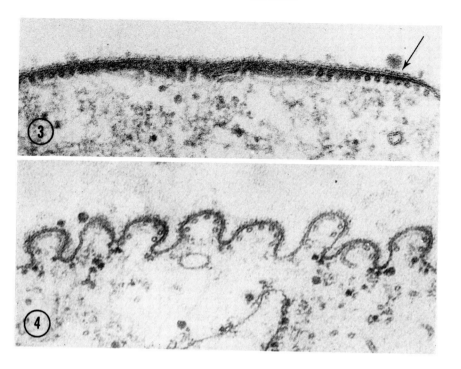

Fig. 3. Nucleocapsids, many in cross section, closely aligned under a long region of the membrane of a SV5 infected monkey kidney cell. A layer of dense material (arrow) corresponding to the viral surface projections is present on the outer surface of the membrane. X 105,000. From Compans et al., 1966 (24).

Fig. 4. A row of eight SV5 virions in the process of budding at the surface of a monkey kidney cell. Many cross sections of nucleocapsids are apparent in the interior of the budding particles. X 80,000. From Compans et al. 1966 (24).

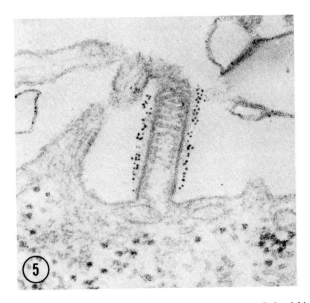

Fig. 5. SV5 filament in the process of budding, tagged with ferritin-conjugated antiviral antibody. The cell surface just adjacent to the filament is devoid of ferritin tagging. X 90,000. From Compans and Choppin, 1971 (22).

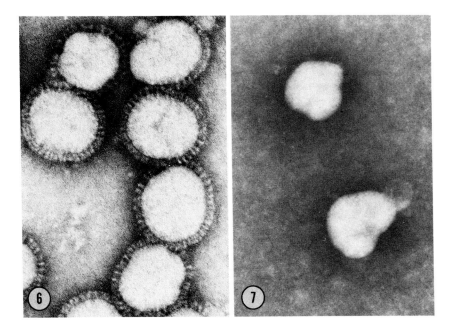

Fig. 6. Influenza virions negatively stained with sodium phosphotungstate. The particles have a diameter of ∿ 1000 A and show clearly the layer of spikes 100-120 A in length on their surfaces. X 220,000. From Compans et al., 1970 (18).

Fig. 7. Influenza virus particles after treatment with bromelain. The spikes have been removed from the surface of the particle, which is bounded by a smooth-surfaced membrane. X 220,000. From Compans et al. 1970 (18).

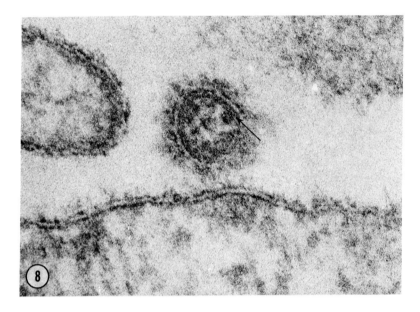

Fig. 8. Thin section of an influenza virion adjacent to the surface of a bovine kidney (MDBK) cell. The viral envelope contains a membrane similar in appearance to that of the host cell, and also an additional electron dense layer (arrow) which is not present in the normal cell membrane. The surface of the virion is covered with a layer of dense material which corresponds to the spikes seen by negative staining. X 300,000. From Compans and Choppin, 1972 (40).

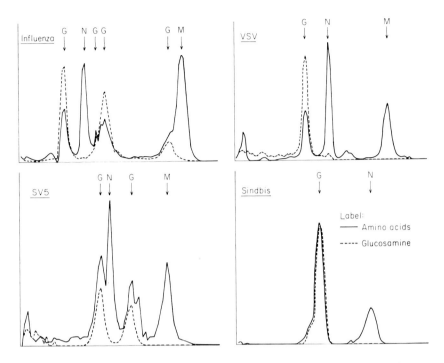

Fig. 9. Polyacrylamide gel electrophoresis of the proteins of influenza, SV5, VSV, and Sindbis virions. Proteins were labeled with ^{14}C-amino acid mixture (solid line) and ^{3}H-glucosamine (dotted line). Glycoproteins are designed, G, nucleocapsid proteins, N, and non-glycosylated membrane proteins, M. Virions were treated with sodium dodecyl sulfate and mercoptoethenal and subjected to electrophoresis processing, and determination of radioactivity as referred to in Methods. The abscissa represents distance migrated and the ordinate radioactivity. Migration was from left to right.

ENVELOPE PROTEINS OF THE AVIAN TUMOR VIRUSES

William S. Robinson and Harriet L. Robinson

Department of Medicine, Stanford University
Stanford, California   94305

ABSTRACT. ATV's[1] appear to contain 9 proteins. Five of
these are nonglycoproteins and these are probably internal components of the virion. Two of the nonglycoproteins
are highly reactive in complement fixation with antiserum
to the ATV gs antigen. The 4 glycoproteins move more
slowly in SDS-gel electrophoresis than the nonglycoproteins and are thus probably larger in size. All 4 glycoproteins and no other virion proteins are iodinated when
intact virions are reacted with I^{125}, lactoperoxidase and
H_2O_2 under conditions which preserve virion structure and
infectivity. Pronase treatment of intact virions results
in cleavage of 3 of the 4 glycoproteins and no other virion proteins. These experiments suggest that the virion
glycoproteins are components of the viral envelope. A
mixture of virion glycoproteins separated from the nonglycoproteins specifically binds virus neutralizing antibody
and interferes with infection of cells with viruses posses-
sing the same envelope antigen.

INTRODUCTION

C-type ATV's have a complex structure, including a
viral envelope or membrane with surface spikes, a second
membrane-like layer inside of the envelope and an electron-
dense core (1,2). These viruses are 1 to 2% RNA by weight,
25 to 30% lipid, and the remainder is protein and carbohydrate (3). The lipid is probably a component of the viral
envelope, because lipid solvents such as ether disrupt the
envelope. These viruses have been shown to contain at
least 9 proteins and 4 of these are glycoproteins (4).
All ATV's contain a common gs antigen (5), which is

[1]Abbreviations: ATV, avian tumor virus; gs, group
specific; anti-gs, antiserum to gs antigen; ts, type specific; RSV(RAV-1), Bryan high titer strain of Rous sarcoma
virus and Rous associated virus; SR-RSV, Schmidt-Ruppin
Rous sarcoma virus; SDS, sodium dodecyl sulfate.

considered to be an internal component of the virion because only dissociated virus and not intact virus reacts
with anti-gs (4). Two of the 9 virion proteins have been
shown to react strongly in complement fixation with anti-
gs, 3 other proteins react weakly and 4 proteins have been
found to give no detectable reaction (4). Subvirion particles or "cores" containing RNA, protein and reverse
transcriptase activity have been isolated after disruption
of the viral envelope with nonionic detergents (6). Such
subviral particles contain the nonglycoproteins including
those which react with anti-gs, again suggesting that the
gs-reacting proteins are internal virion components.

A second virion antigen is a type or strain-specific
(ts) antigen thought to be in the virion envelope (7).
Neutralizing antibody is directed against this antigen, the
antigen appears to determine virus host range by its specificity for genetically controlled chicken cell surface receptors (8) and it is involved in specific viral interference which occurs only between viruses with the same ts
antigen type (9). ATV's have been grouped on the basis of
host range, patterns of viral interference and ts antigens
(10).

The experiments described here were done to determine which virion proteins are components of the viral
membrane or envelope and which are exposed to the virion
surface and may be involved in specific envelope functions
such as viral interference and envelope antigenicity.

METHODS

Viruses. The ATV sub group A (10) RSV(RAV-1) and sub
group D SR-RSV were grown in tissue culture and labelled
with a mixture of 15 C^{14}-or H^3-amino acids or with H^3-glu-
cosamine or H^3-fucose as previously described (4). Virus
was purified from culture medium as previously described
(4).

Gel Electrophoresis. Viral proteins were separated by
electrophoresis in 6% polyacryamide gels containing 0.1%
SDS as previously described (4) after dissociation of virus with 1% SDS, 1% merceptoethanol, and 6 m. urea at 98°C.
for 2 minutes. Gels were sliced as previously described
(4), and each gel slice was extracted in a scintillation
vial with 0.7 ml. water containing 0.1% Triton X-100. Ten
ml. toluene:Triton X-100 (3 vols:1 vol.) was added to each

vial for scintillation counting.
Iodination of Virus. Intact and dissociated viruses
were iodinated with I^{125}, H_2O_2 and lactoperoxidase as described by Phillips and Morrison (11).
Gel Filtration. Components of virus dissociated with
2% Tween 20, 1% mercaptoethanol at 56° for 20 minutes were
layered on a 1.5 X 100 cm. column of Bio-Gel A-5m and
eluted with buffer containing Tris-HCl 0.01 m., NaCl
0.15m., EDTA 0.001 m. and mercaptoethanol 0.001 m. Protein
in aliquots of column fractions was precipitated with TCA
and assayed for radioactivity as previously described (4).

RESULTS

Virion Proteins. Table 1 lists the 9 virion proteins
which have been identified by separating detergent-disso-
ciated virion components first by isoelectric focusing and
then by SDS-gel electrophoresis (4). Proteins P4 and P8
are highly reactive in complement fixation with anti-gs.
The fact that only dissociated virions and not undissoci-
ated virions react with anti-gs indicates that proteins P4
and P8 are probably internal virion components. Proteins
P4-P7 and probably P8 are the main proteins in the RNA-
containing subivral particle isolated after disruption of
the viral envelope with neutral detergent (6), suggesting
that these may all be internal virion components. P4 and
P6 are very basic proteins. The 4 glycoproteins identified in ATV's (P1, P2, P2-a and P3) appear to be larger
in size than the nonglycoproteins, and the 4 contain only
about 18 percent of the radioactive amino acid in virus
labelled with a mixture of 15 radioactive amino acids.
Treatment of Virus with Pronase. Intact virions can
be treated with pronase under conditions which result in
a particle indistinguishable in sedimentation rate, buoyant density and surface charge from untreated virus but
with reproducible alterations in the gel electrophoresis
pattern of the particle proteins. The results of such an
experiment with RSV(RAV-1) labelled with a mixture of 15
H^3-amino acids are shown in Figure 1. Figure 1-A shows
the H^3 proteins of untreated virus and Figure 1-B shows
the proteins of virus treated with pronase (500 µg. per ml.
at 37° for 40 min.) followed by repurification of the virus
in sucrose density gradients. It is clear that the radioactivity in the positions of proteins P1 and P2 is signifi-

cantly reduced by pronase treatment. Figures 2-A and B
show a similar experiment with H^3-fucose labelled RSV(RAV-1)
The fucose-labelled components moving in the positions of
P1, P2 and P2-a are greatly reduced by pronase treatment.
These results indicate the glycoproteins P1, P2 and P2-a
within intact virions are digested by pronase without dis-
rupting the intact particles. When C^{14}-amino acid-RSV
(RAV-1) was dissociated with SDS and digested with pronase,
all 8 virion proteins are susceptible to pronase. When
prolonged digestion ($>$1 hour) of intact particles at
higher pronase concentrations ($>$1 mg/ml) was used, protein P3 almost disappeared in the electropherogram and P4
was decreased somewhat in amount. These experiments with
pronase suggest that at least glycoproteins P1, P2 and P2-a
are on the virion surface and thus can be considered virion
envelope proteins.

    Iodination of Intact Virions. The iodination of pro-
teins catalyzed by lactoperoxidase has been described as a
gentle method which is thought to result in iodination of
only the exposed tyrosine and some histidine residues on
the intact red blood cell surface (11). We have used this
method to iodinate intact virus and under the conditions of
our reaction no measurable viral infectivity was lost
(Table 2). When intact virions were iodinated with I^{125},
followed by repurification of the virus in sucrose density
gradients and analysis of virion proteins by SDS-gel electrophoresis, several labelled components were separated
but clearly not all virion proteins were labelled (Figure3).
The amounts of I^{125} corresponding to the positions of proteins P1, P2, P2-a and P3 are in about the same proportion
as the radioactivity in radioactive amino acid-labelled
virus (Figure 1 and Reference 3). In addition there are
labelled components in the gel between proteins P2 and P2-a
and between P3 and P4. When virus was dissociated with
Tween 20 before iodination, all virion proteins were iodinated. These results, in agreement with the pronase experiments, suggest that at least glycoproteins P1, P2, P2-a
and P3 and possibly other proteins not previously described in ATV's are on the virion surface and are thus
components of the viral envelope.

    Interfering Activity and Immunological Reactivity of
Virion Components. Although virion proteins with gs-anti-
gen reactivity retain their immunological reactivity after
denaturation and separation in SDS and urea (4), the virion

proteins treated in this way react very weakly with virus-neutralizing antibody from chickens (13). In order to clearly identify the protein components of the virion which make up the type specific antigen and those involved in viral interference, we have attempted to dissociate and separate the virion components by methods more gentle than SDS and urea treatment. The experiment in Figure 4 shows the separation of the radioactive components of a mixture of RSV(RAV-1) labelled with C^{14}-amino acids and the same virus labelled with H^3-glucosamine on a column of agarose beads (Bio-Gel A-5m) after dissociation of the virus with 2% Tween 20 and 1% mercaptoethanol. Column fractions were then combined to make 4 pools (a-d). Most of the H^3-labelled material was separated into two major parts, one eluting at the column void volume (pool a) and the second retained by the column and eluting over a broad distribution (pool c). Pool b contained some H^3-labelled material between a and c. Only a small amount of the C^{14}-labelled protein eluted from the column in the positions of a, b and c and most (pool d) was eluted after the H^3 material. The macromolecular material in each pool was then concentrated by ultrafiltration through an Amicon UM-10 Diaflo Membrane. Aliquots of each pool were analyzed by SDS-gel electrophoresis as shown in Figure 5. The material in pools a, b and c contained glycoproteins P1, P2, P2-a and P3 and carbohydrate g4. Proteins P4-P8 were only detected in small amounts in the void volume material (pool a) and not in pools b and c. Pool d contained proteins P4-P8 and almost no detectable glycoprotein. Thus the internal virion proteins (P4-P8) were almost completely separated from the glycoproteins of the virion envelope.

When protein prepared in this way was tested for ability to bind or block the action of neutralizing antibody, all fractions containing the envelope glycoproteins were found to specifically bind antibody to RSV(RAV-1), and no binding was detected with the nonglycoproteins which are internal components of the virion. Similarly, all fractions containing the glycoproteins when placed directly on tissue culture cells specifically interferred with subsequent infection by RSV(RAV-1) but not with viruses with different type-specific antigens. The nonglycoproteins were found not to cause such interference. The details of these experiments will be published elsewhere.

DISCUSSION

ATV's have been shown to contain several glycoproteins which move more slowly in SDS-gel electrophoresis
than the other virion proteins. There appear to be 4 glycoprotein components (proteins Pl, P2, P2-a and P3) which
are regularly found in ATV's, and they have characteristic
electrophoretic mobilities in SDS-gels. There is some variation in relative amount of radioactive glucosamine, fucose or amino acid incorporated into each of the 4 protein
components in different labelled preparations of the same
virus and in different viruses. This could be because one
or more of the 4 components is an aggregate or complex under the conditions of SDS-gel electrophoresis. In addition, other glucosamine and amino acid-labelled components appear in gels in some experiments. Whether this
"background" represents protein aggregates or complexes,
contaminating cellular debris in some virus preparations
or has another explanation is not clear at this time. The
glycoproteins are quite insoluble and difficult to keep in
solution after virus has been dissociated with detergent
and the detergent removed.

The size estimates made for the glycoproteins are
based on the method of Shapiro, et al. (12) for nonglycoproteins. The values given, based on the positions of non-
glycoprotein markers of known size are probably overesti-
mates of the glycoprotein in molecular weights.

Several lines of evidence suggest that the glycoproteins found in the ATV's are all components of the viral
envelope. The iodination of the intact virions using lactoperoxidase under conditions which preserve virion size
and infectivity, revealed I^{125} in the 4 virion glycoproteins and none in the 5 nonglycoproteins. Experiments
with intact red blood cells indicate that only proteins
with tyrosine residues exposed to the cell surface are iodinated by this procedure (11). If lactoperoxidase similarly fails to penetrate the envelope of RSV(RAV-1) then
the glycoproteins iodinated in these experiments must all
have tyrosine residues exposed to the virion surface.
Hung (14) iodinated RSV(RAV-1) using chloramine T which
may be more destructive to the virus than the lactoperoxidase method used here. He found that proteins Pl, P2,
P2-a and P4 were iodinated with chloramine T.

The action of pronase on intact virions leads to a

similar conclusion about the presence on the virion surface of 3 of the glycoproteins. Proteins P1, P2 and P2-a were degraded by pronase action on particles, and the 5 nonglycoproteins remained unchanged. In experiments with the proteolytic enzyme bromelain, Rifkin and Compans (15) have obtained similar but not identical results. In their experiments protein 3 was also cleaved by bromelain. In addition, they showed that bromelain treatment removed the surface spikes from the viral envelope, suggesting that the surface spikes may be composed of one or more of the glycoproteins altered by the enzyme.

There is also evidence that the ATV glycoproteins are involved in determining virus envelope properties such as antigenicity, host range, interference and infecting efficiency. First, viruses with different envelope properties appear to have glycoproteins with different electrophoretic mobilities (16). In addition as described in this study, a mixture of virion glycoproteins, P1, P2, P2-a and P3, separated from the nonglycoproteins (P4-P8) reacts with virus neutralizing antibody and specifically interferes with infection of cells by virus with similar envelope antigen. Duesberg, et al. (17) have also shown that a glycoprotein-rich fraction from RSV(RAV-1) binds virus neutralizing antibody.

All evidence concerning the arrangement of the 5 nonglycoproteins within virions (proteins P4-P8) suggests that they are internal proteins and not exposed to the virion surface.

The presence of one or more glycoproteins exclusively in the virion envelope and the absence of other proteins in the envelope may be a general feature of enveloped viruses. Much evidence available suggests that this is probably true for Sindbis (18,19) vesicular stomatitis (20,22), influenza (23,24) and parainfluenza (25) viruses as well as the ATV's.

ACKNOWLEDGMENTS

This work was supported by U. S. Public Health Service Research Grant CA 10467.

REFERENCES

1. Eckert, E. A., Rott, R., and Schafer, W., _Z_. Natur-

forsch B, 18, 339 (1963).
2. Bonar, R. A., Heine, U., Beard, D., and Beard, J. W., J. Natl. Cancer Inst. 30, 949 (1963).
3. Bonar, R. A. and Beard, J. W., J. Natl. Cancer Inst. 23, 183 (1959).
4. Hung, P., Robinson, W. S., and Robinson, H. L., Virol. 43, 251 (1971).
5. Huebner, R. F., Armstrong, D., Okuyan, M., Sarma, P. S., and Turner, H. C., Proc. Natl. Acad. Sci. U.S. 51, 742 (1964).
6. Robinson, W. S. and Robinson, H. L., Virol. 44, 457 (1971).
7. Hanafusa, H., Hanafusa, T., and Rubin, H., Proc. Natl. Acad, Sci. U.S., 51, 41 (1964).
8. Rubin, H., Virol. 26, 270 (1965).
9. Hanafusa, H., Virol. 25, 248 (1965).
10. Vogt, P. K., Ishizaki, R., and Duff, R., in "Subviral Carcinogenesis: Monograph of the 1st International Aymposium on Tumor Viruses", Aichi Cancer Center, Nagoya, Japan, ed. Y. Ito (1966).
11. Phillips, D. R. and Morrison, M., Biochem. 10, 1766 (1971).
12. Shapiro, A. L., Vinuela, E., and Maizel, J. V., Biochem. Biophys. Res. Comm. 28, 815 (1967).
13. Robinson, W. S. and Robinson, H. L., unpublished data.
14. Hung, P. P. and Straube, L. M., Fed. Proc. 30, 1099 (1971).
15. Rifkin, D. B. and Compans, R. W., Virol. 46, 485 (1971)
16. Robinson, W. S., Hung, P. P., Robinson, H. L., and Ralph, D. D., J. Virol. 6, 695 (1970).
17. Duesberg, P. H., Martin, S., and Vogt, P., Virol. 41, 631 (1970).
18. Straus, J. H., Burge, B. W., and Darnell, J. F., Virol. 37, 307 (1969).
19. Schlesinger, M. J., Schlesinger, S., and Burge, B. W., Virol. 47, 539 (1972).
20. Wagner, R. R., Schnaitman, T. C., Snyder, R. M., and Schnaitman, C. A., J. Virol. 3, 611 (1969).
21. Kang, C. Y. and Prevec, L., J. Virol. 6, 20 (1970).
22. Cartwright, B., Tabot, P., and Brown, F., J. Gen. Virol. 7, 273 (1970).
23. Compans, R. W., Klenk, H. D., Caliguiri, L. A., and Choppin, P. W., Virol. 42, 880 (1970).
24. Shultze, I. T., Virol. 42, 890 (1970).

25. Klenk, H. D., Caliguiri, L. A., and Choppin, P. W.
    Virol. 42, 473 (1970).

TABLE 1

RSV(RAV-1) Protein Components Separated By
SDS Polyacrylamide Gel Electrophoresis

Component in SDS gel electrophoresis	pI[a]	Mol. wt. dal-tons[b] $\times 10^{-3}$	% of total C^{14}-amino acid[c]	% of total H^3-glu-cosa-mine[d]	CF-titer of 1 mg per ml. protein solution
P1 (g1)	4.9	96	1.7	7	<1:4
P2 (g2)	6.5,5.3	75	9.5	52	1:39
P2-a (g2-a)	--	50	1.0	4	--
P3 (g3)	6.3	35	6.0	15	<1:4
P4	8.9	27.5	32.0	0	1:9300
P5	6.3	21	13.4	0	1:53
P6	9.9	19		0	<1:4
P7	4.9	16.5	26.8	0	1:100
P8	7.4	14		0	1:1400
g4	3.5			6	<1:4
CF Controls					
SDS-dissociated Sendai virus					<1:4
SDS-dissociated RSV(RAV-1)					1:1270
Undissociated RSV(RAV-1)					<1:4

[a]Isoelectric points were determined by isoelectric focusing in 6 m. urea (4) after dissociation of virus with Brij 35 1%, mercaptoethanol 1%, and urea 4 M. Proteins P2 and P3 remain complexed under these conditions.
[b]Molecular weight estimates are based on SDS-gel electrophoresis. It is not known how accurately the Shapiro et al. (12) method of molecular weight determination applies to glycoproteins.

[c]The virus was labelled with a mixture of 15 C^{14}- amino acids. The amount of radioactivity in P5 and P6, and P7 and P8 are estimated together since the proteins of these pairs overlap each other in SDS-gel electrophoresis.
[d]The relative amount of radioactive glucosamine in g4 varied from 5 to 20% in different preparations of virus. No radioactive glucosamine is present in P4, P5, P6, P7, and P8.

TABLE 2

Effect of Iodination Reaction On RSV(RAV-1) Infectivity

Virus Concentration in reaction mixture[a] (A$_{280}$)	H$_2$O$_2$ in mμ moles/ml. in reaction mixture					
	0	20	40	80	240	720
1.0	940.[b]	740.	1120.	800.	620.	500.
.20	150.		192.			
.04	27.		24.			
.008	4.8		4.6			

[a]All reaction mixtures (0.25 ml., 1 min.) except H$_2$O$_2$=0 contained NaI 10^{-6} m. and lactoperoxidase 1.2 μg. per ml.
[b]ffu X 10^{-5}/ml. tested at the end of reaction and counted at 6 days.

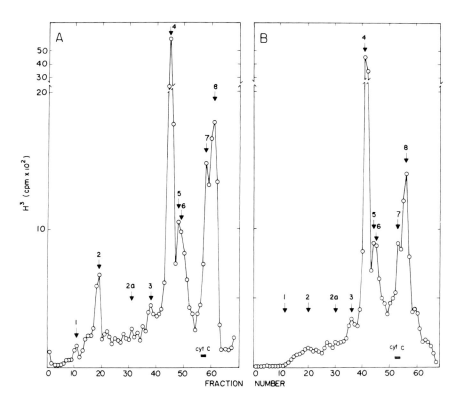

Fig. 1 SDS-gel electrophoresis of H^3-amino acid labelled RSV(RAV-1) proteins before (A) and after (B) treatment with pronase. Purified H^3-amino acid labelled virus was incubated with bovine serum albumen (A) or pronase 500 mg. per ml. (B) at 37°C. for 40 min. Each reaction mixture was then layered over a discontinuous sucrose density gradient (1 ml. 60% and 12 ml. 20% sucrose) in a Spinco SW 40 rotor and centrifuged at 40,000 RPM for 1 hour. The radioactive virus recovered between the 60% and 20% sucrose layers was dissociated with SDS, urea and mercaptoethanol and analyzed by SDS-gel electrophoresis.

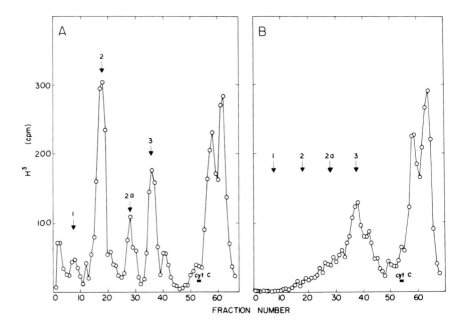

Fig. 2 SDS-gel electrophoresis of H^3-fucose labelled RSV(RAV-1) proteins before (A) and after (B) treatment with pronase. The experiment was carried out as described in Figure 1.

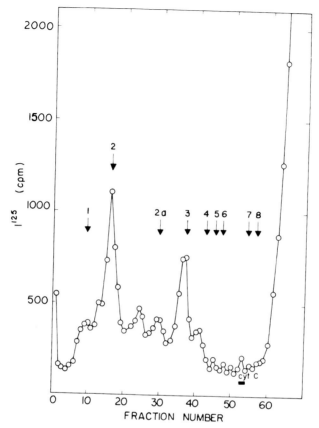

Fig. 3 SDS-gel electrophoresis of proteins from whole RSV(RAV-1) iodinated with I^{125}. Purified RSV(RAV-1) was incubated with NaI^{125}, lactoperoxidase and H_2O_2 as described in Table 2. The reaction was stopped after 1 min. by the addition of excess mercaptoethanol and the virus was repurified and analyzed as described in Figure 1.

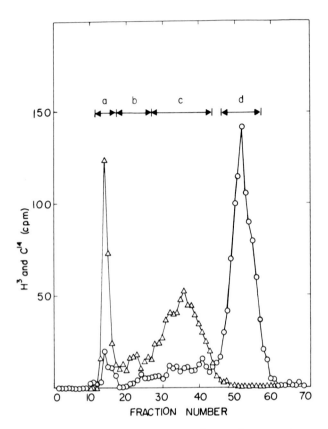

Fig. 4 Gel filtration of RSV(RAV-1) proteins on an agarose (Bio-Gel A-5m) column. RSV(RAV-1) with 11.0 A_{280} and containing RSV(RAV-1) labelled with H^3-glucosamine(-△-) and RSV(RAV-1) labelled with a mixture of 15 C^{14}-amino acids (-O-) was dissociated in 2% Tween 20, 1% mercaptoethanol, 0.15 m. NaCl, 0.01 m. Tris-HCl pH 7.4 and 0.001 m. EDTA at 56° for 30 min. The mixture was then layered on a 1.5 X 100 cm. agarose column and eluted with buffer containing 0.15 m. NaCl, 0.01 m. Tris. HCl pH 7.4, 0.001 EDTA and 0.001 m. mercaptoethanol. One ml. fractions were collected after discarding the first 25 ml.

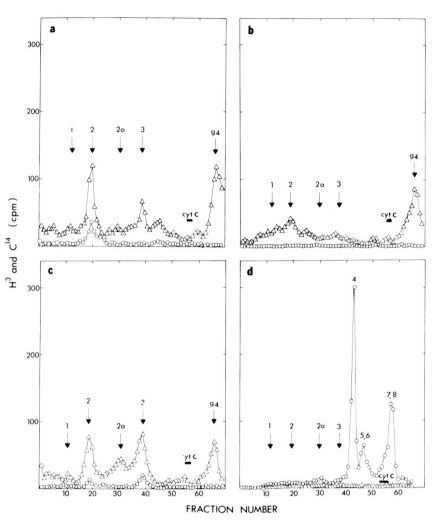

Fig. 5   SDS-gel electrophoresis of RSV(RAV-1) protein fractions from agarose column.   Fractions from the agarose column were combined into pools a-d as shown in Figure 4. The protein in each pool was concentrated with an Amicon UM-10 diaflo membrane filter and analyzed by SDS-gel electrophoresis as described in Figure 1. C^{14} (-O-) and H^3 (-△-).

MODIFICATION OF HUMAN CELL MEMBRANES
BY HERPES VIRUSES

Bernard Roizman and Jochen W. Heine

Departments of Microbiology and Biophysics
The University of Chicago
Chicago, Ill.

Abstract

The growth of herpesviruses in animal cells involves the inner
lamella of the nuclear membrane which envelops the virus and the
cysternae of the endoplasmic reticulum which contain and channel
the infectious virion into the extracellular fluid. Alterations in the
social behavior and immunologic changes have indicated that the
plasma membrane also becomes altered even though it plays no specific role in viral reproduction. Analyses of purified virions have
shown that it contains 24 virus specific proteins and glycoproteins
but no host proteins. Purified plasma membranes of infected cells
contain all of the virus specific proteins plus a small fraction of the
non-glycosylated protein in addition to the full complement of host
proteins. The viral membrane protein synthesis begins early in infection, i.e., before viral DNA synthesis. Viral proteins are glycosylated in membranes; the glycosylation profile is host dependent
and takes place in the absence of viral DNA synthesis and once the
proteins are made, in the absence of protein synthesis. The binding
to membrane is selective, specific and tenacious. Viruses differing
in their effects on the social behavior of cells specify different membrane glycoproteins.

I. Introduction

The studies of the alteration of cellular membranes by herpesviruses began over 10 years ago with the isolation of a herpes simplex

virus subtype 1 (HSV-1) mutant designated MP which caused fusion of human cells in culture (1). The parent strain, designated mP caused human cells to clump but they did not fuse. It was also observed that mouse embryo cells produced some viral progeny, clumped sluggishly, but did not fuse except when the culture was seeded with MP-infected human cells. As a consequence of this and other observations we proposed the term "social behavior" to describe cell-cell interactions and presented (2) five conclusions, i.e., (i) infection alters the social behavior of cells, (ii) ability of the social behavior to be altered is genetically determined by the cell, (iii) the nature of the alteration in social behavior is genetically determined by the virus, (iv) since the change in the social behavior is necessarily mediated by the plasma membrane, a change in the social behavior necessarily reflects structural, functional and immunologic changes in the plasma membrane, and (v) changes in social behavior and in particular, fusion of cells, results from interaction of altered and unaltered membranes at the point of their contact.

The basic conclusion was that the virus somehow altered the plasma membranes, but how? In the years that followed two lines of evidence accumulated independently. On the one hand, studies in our laboratory and in others (3-5) demonstrated that cellular membranes play an important role in viral development (Fig. 1). Thus the viral nucleocapsid is assembled in the nucleus from DNA made in that compartment (5) and proteins made in cytoplasm (6) then transported into the nucleus. The nucleocapsid becomes infectious by being enveloped as it buds through a morphologically modified inner lamella of the nuclear membrane (7,8). Lastly the mature virions accumulate and appear to egress from the infected cell through the endoplasmic reticulum (9). Although the virus development clearly involves the nuclear and the internal cytoplasmic membranes, no functional role has been found for the plasma membrane. On the other hand, studies in our laboratory showed that as expected, changes in immunologic specificity of plasma membranes accompany changes in social behavior of cells and that the altered immunologic specificity of the plasma membrane could not be differentiated from that of the cell (10,11). Similar antigenic identity has subsequently been reported for the herpesvirus associated with certain human proliferative diseases (infectious mononucleosis, Burkitt lymphoma, post nasal carcinoma, etc.) and the cells carrying the genome of the virus (12). In this paper we shall present current data on the relationship between the structure of the virus and the plasma membrane. The basic thesis that we would like to present is that viral structural proteins are incorporated into the plasma membrane, that the binding of the proteins to membranes is selective and specific.

Pertinent to the understanding of the techniques and experimental designs used in these studies are certain characteristics of macromolecular metabolism of cells infected with herpesvirus. Briefly as described elsewhere (13) in greater detail, the outcome of infection of cells with herpesviruses varies: In productive infection the cell produces viral progeny. In the course of the infection host macromolecular metabolism and in particular host protein and DNA synthesis cease within a few hours after infection (5,6,14,15). In nonproductive infection even though some viral functions are expressed, viral progeny is not made and the cell survives. Some non-productive infections can be induced to become productive with subsequent synthesis of viral progeny and cell death. All of the studies described in this paper with HSV and cells of human or simian derivation resulted in production of viral progeny, inhibition of host macromolecular synthesis and, ultimately, in cell death.

II. The Proteins of the Virion

A. Purification of the Virus

Analysis of the number and structure of viral proteins requires highly purified preparations of virions and virion components. Until very recently (16) analysis of the proteins of the herpesvirion was hampered by lack of demonstrably pure viral preparations. Herpesviruses are notoriously difficult to purify for reasons related to the site of maturation and to the complexity of the virion. As indicated earlier in the text, herpesviruses acquire an envelope from the inner lamellae of the nuclear membrane and accumulate in the perinuclear space and in the cisternae of the endoplasmic reticulum (Fig. 1). Virus release from the infected cell is inefficient and acquires momentum only after the infected cell disintegrates spilling both virus and debris (17). In our hands extracellular virus from cells in culture is frequently enclosed in a damaged envelope (permeable to negative stain) and is not a satisfactory source of virus for purification and analysis. On the other hand, separation of intracellular virus from cellular organelles and particularly from membrane vesicles generated during the disruption of the cell presents special problems. The difficulty is compounded by the fact that the membranes of infected cells carry virus-specific proteins (18-21). An additional complication is that infected cells yield not only virions but also naked nucleocapsids in ratios which differ greatly from one cell line to another. Moreover, the envelope of the herpesvirion is relatively unstable.

The procedure for purification of herpesvirions consists of 3 steps outlined in the flow diagram (Fig. 2). All of the steps were carried out at 0-4°C. A brief description of the purpose of each step is essential to an understanding of the final product. The object of the purification procedure was to disrupt the cells and to separate enveloped nucleocapsids in the cytoplasm from cellular organelles and from membrane vesicles generated during cell disruption. Of the various contaminants, membrane vesicles are the most prominent and the most difficult to deal with. The first step consisted of disruption of the cell and separation of the cytoplasm from nuclei. Careful preparation of cytoplasm is essential; isotonic sucrose solutions were used to stabilize the nuclei and prevent leakage of nucleocapsids and of nucleoproteins which tend to aggregate membranous structures. The second step consisted of rate zonal centrifugation of the cytoplasmic extract on Dextran 10 gradients. The centrifugation effectively separates virions from soluble proteins, which remain on top of the gradient, from aggregates of virions, cytoplasmic organelles and large debris which pellet, and from most of the cellular membrane vesicles which remain near the top of the gradient. Dextran specifically is used here because of its low osmolarity; presumably intact membrane vesicles stay distended rather than collapsed and sediment more slowly than the virion which has a high density because of the nucleocapsid. After centrifugation virions were found in a diffuse light-scattering band just above the middle of the tube. This band contained partially purified, mostly enveloped virus and will be referred to in the text as the Dextran 10 virus. The third step in the purification procedure was designed to separate the virions from remaining contaminants, i.e., membrane fragments and small amounts of unenveloped nucleocapsids, by flotation of partially purified virus through a discontinuous sucrose gradient. Membrane contaminants float to a lower density at a higher position in

the tube than the bulk of the virions whereas nucleocapsids, because
of their high density, will pellet. The floated virus band was diluted
and layered over a 10% sucrose, 2 M urea cushion and pelleted by
high speed centrifugation. Viral proteins were quantitatively recovered in the pellets. The band obtained by flotation through the
discontinuous sucrose gradient consisted almost exclusively of enveloped nucleocapsids. However, many of these particles exhibited
loss of integrity of the envelope. Membrane contaminants were extremely rare. An alternative to the last step is to centrifuge partially purified virus (Dextran band) in a potassium tartrate gradient
instead of a discontinuous sucrose gradient; in this way the normal
morphology of the virion envelope is better maintained. However, it
should be emphasized that both alternatives yield equivalent results
in terms of the protein recovered in the final product (16). The purified virions were solubilized and analyzed by a high resolution gel
electrophoresis technique. The presence of proteins in polyacrylamide gels was assayed by one of three techniques, i.e., staining
with Coomassie Brilliant Blue, autoradiography of a stained fixed gel
by x-ray film to detect ^{14}C-labeled amino acids, or by slicing the
gels and assaying the slices by liquid scintillation counting for their
tritium labeled protein content or for mixed ^3H and ^{14}C labeled
protein content. Details of these and other procedures described in
this review were summarized elsewhere (16).

B. Analysis of Purified Virions for Host Proteins

These experiments were prompted by a report (22) that herpesvirus
particles were agglutinated by a serum prepared against uninfected
cells. Analysis for the presence of host proteins in the virion were
made as part of a study of purification of the herpesvirion. In these
studies we took advantage of the fact that host protein synthesis
ceases after infection and that proteins labeled after 4 hours post
infection are probably specified by the virus (6,14). The results may
be summarized as follows:

(i) The degree of purification of herpesvirus virions relative to
host proteins may be calculated directly from data presented in Table 1.
In this   experiment cells were labeled with ^{14}C-amino acids for
48 hours before infection. During and after infection the cells were
incubated in a medium containing excess unlabeled amino acids to
insure that proteins made after infection did not become labeled.
The purification procedure was monitored for both radioactivity and

infectivity. As shown in this Table the p.f.u./cpm for the partially purified virus banded in the Dextran density gradient is nearly 100 times higher than that of the homogenate. The absolute degree of purification achieved after flotation cannot be calculated from the p.f.u./cpm ratio largely because there is considerable loss of infectivity in the last step (16). However the degree of purification may be estimated from another experiment in which the virus was purified from a mixture of infected cells labeled with ^{3}H amino acids after infection and uninfected cells labeled with ^{14}C amino acids. In this experiment as well as in several others the ratio of ^{3}H to ^{14}C improved approximately 2-fold in the last step of the purification procedure. On the basis of these two experiments as well as several others cited in greater detail elsewhere (16), we may conclude that the purification of virus is approximately 200-fold.

(ii) Analysis of the number, distribution and origin of the residual host proteins in the virion indicate that these are contained in residual membrane contaminants and are not structural proteins of the virion. The conclusion is based on several experiments. First, electropherograms, shown in Fig. 3, of partially purified and purified virus prepared from cells which had been incubated with ^{14}C amino acids 48 hours prior to infection indicate that both contain small residual amounts of host (prelabeled) proteins but in relatively different amounts. Specifically, while the virus-specified proteins remain in relatively constant ratios with respect to one another during the last step of purification, the host proteins non-selectively diminish in amount and appear to be separable from the virion proteins.

The second finding is that labeled host proteins diminish in quantity to about the same extent as a result of flotation through sucrose gradients regardless of whether the label is incorporated into the proteins of the same cells in which the virus was grown or whether the label is from uninfected cells artificially mixed with the infected cells and therefore not incorporated into the virion during morphogenesis (16). Lastly, as discussed in a subsequent section in detail, purified plasma membranes and other membranes of the infected cells contain not only virus-specific proteins but also host proteins. Comparison of the electrophoretic mobilities of the major host proteins present in the virus preparations in the membrane contaminants separated from the virus, and in the plasma membrane indicate that they are very similar or identical (16,23). These data indicate that host proteins are not structural proteins of the virion.

C. The Number and Molecular Weights of the Virion Proteins

The number and properties of the virion proteins were determined in a series of experiments designed as follows: Infected cells were labeled with ^{14}C amino acids or with ^{14}C glucosamine from 5 to 20 hours post infection. Purified virus was prepared, solubilized, and subjected to electrophoresis on gels containing 6, 7, 8.5, 9 and 14% acrylamide. The gels were then fixed, stained with Coomassie Brilliant Blue, and scanned in a Gilford or a Joyce Loebel spectrophotometer. The gels were then sectioned longitudinally and a thin slice of the gel was dried on filter paper and exposed to x-ray film. The autoradiographic image on x-ray film was developed and then scanned, again with the Gilford or Joyce Loebel spectrophotometers. The results of these analyses may be summarized as follows:

(i) The 8.5% acrylamide gels (Fig. 4) gave the best overall separation. In general the ^{14}C amino acids autoradiograms and the absorbance profiles of the stained proteins agree remarkably well except in the right 1/3 of the gel which still showed traces of the host proteins identified earlier (Fig. 3). The light radioisotopic labeling of the band No. 22 is reproducible. In recent experiments (Roizman and Terlizzi, unpublished data), it was found that the proteins in this band are synthesized largely during the first 4 hours of infection.

(ii) We have detected 24 virus-specific proteins in the virion. To distinguish these from other proteins specified by the virus we have designated the virion proteins as VP1 through VP 24. Some bands were assigned two numbers because evidence exists that they contain at least two proteins. The two bands numbered VP1-3 can be resolved into three bands on 6% gels. The band numbered VP7-8 is known to contain two proteins because glucosamine label is distributed asymmetrically toward the leading edge. VP7 may not be glycosylated or may be glycosylated to a lesser extent than VP8. Visual inspection of 6% and some 8.5% gels revealed that the bands labeled VP13-14 and VP17-18 consist of at least two proteins each. Electrophoresis of virion proteins in 14% acrylamide gels failed to reveal the presence of additional small molecular weight proteins in the virion. The bands given numerical designations are those which were highly reproducible in appropriate gel concentrations. The small blips between bands 3 and 4 were not reproducible whereas bands 4, 6, 9, 10, 20 and 24 were reproducible in appripriate gels both by Coomassie Blue staining and by autoradiography.

(iii) The glycoprotein bands were reproducibly broad by visual inspection of stained gels and of the autoradiographic tracings. We cannot say with any degree of certainty how many glycoproteins with distinct protein moieties are present in the virion nor precisely how many of the protein bands are glycosylated. Recent studies, described elsewhere in detail (24), indicate that all of the proteins from VP11 through VP19 are glycosylated and that some are present in larger amounts in the virion than in the infected cell membrane.

(iv) It has been repeatedly demonstrated that the mobilities of proteins in SDS-acrylamide gels are proportional to the logarithm of the molecular weights of the proteins (25, 26). An apparent exception to this is an erythrocyte glycoprotein observed by Bretscher (27) to yield a range of molecular weight estimates which depended on the concentration of acrylamide used and none of which agreed with the value obtained by another method. The electrophoresis of virion proteins through polyacrylamide gel concentrations ranging from 6% to 14% were done specifically to determine whether the mobility of the glycoproteins through gels were a function of gel concentration. In each run we included proteins of known molecular weight spanning the range of molecular weights which could be resolved.

The electrophoretic mobilities of the protein standards and of the viral proteins in these gels are shown in Fig. 5 along with the molecular weights calculated from mobilities of the viral proteins in each gel. The data show a remarkable agreement between the molecular weights calculated from the migration of the proteins in different gels. The average molecular weights of viral proteins range from 25,000 for VP24 to 275,000 daltons for VP1. The calculations summarized elsewhere (16) indicate that the sum of molecular weights of virus specific structural proteins is approximately 2,580,000 daltons. Assuming that viral messenger RNA is transcribed asymmetrically and is not complementary, it can be calculated that HSV DNA has sufficient genetic information to specify the sequence of 55,000 amino acids. The data presented in this study indicate that approximately 25,800 amino acids, or 47% of the genetic information (23% of viral DNA), is concerned with the structural proteins of the virus.

III. Viral Proteins in the Plasma Membrane

A. Purification of Plasma Membranes

The procedure for the purification of the plasma membranes used in these studies was selected for several reasons. First, the procedure is very gentle in that it permits rupturing the cells with minimal discharge of lysosomes. Second, the major constituents of the washed microsomal fraction were cytoplasmic membranes and plasma membranes which allowed us to monitor the purification procedure with specific enzyme markers. Third, the most important advantage of this procedure is that it allows the separation of the two smooth membrane populations which are similar in many physical properties.

The purification procedure involved two steps and is outlined in the Flow Diagram (Fig. 6). Details of this and other procedures are described in Heine et al. (23). In the first step the cells were ruptured by microcavitation (28) which caused the formation of micro-vesicles from surface membranes. These vesicles were separated from the nucleus, endoplasmic reticulum complex, mitochondria and large cytoplasmic membrane fragments by low speed centrifugation. The conditions for the disruption of HEp-2 cells was determined in the same fashion as that reported for L cells by Heine and Schnaitman (29) using glucose-6-phosphate dehydrogenase and acid phosphatase as markers for ruptured cells and discharged lysosomes respectively.

The supernatant from the low speed centrifugation, i.e., the microsomal fraction, was enriched in plasma membrane and contained as its major contaminant endoplasmic reticulum. This fraction appropriately treated (23) was centrifuged through a Dextran 110 barrier. After centrifugation the plasma membrane remained at the buffer Dextran interphase while most of the smooth cytoplasmic membranes along with the other microsomal components were found in the pellet (28,30).

The second step in the purification of plasma membranes involved a fractionation of the membranes banded in Dextran 110A according to their physical properties and served to remove residual virus particles still associated with the membranes. In this step the partially purified plasma membranes suspended in tris-EDTA buffer were floated through a discontinuous sucrose gradient as described by Spear et al. (18). The purified plasma membrane banded at the 25-30% w/w sucrose interphase. The purification of the plasma membrane was monitored with three markers, i.e., 5'-nucleotidase,

NADH-diaphorase and fucose. The markers were selected on the basis of a report (31) of purification of HeLa plasma cell membranes and in part on the basis of conclusions by Steck et al. (32) that the most useful markers for measuring the extent of separation of plasma membranes and of cytoplasmic membranes are the 5'-nucleotidase associated with the plasma membrane and the NADH-diaphorase associated with the endoplasmic reticulum which constitutes the bulk of cytoplasmic membranes. The results obtained in two experiments described in detail elsewhere may be summarized (Table 2) as follows: (i) The degree of purification indicated by the ratios of the specific activities of 5'-nucleotidase to diaphorase is approximately 26-fold, i.e., as good or better than that reported for other kinds of cells (28,33,34). (ii) The degree of purification as measured by monitoring fucose was only 14-fold, i.e., somewhat less than that reported by Atkinson and Summers (31). It should be noted here that the discrepancy could be expected. The experimental design required that cells be labeled with fucose before infection since viral glycoproteins contain fucose (20,21). Although the cells were incubated in unlabeled fucose for 12 hours before infection and acrylamide gel electrophoresis of the doubly labeled plasma membranes failed to show appreciable amounts of fucose in the viral glycoproteins, we cannot exclude the possibility that (a) some labeled fucose remained in the nucleotide sugar pool and was available to the viral glycoproteins at all membranes or that (b) fucose was removed from host glycoproteins and reutilized. The extent of purification of plasma membranes for infected cells was similar to that of uninfected cells suggesting that (a) the physical properties of the infected cell membranes do not change and (b) there is no indiscriminate re-distribution of the enzymes or outright exchange of membrane material between the two membrane populations. (iii) Electron microscopic studies on thin sections indicated that the purified plasma membranes were free of naked and enveloped nucleocapsids. The absence of virus particles was also evident from the absence of the major capsid protein from the acrylamide gel electrophreograms of membrane proteins reported in the next section of this paper.

B. The Fate of Host Proteins

The major objectives of these experiments were to determine (a) the protein composition of plasma membranes extracted from infected and uninfected cells and (b) the fate of host membrane proteins after infection. In the first series of experiments 4 sets of cell cultures each containing 4×10^8 cells were treated as follows: The first set of cell cultures was replenished with labeling medium containing ^{14}C amino acids at time 0. After 24 hours of incubation, the cells were washed and incubated for an additional 22 hours in medium containing unlabeled amino acids. The second set of cell cultures was treated from time 0 to 24 hours exactly as the first set. At that time the labeling medium was removed and the cells were incubated for 4 hours in unlabeled medium, then infected and reincubated in unlabeled medium for an additional 18 hours. The cells in the third set were labeled with ^{14}C amino acids between 32 and 46 hours post 0 time. The cells in the fourth set were infected at 28 hours and labeled between 32 and 46 hours post 0 time, i.e., between 4 and 18 hours post infection. At 46 hours post 0 time all cell cultures were harvested and the plasma membranes were purified, solubilized and subjected to electrophoresis on a flat acrylamide gel slab (23). Fig. 7 shows the absorbance of the bands stained with Coomassie Brilliant Blue and of the autoradiograms developed in the x-ray film. The results may be summarized as follows:

(i) The proteins in the plasma membranes of uninfected cells differ with respect to turnover rate. This conclusion is based on two sets of observations. On one hand, the absorbance of the stained protein bands (Fig. 7A,C) from membranes of uninfected cells labeled between 0 and 24 hours (set 1) and after mock-infection between 32 and 46 hours post 0 time (set 3) appear identical. However, minor differences are apparent in the absorbance tracings of their autoradiograms (Fig. 7B,D). Thus the membranes of the cells labeled after mock-infection (set 3), i.e., immediately before harvesting, contain three prominent bands of labeled proteins (Fig. 7D) which are present in reduced amounts relative to others in the autoradiograms of membranes labeled from 0 to 24 hours (set 1).

(ii) The plasma membranes of infected cells contain two sets of protein bands, i.e., proteins characteristically present in uninfected cells and new proteins, absent from uninfected cells. This conclusion is based on the comparison of the absorbance tracings of the stained bands of plasma membrane proteins from infected cells (Fig. 7E,G) with those of uninfected cells (Fig. 7A,C).

(iii) In infected cells, the synthesis of host plasma membrane proteins ceases and only the new membrane proteins are made. This emerges from comparisons of the autoradiographs of the gels of plasma membranes from uninfected cells and infected cells labeled before infection (Fig. 7B,D,F, sets 1,2,3) and infected cells labeled after infection (Fig. 7H, set 4). Direct comparison of the autoradiographic absorbance tracings from the gels of infected and uninfected cell membranes indicates that infected cell membrane proteins differ in number and electrophoretic mobilities from those of uninfected cell membranes and resemble the electropherograms of the membrane proteins of smooth membranes as published previously (18,19,20). In the earlier studies it was demonstrated that the number and electrophoretic mobility of the membrane proteins made after infection were genetically determined by the virus and hence were virus-specific (18,19).

(iv) Host proteins are not selectively removed or expelled from the membranes of infected cells. This conclusion is based on comparisons of the autoradiograms from gels of membranes from infected cells labeled before infection, i.e., between 0 and 24 hours (set 2) and uninfected cells labeled during the same interval (set 1). On the contrary, the data suggest that infection may slightly decrease the rate of cellular protein turnover in the plasma membrane. This conclusion is based on the observation that the Coomassie Brilliant Blue stain profiles of infected cell membranes (Fig. 7E,G) reveal that the virus specified proteins are a significant fraction of the total protein in the membranes (approximately 10%) and analyses of the specific activities of membranes infected or mock-infected after labeling (23).

C. Characterization of Viral
   Membrane Proteins
   The purpose of these experiments was to determine how many virus-specific membrane proteins are made and bound to the plasma membrane, how many are glycosylated, and how are they related to the structural proteins and glycoproteins of the virion. Approximately 4×10^8 cells were labeled either with ^{14}C amino acids and with ^{14}C glucosamine, respectively, between 4 and 18 hours post infection. The plasma membranes were then extracted, purified and subjected to electrophoresis concurrently with solubilized proteins from highly purified preparations of enveloped virus labeled with ^{14}C glucosamine. Comparison of the autoradiographic tracing of the amino acid-labeled proteins in the plasma membrane (Fig. 8C) with

the absorbance tracings of the stained bands of the virion proteins (Fig. 8D) suggests that the proteins made after infection and binding to plasma membrane are also present in the virion, albeit in somewhat different proportions. This conclusion is reinforced by comparisons of the autoradiograms of the ^{14}C glucosamine-labeled proteins in the virion and in the plasma membrane (Fig. 8A, B). The electropherograms presented in Fig. 8 show an additional band of glycosylated protein which migrates slightly slower than No. 22. This band, however, has not been found consistently and we are not certain whether it should be included among the invariant population of virus-specific membrane proteins. Several points should be made in connection with these data.

First, the differentiation of proteins 7-8, 11-14 and 17-19 is based on visual examination of both the autoradiograms and the stained gels. Second it seems very likely that 8, 17, and 18 are extensively glycosylated and that glucosamine is present in at least 3 of the proteins in the 11-14 group.

Lastly, it seems pertinent to note that proteins 13,14, 19 and 22 are present in larger amounts in the virion than in the plasma membrane.

IV. The Entry and Function of Viral Proteins in Membranes

A. Time of Synthesis of Viral
Membrane Proteins
These experiments were prompted by the observation that some human cells non-productively infected with a herpesvirus associated with lympho-proliferative diseases exhibit the same immunologic reactivity as the virus infecting them yet infectious progeny is not made and the cell survives. These (35,36) and others (37,38) observations led to the conclusion that the membrane specific antigens were "early" antigens, i.e., virus products made before and independently of viral DNA synthesis. Interest in "early" viral antigens stems from the fact that such products are frequently present in non-productively infected, virus-transformed cells and might be responsible for the malignant properties of the cell. As reviewed elsewhere in greater detail (10,11,39), several lines of evidence indicate that the immunologic reactivity of the antigens on the surface of the virion and on the membrane are identical. The most direct evidence is that purified membranes compete with the virion

for antibody in neutralization tests (Fig. 9) and, as shown earlier in the text, both membranes and virions share new, virus specific proteins which cannot be differentiated with respect to glycosylation or electrophoretic mobility (23). The question therefore arises whether viral membrane proteins are "early" products.

Several lines of evidence indicate that a large fraction of viral structural proteins including the membrane glycoproteins are "early functions" of the virus, i.e., expressed before and independently of viral DNA synthesis. The data briefly, is as follows: (i) Extensive analyses of the transcription of viral DNA (40 , Frenkel and Roizman manuscript in preparation) indicate the existence of 2 classes of viral RNA transcripts differing in molar ratios. The total DNA transcribed is 44 and 48 per cent at 2 hours and 8 hours post infection, respectively. At 2 hours post infection, i.e., before the onset of DNA synthesis, most abundant RNA class is transcribed from 14 per cent of the DNA and corresponds to nearly 98 per cent of total viral RNA. At 8 hours post infection, i.e., after viral progeny begins to accumulate, the abundant viral RNA comprises 96 per cent of total and is derived from 19 per cent of the DNA. Abundance competition experiments indicated that the most abundant RNA class at 2 hours after infection is a subset of the most abundant RNA class at 8 hours post infection. A priori it could be predicted that the most abundant RNA class would specify structural proteins of the virus. Indeed, the abundant RNA class is transcribed from 19 per cent of the DNA, i.e., roughly from the same amount of DNA as that calculated to contain the genetic information for the structural proteins in the virion. The synthesis of viral structural proteins early in infection was then confirmed directly by pulse labeling the cells from 0.5 to 2 hours post infection. The cells were then incubated in medium containing excess unlabeled amino acids. Comparison of the electropherograms of viral protein labeled between 0.5 to 2 hours with viral proteins labeled between 4 and 20 hours post infection clearly indicates that the bulk of viral structural proteins including the glycoproteins are already made between 0.5 and 2 hours post infection (Fig. 10). The exceptions are some high molecular weight structural proteins (Nos. 1-4) which together constitute some 35 per cent of the genetic information of the virus, i.e., roughly in accord with the difference between the total DNA template for high abundance RNA class at 8 and 2 hours post infection. (ii) Inhibition of DNA

synthesis by cytosine arabinoside does not block the synthesis or glycosylation of viral membrane proteins in infected cells (Fig. 11). It
is noteworthy that in the presence of the inhibitor of DNA synthesis
the glycosylation profile becomes altered in that some proteins become more extensively glycosylated than others. The reason for this
is not known.

Parenthetically, the finding that viral structural proteins including viral membrane glycoproteins are made early is unusual and not
readily explained. In bacteria infected with phages and in animal
cells infected with some animal DNA virus the early products are
usually nonstructural protein. The question arises as to whether the
synthesis of these products early in infection is mandatory and what
function they have other than that of structural proteins of the virus.
The answer is not clear. It is conceivable that initiation of viral
DNA synthesis requires modified cell membranes. Equally if not
more interesting are the consequences of midification of cellular
membranes in non-productive infection in which viral progeny is not
made and the cell survives. The data presented here gives credence
to the conclusions of Gergely et al. (37,38) that viral membrane
antigens in Burkitt lymphoma cells are probably "early" products and
lends support to the suggestion made earlier (13) that the changes in
the plasma membrane resulting from the entry of viral glycoproteins
might well be responsible for the neoplastic characteristics of the
non-productively infected cells.

B. Glycosylation of Viral
   Membrane Proteins

We have very little information concerning the sugars present in
the membrane glycoproteins, the site and specificity of glycosylation. The available information may be summarized as follows:
(i) to date incomplete analyses of the sugars present in membrane
glycoproteins have revealed the presence of galactosamine, mannose,
fucose, galactose in addition to glucosamine. Of these, fucose is
quantitatively incorporated into membranes as fucose. Most of the
glucosamine is incorporated into the glycoproteins as glucosamine
and galactosamine (Keller, unpublished observations). (ii) Experiments designed to determine the site of glycosylation of these proteins have shown that the glucosamine is not incorporated into nascent peptides on polyribosomes, that glucosamine is incorporated
into acid-soluble macromolecules in the presence of puromycin, and,

finally, that membrane proteins are largely glycosylated in membranes. In these experiments we pulse-labeled cells for 1 hour intervals at different times after infection. A portion of the infected
cell population was then harvested and a smooth membrane fraction
was purified free of ribosomes, soluble proteins and virus (18).
Another portion of the cell population was incubated in medium containing unlabeled amino acids, then processed in the same fashion.
As shown in Fig. 12, concurrently with glycosylation (amino acid
label is conserved during the chase while glucosamine label increases) membrane proteins increase in apparent molecular weight
and homogeneity. This is evidenced by the observation that after
the chase some of the proteins migrate in acrylamide more slowly and
in sharper bands with complete coincidence of amino acid and glucosamine label (41). (iii) We have previously reported (19) that
glycosylation of viral proteins appears to be at least in part a host
function. Thus, the minor glycoproteins appear to be more extensively glycosylated in simian (Vero) cells than in human (HEp-2)
cells. (iv) As cited earlier in the text, cytosine arabinoside in concentrations sufficient to block the synthesis of viral DNA also modi-
fies the pattern of glycosylation of viral membrane proteins.

C. The Selectivity, Specificity and
    Tenacity of the Binding of Mem-
    brane Proteins to the Plasma
    Membrane
    Several lines of evidence indicate that the binding of viral pro-
teins to membranes is not random but rather, it shows some degree
of selectivity and specificity.
    Selectivity is demonstrable from the fact that only a fraction of
all the virus-specific proteins present in the cell bind to membranes.
All of the virus-specific proteins binding to membranes are also
structural proteins of the virus.
    Specificity is apparent from comparisons of the membrane glyco-
proteins in the plasma membrane and in the whole cell or in the
virion. Thus NP-40 (2.5%) extract of cells infected with HSV-1
contains the same population of glycoproteins as the whole cell solubilized and subjected to electrophoresis on polyacrylamide gels
(Fig. 13). However, as shown in the same Fig. the population of
glycoproteins in the purified plasma membrane differs from that of
the extract with respect to relative amounts of several glycoproteins.

In a similar vein, the glycoprotein content of plasma membranes from cells infected with the mP mutant suffers from that present in the purified virion, again with respect to relative amounts of glycoproteins. These data indicate that the glycoproteins do not bind randomly to all the membranes of the infected cell. The tenacity of the binding emerges from observation that membranes of infected cells withstand considerable amounts of manipulation without evidence of selective loss of virus-specific membrane proteins (18,19). Perhaps more impressive, the virus-protein--membrane complex withstands considerable hydrodynamic stress augmented by the presence of the antibody to the glycoproteins on the membranes (42).

D. Viral Membrane Glycoproteins
   and the Social Behavior of
   Infected Cells

As indicated in the Introduction, much of the initial impetus to study the membranes of infected cells was based on the observations that herpesviruses differ with respect to their effects on the social behavior of infected cells (2,20,21). The question arises therefore whether the changes in the membranes described earlier in the text related in any fashion to the changes in the behavior of cells.

Much of the available data suggest that there is a relationship between the virus-specific membrane glycoproteins and the social behavior of infected cells. Briefly as reported elsewhere (20,21) in greater detail: (i) virus strains differing with respect to their effects on the social behavior of cells (43) also differ with respect to the membrane and envelope glycoproteins they specify (19-21) (Fig. 14). (ii) Fusion of cells caused by MP virus is recessive to clumping of cells caused by the parent, HSV-1 virus (2). Membranes from MP-infected cells appear to lack one major glycoprotein (protein 7 or 8). Moreover, MP virus appears to be a deletion mutant lacking 4×10^6 daltons in molecular weight. In cells doubly infected with both HSV-1 and MP the cells clump and both membranes of virions contain the full complement of virus-specific proteins characteristic of the HSV-1 virus (20,21).

The data are significant, but, alas, they are not sufficient evidence that glycoproteins by themselves or in combination with other membrane constituents are responsible for the change in the behavior of the cell. Two types of additional evidence are necessary. The first and perhaps most important is an analysis of the topology of the

viral proteins in the plasma membranes and their effect on the topology and composition of host membrane constituents. The second
and perhaps equally important are evidence from reconstruction
experiments that the social behavior of cells can be altered by
appropriate addition of viral membrane constituents to the membranes
in the absence of infection. Analyses of the topology and composition of plasma membranes from infected cells are now in progress.

Acknowledgments

These studies were aided by grants from the National Cancer Institute, United States Public Health Service (CA-08494), the American Cancer Society (NP-15G), and the National Science Foundation (GB-27356X). J.W.H. is a postdoctoral trainee of the
United States Public Health Service (AI-00238).

References

1. M. D. Hoggan and B. Roizman, Am. J. Hyg. 70, 208 (1959).
2. B. Roizman, Cold Spring Harbor Symp. Quant. Biol. 27, 327 (1962).
3. B. Roizman and P. G. Spear, in Atlas of Viruses, Academic Press, New York, 1971, in press.
4. R. W. Darlington and H. L. Moss, III, Progr. Med. Virol. 11, 16 (1969).
5. B. Roizman, in Current Topics in Microbiology and Immunology, Springer-Verlag, Heidelberg, 1969, Vol. 49, p. 1.
6. P. G. Spear and B. Roizman, Virology 36, 545 (1968).
7. R. W. Darlington and L. H. Moss, III, J. Virol. 2, 48 (1968).
8. J. Schwartz and B. Roizman, J. Virol. 4, 879 (1969).
9. J. Schwartz and B. Roizman, Virology 38, 42 (1969).
10. P. R. Roane, Jr. and B. Roizman, Virology 22, 1 (1964).
11. B. Roizman and S. B. Spring, Proc. Conf. on Cross Reacting Antigens and Neoantigens, Williams & Wilkins Co., Baltimore, Md., 1967, p. 85.
12. G. Pearson, F. Deney, G. Klein, G. Henle and W. Henle, J. Nat. Cancer Inst. 45, 989 (1970).
13. B. Roizman, Proc. Symp. Oncogenesis and Herpes-Type Viruses, Cambridge, England, 1971.
14. R. J. Sydiskis and B. Roizman, Virology 32, 678 (1967).

15. B. Roizman and P. R. Roane, Jr., Virology 22, 262 (1964).
16. P. G. Spear and B. Roizman, J. Virol. 9, 143 (1972).
17. M. D. Hoggan and B. Roizman, Virology 8, 508 (1959).
18. P. G. Spear, J. M. Keller and B. Roizman, J. Virol. 5, 123 (1970).
19. J. M. Keller, P. G. Spear and B. Roizman, Proc. Nat. Acad. Sci. U.S. 65, 865 (1970).
20. B. Roizman, in Proc. 3rd Int. Symp. Applied and Medical Virology, Warren Green Publ., St. Louis, Mo., 1971, p. 37.
21. B. Roizman and P. G. Spear, in Nucleic Acid-Protein Inter-actions and Nucleic Acid Synthesis in Viral Infection, North Holland Publ. Co., Amsterdam, 1971, Vol. 2, p. 435.
22. D. H. Watson and P. Wildy, Virology 21, 100 (1963).
23. J. W. Heine, P. G. Spear and B. Roizman, J. Virol.9,(March) (1972).
24. T. Savage, B. Roizman and J. W. Heine, J. Gen. Virol. (1972) in press.
25. A. L. Shapiro, E. Vinuela and J. V. Maizel, Jr., Biochem. Biophys. Res. Commun. 23, 815 (1967).
26. K. Weber and M. Osborn, J. Biol. Chem. 244, 4406 (1969).
27. M. S. Bretscher, Nature 231, 229 (1971).
28. V. B. Kamat and D. F. H. Wallach, Science 148, 1343 (1965).
29. J. W. Heine and C. A. Schnaitman, J. Virol. 8, 786 (1971).
30. D. F. H. Wallach and V. B. Kamat, Methods in Enzymol. 8, 164 (1966).
31. P. H. Atkinson and D. F. Summers, J. Biol. Chem. 246, 5162 (1971).
32. T. L. Steck and D. F. H. Wallach, Methods in Cancer Res. 5, 137 (1970).
33. R. M. Burger and J. M. Lowenstein, J. Biol. Chem. 245, 6274 (1970).
34. R. A. Weaver and W. Boyle, Biochim. Biophys. Acta 173, 377 (1969).
35. G. Klein, Advan. Immunol. 14, 187 (1971).
36. J. Nadkarni, P. Clifford, G. Manolov, E. Fenyo and E. Klein, Cancer 23, 64 (1969).
37. L. Gergely, G. Klein and I. Ernberg, Int. J. Cancer 7, 293 (1971).
38. L. Gergely, G. Klein and I. Ernberg, Virology 45, 10 (1971).

39. B. Roizman, Transplantation Proc. 3, 1179 (1971).
40. B. Roizman and N. Frenkel, in Molecular Studies in Viral Neo-plasia, William & Wilkins Co, Baltimore, Md., 1972, in press.
41. P. G. Spear and B. Roizman, Proc. Nat. Acad. Sci. U.S. 66, 730 (1970).
42. B. Roizman and P. G. Spear, Science 171, 298 (1971).
43. P. M. Ejercito, E. D. Kieff and B. Roizman, J. Gen. Virol. 3, 357 (1968).

TABLE 1

Purification of Infectious Herpesvirions
with Respect to Labeled Host Proteins

Fraction	Cellular Protein $cpm \times 10^{-3}$	$p.f.u. \times 10^{-6}$	$\dfrac{p.f.u.}{cpm} \times 10^{-3}$
Homogenate	23,960	21,030	0.9
Cytoplasm	7,550	20,020	2.7
Dextran Virus Band	51.8	4,130	80.0

Data from Ref. 16.

TABLE 2

Purification of Plasma Membrane

Materials tested	Exp. 1			Exp. 2
	NADH-diaphorase $\Delta OD/min/mg$	5'-Nucleo-tidase $\mu moles$ $PO_4/hr/mg$ released	Ratio of Specific Activities[*]	^3H-Fucose cpm/mg protein $\times 10^4$
Microsomal fraction	0.403	36.92	1.0	11
Dextran 110 band	0.098	86.20	9.6	90
Plasma membrane	0.038	92.80	29.6	151

Data from Ref. 23.

*Ratio of 5'-nucleotidase (sp. act.) to NADH-diaphorase (sp. act) Normalized with respect to that of the microsomal fraction.

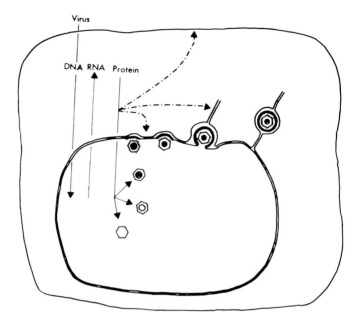

Fig. 1 Schematic diagram of herpesvirus reproduction in animal cells in culture. The sequence of events is depicted from left to right, i.e., reproduction is initiated by the entry of the virus into the cell. The DNA becomes uncoated and transported into the nucleus where it is transcribed, etc. The reproductive cycle lasts 18–20 hours. Viral DNA synthesis cannot be detected before 3 hours but reaches peak levels between 4 and 8 hours post infection. Infectious progeny first appears about 6 hours post infection.

PURIFICATION OF HERPES VIRIONS

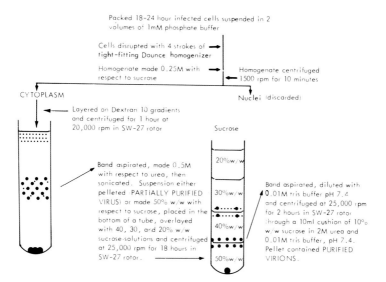

Fig. 2   Flow diagram for purification of herpes simplex virus from infected HE--2 cells.   Adapted from procedures described in Ref. 16.

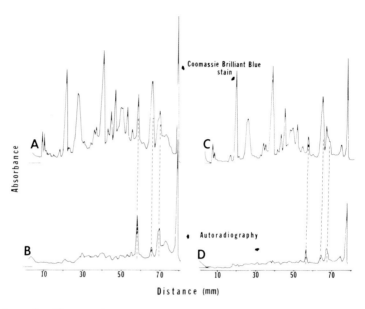

Fig. 3   Electropherograms of partially purified (A & B) and purified (C & D) virus prepared from cells which had been incubated with ^{14}C amino acids for 48 hours prior to infection. The proteins were analyzed in cylinders of 8.5% acrylamide. Absorbance profiles of Coomassie Brilliant Blue stained gels and of autoradiograms prepared from the same gels are shown. The dashed lines mark the positions of the most prominent cellular proteins. In this and all following electropherograms the position of the base line is indicated by a short line segment at the extreme left of the profile. Data from Spear and Roizman (ref. 16).

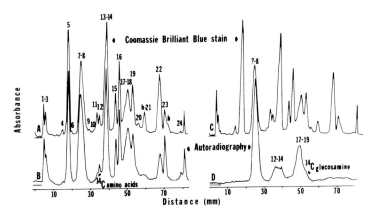

Fig. 4   Electropherograms of purified virion proteins labeled
either with ^{14}C amino acids (A & B) or ^{14}C glucosamine (C & D).
The absorbance profiles of Coomassie Brilliant Blue stained gels
(8.5% acrylamide gel cylinders) and of autoradiograms developed
from those same gels are shown.   The virus-specific protein bands
have been assigned numbers from 1 to 24. The letter h designates the
positions of residual cellular proteins.   Data from Spear and Roizman
(ref. 16).

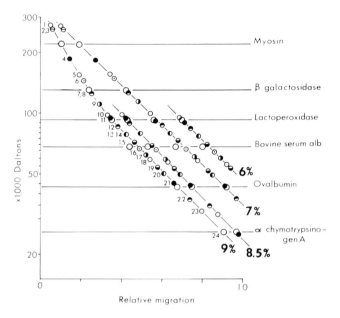

Fig. 5 The molecular weights of HSV-1 virion proteins graphically determined by their migration rates relative to known protein standards in 6, 7, 8.5 and 9% polyacrylamide gels. The large open circles represent the protein standards of known molecular weight. The small circles represent viral proteins as numbered on the left. Data from Spear and Roizman (16).

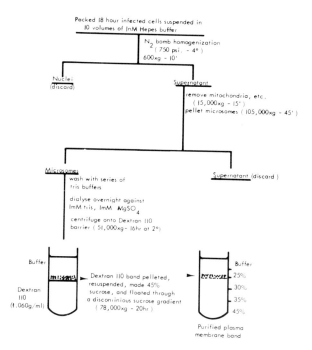

Fig. 6   Flow diagram for purification of plasma membranes from HEp-2 cells infected with herpes simplex virus. Adapted from procedures described in Ref. 23.

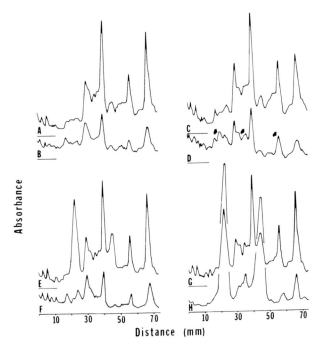

Fig. 7   Electropherograms of purified plasma membrane proteins of infected and uninfected cells labeled with ^{14}C amino acids before and after infection.   Cells were either infected or mock-infected at 28 hours and harvested at 46 hours post 0 time. The absorbance of profiles of a Coomassie Brilliant Blue stained flat gel (A,C,E,G) and of autoradiograms (B,D,F,H) developed from the same gel are shown.

A. Absorbance of stained bands of plasma membrane proteins of cells incubated from 0-24 hours in radioactive medium followed by incubation in non-radioactive medium until 46 hours post 0 time.

B. Autoradiographic tracing of labeled proteins in A.

C. Absorbance of stained bands of plasma membrane proteins of cells incubated from 32-46 hours post 0 time in radioactive medium.

D. Autoradiographic tracing of labeled proteins in C.   The arrows point out the proteins reduced in B.

E. Absorbance of stained bands of plasma membrane proteins of cells incubated from 0-24 hours in radioactive medium followed by infection at 28 hours and incubation in non-radioactive medium from 24-46 hours after 0 time.

F. Autoradiographic tracing of labeled proteins in E.

G. Absorbance of stained bands of plasma membrane proteins of
   cells infected at 28 hours after 0 time and incubated in
   radioactive medium from 32–46 hours.
H. Autoradiographic tracing of labeled proteins in G.
   Data from Heine et al. (23).

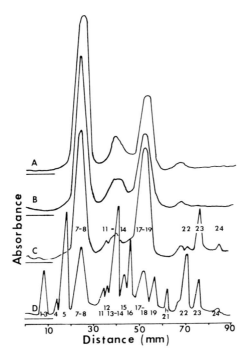

Fig. 8   Electropherograms of purified plasma membrane proteins
of infected cells labeled with ^{14}C amino acids or ^{14}C glucosamine
and of purified enveloped nucleocapsids of herpes simplex virus pro-
totype F labeled with ^{14}C glucosamine. Note that the resolution obtained in the flat gel slab as shown here is not as good as that obtained in cylindrical gels. In this instance some of the minor virion
proteins are unresolved. The absorbance profiles of Coomassie Brilliant Blue stained protein bands of the purified virion (D) and of autoradiograms of labeled proteins (A, B, C) are shown.
   A. Purified plasma membrane proteins labeled with ^{14}C-glucos-
      amine.
   B. Purified enveloped nucleocapsid proteins labeled with ^{14}C-
      glucosamine.
   C. Purified plasma membrane protein labeled with ^{14}C amino
      acids.
   D. Stained protein profile of purified enveloped nucleocapsids.
      Data from Heine et al. (23).

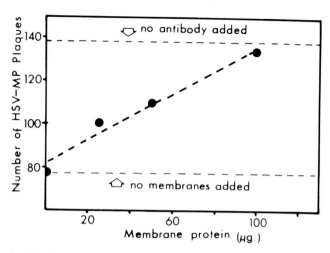

Fig. 9   Evidence for the identity of surface antigens on virions and on the smooth membranes of infected cells.   Smooth membranes (18) from HEp-2 cells infected with HSV-1 were purified according to the procedure of Spear et al. (16).   The test system consisted of 50 μl of challenge virus (MP strain) containing 500 p.f.u., 5 μl of anti HSV-1 serum (No. 39) and 200 μl of variable amounts of puri-fied membranes as shown.   In the absence of membranes 5 μl of serum reduced challenge virus plaque count by 50 per cent.   The membrane antigen competed with the virus for neutralizing antibody thereby sparing the infectivity.   The plaques formed by the chal-lenge virus were polykaryocytes, i.e., readily differentiated from the plaques produced by HSV-1 virus which might have contaminated the smooth membrane preparation.   In this instance the smooth mem-brane was free of demonstrable virus (Brown, Spear and Roizman, unpublished data).

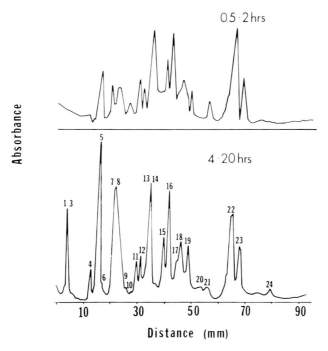

Fig. 10 Autoradiographic tracing of HSV-1 structural proteins
subjected to electrophoresis in 8.5% polyacrylamide gels. Top panel:
4×10^8 cells were exposed to 30 p.f.u./cell for 30 minutes, (effective multiplicity 6 p.f.u./cell) then overlayed with 40 ml of Eagle's
minimal essential medium lacking the amino acid mixture but supplemented with 10 μCi of ^{14}C reconstituted protein hydrolysate
(New England Nuclear, Boston, Mass.) and 1% dialyzed calf serum.
At 2 hours post infection the labeling medium was removed, the culture rinsed with unlabeled medium and replenished with mixture 199
supplemented with 1% calf serum. The cells were harvested 29 hours
later and the virus purified, solubilized and subjected to electrophoresis as described in ref. 16. Bottom panels approximately 0.5 ×
10^8 cells were infected at a multiplicity of 6 p.f.u./cells and incubated between 4 and 20 hours post infection in medium consisting of
Eagle's minimal essential medium containing only 1/10 the usual
concentration of serum but supplemented with 0.5 μCi of ^{14}C reconstituted protein hydrolysate per ml of medium and 1% dialyzed calf
serum. All other procedures were the same (Spear and Roizman, 1972).

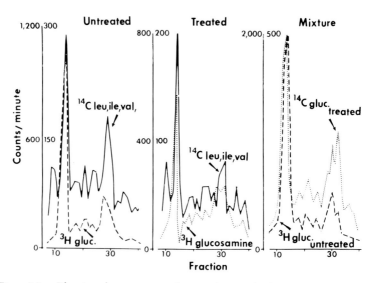

Fig. 11   Electropherograms of proteins and glycoproteins extracted from purified smooth membranes of HEp-2 cells infected with HSV-1 and either treated or untreated with cytosine arabinoside (20 μg/ml). The drug was added at the time of exposure of the cells to virus. Radioactive precursors were added at 4 hours post infection. The left and middle panels show the electropherograms of membrane proteins from cells labeled simultaneously with ^{14}C-amino acids (leucine, isoleucine, and valine) and ^{3}H-glucosamine. The right panel shows the electropherograms of a mixture of the membrane preparations of cells treated and labeled with ^{14}C-glucosamine and untreated and labeled with ^{3}H-glucosamine. Data from Roizman (13).

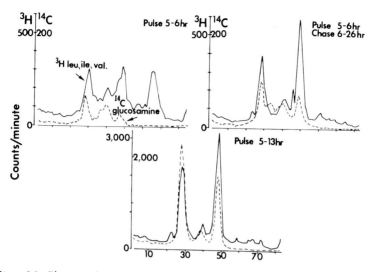

Fig. 12 Electropherograms of radioactive proteins present in smooth membranes purified by flotation through a discontinuous sucrose gradient (18) immediately after a 1 hour pulse from 5 to 6 hours post infection or after further incubation in non-radioactive medium. Conditions for solubilization of proteins and electrophoresis were as described in Ref. 18. Direction of migration was from left to right and the numbers of the abscissa represent gel slices. The cells were labeled with ^{3}H-amino acids(leu, isoleu, val) and ^{14}C-glucosamine. Amino acid label was conserved during the chase while glucosamine label increased. Amino acid label from the non-glycosylated band present after the pulse apparently distributed among the glycosylated bands during the chase. That further glycosylation of the membrane bound proteins occurs during the chase is indicated by the observations that (1) glucosamine peaks trail the broad amino acid peaks after the pulse, (2) after the chase the amino acid peaks are sharper indicating increased homogeneity, and migrate more slowly in coincidence with the glucosamine peaks. From Spear and Roizman (42).

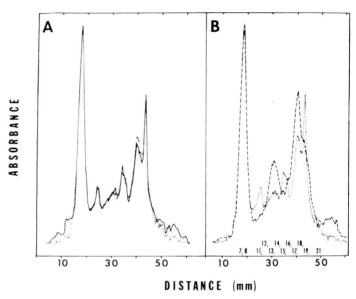

DISTANCE (mm)

Fig. 13 Comparison of HSV-1 glycoproteins in whole infected cells, purified plasma membranes and NP-40 extract of infected cells. The cells were labeled with ^{14}C glucosamine from 4 to 24 hrs post infection, then extracted with NP-40 as described in the text. The purified plasma membranes were prepared as described by Heine et al (23). The absorbance profiles of the autoradiographic image of viral glycoproteins subjected to electrophoresis in polyacrylamide gels were normalized with respect to the glycoprotein, 7-8. The assigned numbers are based on the designations given to structural proteins of the virion and virus-specific membrane proteins stated in the text.

A. Whole infected cells and NP-40 extract.

B. NP-40 extract and purified plasma membranes.

Solid line (———) solubilized whole infected cells.

Dotted line (° ° ° °) NP-40 extract of the same batch of infected cells.

Dashed line (----) purified plasma membranes.

Data from Savage et al. (24)

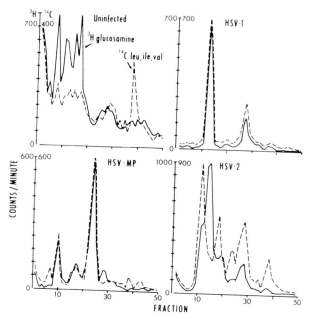

Fig. 14 Electropherograms of proteins and glycoproteins present in smooth membranes of infected and uninfected cells. The cells were simultaneously labeled with D-glucosamine and L-amino acids (leucine, isoleucine and valine) from 4 to 24 hrs post infection. The smooth membranes were prepared as described (18). Electrophoresis was carried out in 8.5% acrylamide gels (16). In this and subsequent electropherograms the direction of migration of the proteins is from left to right.

V

MEMBRANES OF TRANSFORMED
CELLS

Mitotic Cell Surface Changes

Max M. Burger

Department of Biochemical Sciences
Princeton University
Princeton, New Jersey

Most cells go through a period of drastic morphological alteration at cell division. An increased kinetic activity of the whole cell is also evident in both fibroblasts and epithelial cells and includes the cell surface.
Using plant agglutinins as probes, we observed earlier alterations in the surface of tumor cells.[1] Since such cells
have a high mitotic rate, we were interested to find some
plant agglutinin which may detect plasma membrane or surface coat alteration in dividing cells.

First we will discuss and introduce briefly some of
the agglutinins used. Then the permanent alterations found
in transformed cell surfaces will be compared to those found
only in mitosis in normal cells. A possible functional
significance of such a surface alteration will be discussed
subsequently, particularly in view of the fact that concomitantly with the surface alteration, changes in the level of
intracellular cyclic-AMP could be observed.

Agglutinins that react preferentially with transformed cells.

Based on earlier work from Ambrose's group,[2] Aub and
his coworkers[3,4] found that wheat germ lipase agglutinated
tumor cells preferentially even though some normal cells
seemed also to react.[5] We isolated and purified the agglutinin which was an impurity in the lipase preparation
(WGA),[1,6] and concluded from comparisons between pairs of
virally, chemically and spontaneously transformed tissue
culture cells that the untransformed parent cell agglutinated with very few exceptions to a negligible degree if
compared with the transformed cell line.

These differences were subsequently shown also for other agglutinins,[8,9] DNA virus and RNA virus transformed cells,[10,11,12] as well as[13] for epithelial cells and their transformed derivatives. We do not know at the present time, however, whether increased agglutinability is a cell surface property that follows altered growth control _in vivo_, or _in vitro_ (contact inhibition of growth) or whether it is an irrelevant by-product of transformation. We do not know either what exactly the biochemical surface modification consists of that leads to the increased agglutinability except that any explanation will have to take into account that after a brief proteolytic treatment, nonagglutinating normal cells can be brought into the same agglutinable state seen in virally and otherwise transformed cells.

Even though most transformed cells tested so far (62) agglutinated better with WGA, it is not very likely that WGA would detect all tumor cells without exception. Indeed, Dr. Phillips W. Robbins and Dr. Friberg's laboratory have recently found such exceptions. On the other hand, similar exceptions have been found for the other purified agglutinin, Concanavalin A. Many leukemia cells[15] do hardly agglutinate with Con A, like e. g. L1210 cells. I would suggest therefore that in every transformed cell a surface alteration took place which can be detected with one of the many plant agglutinins available.

Binding Fluorescein-labelled Agglutinin to Untransformed Mitotic Cells

Fluorescein labelled wheat germ agglutinin (Fl-WGA) binds preferentially to mitotic mouse fibroblasts and hardly[16] to the interphase cells under identical conditions. This can be seen in Fig. 1 where mitotic indices are perfectly overlapping with fluorescence indices in synchronized 3T3 mouse fibroblasts. Using again fluorescence to detect the binding of another agglutinin that reacts preferentially with transformed cells (Concanavalin A), Dr. R. S. Turner was able to confirm the observations with Fl-WGA. Furthermore, Dr. K. D. Noonan has found that mitotic cells bind[3]H-labeled Concanavalin A to about the same degree as non mitotic transformed cells.

From control experiments these reactions were shown to

be due to specific binding to carbohydrate receptors on the mitotic cell surface. They were not due to the rounding[16] up _per_ _se_ which occurs also during mitosis.

A Drop in Cyclic-AMP Levels During Mitosis

Since transformed cells seemed to show a cell surface alteration throughout the cell cycle that was similar to the one found during mitosis of normal cells, we suggested that the transformed surface might be in a permanently "locked-in" mitotic cell surface conformation and that it would be the virus which would fix this surface modification.[17] Whether the agglutinating state is characterized[18] by exposed agglutinin sites as we originally suggested in Fig. 2 is irrelevant for the argument here.

Sheppard[19] and Otten[20] et al. found a decrease in cyclic-AMP in transformed cells when comparing them to those in untransformed cells. Even though these two laboratories obtained different results when comparing growing cells, they agree when comparing confluent cells. We wondered therefore whether the correlation which we found between the cell surfaces of normal mitotic cells and transformed interphase cells might also hold regarding intracellular levels of cyclic-AMP. In other words, we thought that agglutinability may be equated in Fig.2 with low levels of cyclic-AMP and non agglutinability with high levels.

Fig. 3, taken from a collaborative effort with two other laboratories, shows that cyclic-AMP levels must drop precipitously during mitosis.[21] Since the total level of cyclic-AMP in a whole culture drops by a good 50% at a point where at best about 20% of the cells are in mitosis, we have to assume that mitotic cells have low levels for more than the hour during which they go through the classical phases of cell division (mitosis). A slight shift of the cyclic-AMP minimum in Fig. 3 towards the next following part of the cell cycle (G-1) and away from the peak of the mitotic index, may indicate that the low cyclic-AMP levels rather than beginning before mitosis, persist after mitosis, and continue into the G-1 phase. Our techniques used in this study do not permit us, however, to make a reliable statement about the duration of the low cyclic-AMP during division.

Inhibition of the Escape from Growth Control with the Cyclic Nucleotide

A brief treatment of untransformed 3T3 mouse fibroblasts with a proteolytic enzyme did not only alter the surface architecture to the same degree as transformation by a virus but growth control, i. e. density dependent inhibition of growth typical for untransformed cells could also be overcome as in transformed cells.[22,23] In other words, resting untransformed cells in confluency could be stimulated to grow like transformed cells, i. e. they escaped density dependent inhibition of growth. This uncontrolled growth persisted only for one generation since these cells repair the surface alteration within one cell generation. Without rounding off the cells or changing their gross morphology, a 5 minute treatment with pronase or trypsin results in the stimulation of the majority of cells which go through one round of cell division 24 to 36 hours later. (Fig. 4). We could show that the brief proteolytic treatment can be restricted only to the cell surface by using trypsin or pronase covalently linked to beads that were not phagocytosed (Fig. 5).

If the drop in cyclic-AMP found during mitosis would be the signal activating a new round of a cell cycle, two predictions could be made: 1) one should see a drop in cyclic-AMP after the brief treatment with proteases as well as other stimuli (e. g. serum) that lead to an escape from growth control. Sheppard found such a drop both after protease and serum. 2) Prevention of the cyclic-AMP drop by the addition of easily penetrating dibutyryl-cyclic-AMP to the medium should prevent also the escape from growth control. Whether dibutyryl-cyclic-AMP mimicks cyclic-AMP and "replenishes" therefore the intracellular pool of the cyclic nucleotide or whether it inhibits the phosphodiesterase that usually degrades cyclic-AMP is presently debated but irrelevant in our context.

As we showed earlier,[24] and as can be seen from Fig. 4, 10^{-5} M of the cyclic nucleotide can already inhibit the protease induced growth response. We found that the protease induced "signal" was sensitive to the cyclic nucleotide only for a very short time and that it was irreversibly established 3 minutes after the protease treatment.[21] Incubation of cells with dibutyryl-cyclic-AMP and then

removal of the nucleotide prior to stimulation with protease did not inhibit the "signal" from being triggered.

Since the protease still alters the surface and brings
about the agglutinable state in the presence of the cyclic
nucleotide,[21] we think it more likely that the surface alteration leads to a change in cyclic-AMP content than vice
versa.

Our observations led us to the working model shown in
Fig. 6 which is clearly an oversimplification and will require modification later on as probably many other hypotheses made for the function of the panacea cyclic-AMP. I
would therefore like to conclude with the open questions:

1. We do not know yet whether the alteration of the
cell surface is a requirement for the alteration in cyclic-
AMP levels and whether the drop in cyclic-AMP levels is the
message required to promote the cell into the next cell
cycle.

2. We do not know whether it will be the low level of
cyclic-AMP reached during mitosis or the degree by which
the level drops which contributes the "signal".

3. We do not know by what mechanism the drop of intracellular cyclic-AMP is achieved: a drop in the enzyme that
seems to be in the surface membrane and synthesizes cyclic-
AMP, i. e. the cyclase, an increase in the enzymatic activity which degrades cyclic-AMP, i. e. phosphodiesterase or
-- seldom considered -- simply a release of cyclic-AMP from
the cell. A decrease in cyclase activity seems to be unlikely at the present time since high cyclase values were
found in mitotic cells.[25] This possibility is not rigor-
ously excluded, however, since a slight shift in the cyclase increase beyond the mitotic peak which might be a
response to the low levels of cyclic-AMP during mitosis
would not have been detected by the sparse observation
points distributed over the cell cycle in that particular
study.

The hypothetical model is presently put to several
tests and others as well as this laboratory are involved in
an effort to answer the open questions discussed.

References

1. M. M. Burger and A. R. Goldberg, Proc. US Nat. Acad. Sci. 57, 359 (1967).
2. E. J. Ambros, A. I. Dudgeon, D. M. Easty, and G. C. Easty, Exptl. Cell Res. 24, 220 (1961).
3. J. C. Aub, C. Tieslan, and A. Lankester, Proc. US Nat. Acad. Sci. 50, 613 (1963).
4. J. C. Aub, B. H. Sanford, and M. N. Cote, Proc. US Nat. Acad. Sci. 54, 396 (1965).
5. J. C. Aub, B. H. Sanford, and L. H. Wang, Proc. US Nat. Acad. Sci. 54, 400 (1965).
6. Y. Nagata, and M. M. Burger, J. Biol. Chem. 247, 2248 (1972).
7. M. M. Burger, in Permeability and Function of Biological Membranes (edit. L. Bolis et al.) p. 107 (North Holland, Amsterdam, 1970).
8. M. Inbar, and L. Sachs, Nature 223, 710 (1969).
9. B. Sela, H. Lis, N. Sharon, and L. Sachs, J. Membr. Biol. 3, 267 (1970).
10. J. M. Biquard, and P. Vigier, CR Acad. Sci. 274, 144 (1972).
11. M. Kapeller, and F. Doljanski, Nature 235, 184 (1972).
12. M. M. Burger, and G. S. Martin, Nature 237, 9 (1972).
13. C. Borek, M. Grob, and M. M. Burger, in preparation.
14. M. M. Burger, Proc. US Nat. Acad. Sci., 62, 994 (1969).
15. M. M. Burger, unpublished observation.
16. T. O. Fox, J. R. Sheppard, and M. M. Burger, Proc. US Nat. Acad. Sci. 68, 244 (1971).
17. M. M. Burger, in Ciba Foundation Symposium on Growth Control in Cell Culture (edit. Wolstenholme and Knight) p. 45 (Churchill Livingstone, London, 1971).
18. M. M. Burger, K. D. Noonan, J. R. Sheppard, T. O. Fox, and A. J. Levine, 2nd Lepetit Colloquium on the Biology of Oncogenic Viruses, (edit. L. G. Silvestri) p. 258 (North Holland, Amsterdam, 1971).
19. J. R. Sheppard, Nature 236, 14 (1972).
20. J. Otten, G. S. Johnson, and I. Pastan, Biochem. Biophys. Res. Commun. 44, 1192 (1971).
21. M. M. Burger, B. M. Bombik, B. M. Breckenridge, and J. R. Sheppard, Nature, submitted.
22. M. M. Burger, Nature 227, 170 (1970).
23. B. M. Sefton, and H. Rubin, Nature 227, 843 (1970).
24. M. M. Burger, B. M. Bombik and K. D. Noonan, Invest.

Dermatology, in press.
25. M. H. Makman, and M. I. Klein, Proc. US Nat. Acad. Sci. 69, 456 (1972).
26. G. Todaro, Y. Matsuya, S. Bloom, A. Robbins, and H. Green, in Growth Regulating Substances for Animal Cells in Culture (edit. V. Defendi, and M. Stoker) No. 7. p. 87 (The Wistar Inst. Press, Philadelphia, 1967).
27. A. G. Gilman, and M. Nirenberg, Proc. US Nat. Acad.Sci. 68, 2165 (1971).
28. G. Krishna, B. Weiss, and B. B. Brodie, J. Pharmacol. Exp. Ther. 163, 379 (1968).

Acknowledgments

This work was supported by grants from the NCI (CA-10151 and CA- 16765) as well as a contract with the SVCP of the National Cancer Institute, 71-2372.

Table 1
Effect of hyaluronidase treatment on the agglutinability
of RSV-infected and uninfected chick embryo fibroblasts

Infection	Temperature of growth +	μg agglutinin per ml required for half maximal agglutination*			
		WGA		Con A	
		without hyaluronidase treatment	with hyaluronidase treatment	without hyaluronidase treatment	with hyaluronidase treatment
None	36°		350±50		300±100
	41°	625±125	350±50	300±50	250±50
SR-RSV	36°		19±9		47±13
	41°	145±31	18±7	176±24	44±10
T5 + +	36°	177±24	30±10	221±31	51±11
	41°	410±60	187±38	380±70	194±31

* Data for uninfected cells were averaged from two experiments, for SR-RSV infected from 11, and for T5 infected from 8.

+ Cells were infected and grown at 41° C, subdivided with trypsin, and incubated at 41° C for 36 hrs. Some plates were then shifted to 36°C. The cells were harvested 36 hrs. later.

+ + Chick embryo fibroblasts transformed with a temperature sensitive Rous Sarcoma Virus which is nonpermissive for the transformed state at 41° C. (Taken from reference 12).

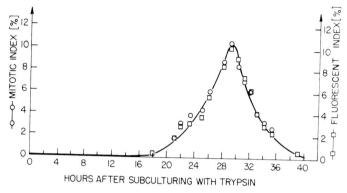

Figure 1. Mitotic and fluorescent indices of synchronized 3T3 cells. Cells were exposed to fluorescein isothiocyanate treated WGA, fixed with ethanol, stained with Evans blue and mounted in Elvanol. In control experiments, cells were exposed to FITC-WGA and counted without fixing and staining, and the fluorescent index was identical to the data reported here. Blind counts of several hundred cells were made by two investigators and were in good agreement. (From reference 16).

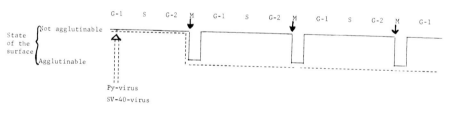

Figure 2. Hypothetical model for a permanent "fixation" of the mitotic cell surface structure in transformed cells. (Modified from reference 18).

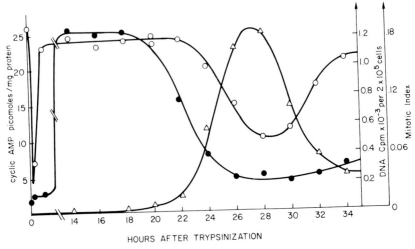

HOURS AFTER TRYPSINIZATION

Figure 3. 3T3 cells were synchronized by trypsinization (0.05% trypsin for 2 min.) of 4 day old monolayers and plating them at 5 times lower densities with 20% calf serum in the Dulbecco-Vogt modification of Eagle's medium (Gibco). The ratio of cells in mitosis to total cells (mitotic index) was determined with the phase microscope. For the measurement of DNA synthesis, 2 μ Curies of ^{3}H-thymidine (5.95 Curies per millimole) were added in 2 ml per 75 cm^2 Petri dish for 20 minutes. The cells were washed, removed with trypsin, extracted on millipore filters with 5% TCA and washed with ethanol before counting.[26] Cyclic-AMP was measured with Gilman's assay.[27] The cells were washed with Eagle's minimum essential medium, fixed with 5% TCA, removed, homogenized and the 20,000 r.p.m. supernatant treated with ether, lyophilized and assayed. Assays on one complete experiment with synchronized cells were carried out after separating cyclic-AMP from other nucleotides by column chromatography.[28] All points are averages of 3 independent experiments.
0——0 Cyclic-AMP level. ●——● DNA synthesis. △——△ Mitotic index. (From reference 21).

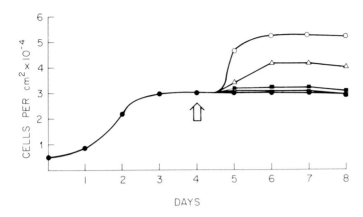

Figure 4. Inhibition of protease induced growth stimulation by dibut-cyclic-AMP: Dependence on cyclic-AMP concentration.

At the arrow, the medium was removed from 3.5 cm Fal-
con dishes containing confluent 3T3 fibroblasts and various
concentrations of dibut-cyclic-AMP were added together with
10 μg pronase (Calbiochem) per ml phosphate buffered saline
and filled with the conditioned medium removed before the
treatment. Cells were counted daily after removal with
0.025% trypsin. The average from 6 determination is given
in the graph.

● — ● Control cultures incubated for 10 min. with phosphate
buffered saline and subsequently washed with the
same buffer.

0 — 0 10 μg/ml Pronase for 10 min.

△ — △ 10 μg/ml Pronase and 5×10^{-6} M Dib-cyclic-AMP for 10
min.

■ — ■ 10 μg/ml Pronase and 10^{-5} M Dib-cyclic-AMP for 10 min.

□ — □ 10 μg/ml Pronase and 5×10^{-5} M Dib-cyclic-AMP for 10
min. (From reference 21).

251

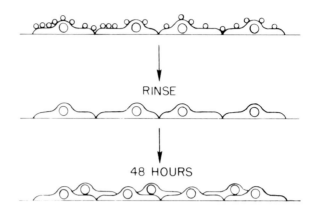

RINSE

48 HOURS

Figure 5. To confluent 3T3 cells which were resting, Sephadex beads with a diameter of 5 micron were added for 5 min. Trypsin or Pronase were coupled to the beads in advance and the beads were rinsed off the monolayer of 3T3 cells to an extent that less than one bead was left on the culture as verified with ^{3}H-labeled beads. The growth response (80-90% as many cells after 48 hours) did not occur with untreated beads. If any protease leaked off the beads during the incubation, it could not have been enough since on an activity basis, the beads did not lose more than 3% of their protease during incubation, neither could any substantial protease activity be found in the soluble supernatant after the incubation (see references 17 and 22).

GLYCOLIPIDS OF NORMAL AND TRANSFORMED CELLS
A DIFFERENCE IN STRUCTURE AND DYNAMIC BEHAVIOR*

By
Sen-itiroh Hakomori, Shigeko Kijimoto,
and Bader Siddiqui

Departments of Pathobiology and Microbiology
School of Public Health and School of Medicine,
University of Washington, Seattle, Washington 98105

Introduction and a Brief Survey

The regulatory mechanism of cellular growth is be-lieved to be mediated through cell surface membranes, and defects of the mechanism in tumor cells can be ascribed to a change of cell surface membranes (1).

The common properties of tumor cell surfaces can be described as the absence of the functions that are exhi-bited by normal cells, such as "contact inhibition" (2), intercellular association (3), and "contact communication" (4).

It is now obvious that plasma membranes of animal cells are characterized by a much higher quantity of gly-cosphingolipids than those in intracellular membranes (5-8), and some glycolipids on cell surfaces serve as potential antigen sites (9) and intercellular recognition sites.

Human tumor cells have been characterized by the al-tered status of blood group isoantigens as shown in Table 1. The chemical structures of the accumulated glycolipid hap-ten that showed Le^a and Le^b activity (17) and that showed a still unidentified specificity (20) have been described.

During the past few years, a great deal of interest as to the transformation-dependent changes of cell surface

---
*Supported by the National Cancer Institute Research Grants CA 12710 and CA 10909 and by the American Cancer Society Grant BC9-B.

glycoproteins and glycolipids has been aroused (21-34).
The structures of membrane-bound heteroglycans and
their possible changes associated with malignant transfor-
mations can be summarized as shown in Table 2. The most
obvious and well-established structural change of heteroglycans is the simplification of carbohydrate chains by
deletion of the terminal residue, thus creating a new carbohydrate residue, which was originally located at the
penultimate position (21).
The organizational changes of membrane heteroglycans
have been studied rather extensively by means of phytoagglutinin and anti-glycolipid antisera (22, 35-39). The
basis of the enhanced cytoagglutination of tumor cells is
now partially known to be the relocation and redistribu-
tion of the reactive sites, rather than actual exposure of
the reactive sites (40, 52-55).
In this article we would like to present 1) chemical
changes of membrane glycolipids, 2) dynamic behavior of
glycolipid synthesis and degradation in relation to contact inhibition, and 3) organization of glycolipids in
membranes and possible intercellular linkages.

Chemical Changes of Membrane Glycolipids

Glycolipids of a number of "normal" and malignant
transformed cells have been characterized, and their
chemical quantities were determined. The results indicated that a majority of glycolipids of "normal" and
"transformed" cells were structurally related or identical.
A significant difference was found, however, in the proportions of more than two glycolipids which were structurally related. In general, glycolipids with more complete carbohydrate residues were deleted in transformed
cells, suggesting an incomplete synthesis of the carbohydrate chain (21). Occasionally incomplete synthesis
was coupled with accumulation of a glycolipid having a
precursor residue (21,22).
This phenomenon was observed not only in in vitro
transformed cell systems, but also in in vivo tumor cells
(26). The same phenomenon could apply to human tumors as
well. Thus, the deletion of blood group A or B and the
accumulation of Le^a, Le^b and H-hapten were observed (see
Table 1). Further strong evidence of this phenomenon for

the human tumor will be extremely difficult to obtain, as we do not have reliable "normal controls" for human tumor tissue.

Various examples of incomplete synthesis of glyco-lipids are shown in Table 3.

Contact Inhibition and Glycolipid Synthesis

The data described in the previous section suggests that altered membrane heteroglycans may be linked to un-controlled cell division. It should be noted, however, that the total amount of cellular heteroglycans was higher in cells grown on monolayer than in those grown in a sus-pension culture (43) where cell-to-cell contact did not take place. A related observation is that one of the mem-brane glycopeptides of mouse fibroblasts increased on cell confluence (31). Further evidence for the cell-density-dependent change of the membrane glycolipid has been fur-nished more recently (28,44).

1. Glucosylceramide, β-galactosylglucosylceramide, α-galactosylgalactosylglucosylceramide (CTH), and hema-toside (N-acetylneuraminylgalactosylglucosylceramide) were found in some BHK cells where saturation density was not very high. Only the concentration of α-galactosylgalacto-sylglucosylceramide increased when cells were crowded and contact-inhibited. Polyoma-transformed BHK cells lacked in α-galactosylgalactosylglucosylceramide, which was in-creased in contact-inhibited normal BHK cells (28).

2. β-glucosylceramide, hematoside (N-acetylneuraminyl-galactosylglucosylceramide), disialohematoside (N-acetyl-neuraminyl-N-acetylneuraminylgalactosylglucosylceramide), and monosialoganglioside (GM_1-ganglioside) were found in human 8166 diploid cells. The quantities of hematoside, disialohematoside, and monosialoganglioside increased when cells were contact-inhibited. The increasing rate of di-sialohematoside was more obvious than that of monosialo-hematoside. The SV_{α_0}-transformed human heteroploid cells had very low levels of disialohematoside and monosialo-ganglioside (28).

Hematoside, disialohematoside, and monosialoganglio-side were also found in chick embryonic fibroblasts. The quantity of disialohematoside increased when cells were confluent. The glycolipids that increased on confluency were greatly decreased when these cells were transformed with Rous sarcoma viruses (41).

4. In NIL cells (hamster fibroblasts), glucosylceramide (CTH), globoside (N-acetylgalactosaminosylgalactosyl-
galactosylglucosylceramide), and Forssman glycolipid antigen (N-acetylgalactosaminosyl) ($\alpha 1 \rightarrow 3$)-N-acetylgalacto-
saminosyl ($\beta 1 \rightarrow 3$) galactosyl ($\alpha 1 \rightarrow 4$) galactosyl ($\beta 1 \rightarrow 4$)
glucosylceramide have been found and identified (45). In
confluent cell sheets, the quantities of CTH and Forssman
antigen increased as compared to the growing cell cultures.
Transformed cells with polyoma virus, CTH, globoside, and
Forssman glycolipid antigen were almost absent (42,44).

The results of these studies suggest that the concentration of certain glycolipids increases on cell-to-cell
contact of normal cells. This is presumably due to the
attachment of sugar residue to precursor lipids. The most
typical and sensitive extension response was demonstrated
by the terminal residue of CTH in a contact-sensitive BHK
and NIL cells. The activity of UDP-galactose:lactosyl-
ceramide α-galactosyl transferase was determined in comparison with the activity of UDP-galactose:glucosylcera-
mide β-galactosyl transferase. (See Figure 1). On the
other hand, α-galactosidase activity was also determined
using ceramide trihexoside that had a radioactive label at
the terminal of α-galactose moiety.

The activity of CTH synthesis in BHK and NIL cells
was greatly enhanced when the cell population density
reached over $10^5/cm^2$. Such a sensitive response was not
observed in CDH synthesis. The CTH synthesis in either
polyoma-transformed BHK or NIL cells was extremely low and
did not respond to their cell population densities (See
Figure 1).

The α-galactosidase for CTH ("ceramide trihexosidase")
was not enhanced when the cell population increased; however, it did increase in polyoma-transformed cells (46).

A number of trials have been made to demonstrate the
synthesis of globoside and Forssman glycolipid hapten in
cell-free systems. So far the trials have been unsuccess-
ful. However, pulse-labelling experiments of NIL cells and
of their polyoma transformants with ^{14}C-galactose have been
carried out, and the results are shown in Table 5.

A considerable amount of all the glycolipids appeared
to be synthesized during a short period (one hour). These
numbers do not represent the chemical quantity of glycolipids present in cells, but indicate the cell activity for

net synthesis of individual glycolipids under varying degrees of cell-to-cell contact. While synthesis of CDH in normal NIL cells was independent of cell population density, the synthesis of CTH, globoside, Forssman glycolipid, and hematoside was significantly enhanced when the cells were crowded and contacted each other. The enhanced synthesis was particularly pronounced in Forssman glycolipid and hematoside.

It is also noteworthy that the enhanced synthesis was most remarkable at the early stage of cell confluency (cell population density $1.4 \times 10^5/\text{cm}^2$) than at the later stage of confluency (cell population density $2.5 \times 10^5/\text{cm}^2$).

The synthesis of CDH was rather high in py-transformed cells, but very low synthesis of CTH, globoside and Forssman glycolipid was demonstrated in py-transformants. Because the synthesis of higher glycolipids in py-transformants was low, density-dependent change was difficult to recognize.

The structures of CTH, globoside and Forssman glycolipids have been established (45,47), and they interrelated as shown in Figure 2. The metabolic behavior of these three glycolipids on cell-to-cell contact and their changes on malignant transformation were similar, as they are structurally related. However, synthesis of hematoside was enhanced on cell-to-cell contact but did not decrease on malignant transformation.

The enhanced synthesis of Forssman glycolipids on cell-to-cell contact and its deletion in malignant cells were further assessed by incorporation of "C-galactose" radioactivity into the precipitin line, demonstrated on an immunodiffusion plate (see Figure 3).

Organization of Glycolipids in Surface Membranes, Intercellular Linkage, and Contact Inhibition

Glycolipids are supposed to bind with membrane proteins and lipids in hydrophobic-hydrophobic interactions, and the carbohydrate moiety can be stretched outwards. It is suspected, however, that not all of the carbohydrate moieties of glycolipids are exposed to extracellular environments, based on the following experimental findings: 1) a purified Jack-bean β-N-acetylgalactosaminidase can hydrolyze only a very small portion (a few %) of total β-N-acetylgalactosaminosyl residue of globoside in the mem-

brane, while the same enzyme can hydrolyze β-N-acetylgalac-
tosaminosyl residue of an isolated globoside very easily
(B. Siddiqui, unpublished data); 2) sialidase cannot hydrolyze sialyl residue of hematoside in membranes, while
the same enzyme can hydrolyze sialyl residue of hematoside
very easily (7,48); 3) only a part of ganglioside in membranes can be extracted by ether-methanol or aqueous ethanol; after treatment with protease, extraction can be quantitative (7,49); 4) antibodies against glycolipid (e.g.
anti-globoside and anti-hematoside) can react with human
erythrocyte membranes to a small extent, while trypsin
treatment of erythrocytes enhances the reactivity to a
great extent (22,50,51).

Non-susceptibility of glycolipids to various glycosidases contrasts to the higher susceptibility of glycoproteins by glycosylhydrolases in general. It is possible
that a majority of glycoprotein carbohydrate chains can be
located at the external part of "glycocalyx" of the membrane by being attached to a stretching protein core, while
glycolipids are directly imbedded in the membrane matrix
(see Figure 4).

The enhanced reactivity of cells with phytoagglutinin
and with anti-glycolipid antiserum was observed after treatment with protease (35-37,22,50-51).

Recent studies by Nicolson and Singer have pointed out
that trypsin-dependent enhancement of agglutination of cells
with phytoagglutinin is not due to an exposure of non-
cryptic carbohydrate sites on cells, but rather to a dis-
location and re-distribution of the reactive sites on
fluidy cell membranes (52,53).

Unmasking of "cryptic" antigen sites, however, does
not explain the fact that untreated cells bind radioacti-
vity-labelled agglutinins. Radioisotope-labelled agglutinin was bound approximately to the same extent or even
less to untreated cells or to cells with proteolytic enzyme treatment (52-60). In view of these recent develop-
ments, our previous study on the enhanced reactivity of
erythrocytes to anti-globoside antiserum by trypsinization
(50,51) was thoroughly reinvestigated.

The trypsinized and non-trypsinized membranes were
reacted with anti-globoside antiserum, washed with saline,
and the absorbed antibody was eluted with glycine-HCl buffer pH 2.4. The eluted antibody fraction was dialyzed and

the reactivity was tested by precipitin reaction. As shown in Table 6, the precipitin value of membrane globoside increased four times with proteolysis of membranes, whereas agglutination titer and hemolysis titer increased 50-100 times.

It is assumed, therefore, that the increased capability of membranes for absorption of anti-globoside can be ascribed to an increased number of globoside molecules on cell surfaces. However, the degree of agglutination or hemolysis can be affected by another organizational factor, namely re-location of reactive sites, as pointed out by Singer and Nicolson (52). Further ultrastructural study is necessary to evaluate this possibility.

The reactivity of fetal, cord and adult erythrocytes to anti-globoside serum and to anti-lactosylceramide was tested. As is shown in Table 3, there was a striking difference in the reactivity of erythrocytes with anti-globoside in absorption capacity, agglutination, and hemolysis; no difference in reactivity was observed with anti-lactosylceramide serum.

The results indicate that the organizational state of globoside and lactosylceramide on cell surfaces are different, and that the organizational state of globoside in erythrocyte membranes can be greatly changed during the early stage of development. Most probably, the globoside was in a highly exposed state in the fetal stage and became "cryptic" during the gestation period. The possibility of a change of topological distribution may also be considered according to the Singer-Nicolson model (52,53). Further ultrastructural study is waiting to be developed.

The reactivity of glycolipids on cell surfaces of fibroblastic cells at different stages of cell growth and their malignant transformants have been studied. The results of earlier studies of the reactivities of antihematoside antiserum to "normal" and "transformed" fibroblastic cells has been summarized briefly in Table 8.

Increased reactivity was observed in "transformed" BHK and 3T3 cells, and the reactivity of the transformed cells did not further increase on trypsinization. Reactivity of "normal" cells did increase on trypsinization. These results were very similar to those described by Burger on the agglutinability of "normal" and "transformed" cells to the wheat germ agglutinin (35).

259

In view of a significant increase of various glyco-lipid antigens in NIL cells at higher population densities, the reactivities of NIL cells and their polyoma transfor-mants to the antisera against glycolipids have been inves-tigated. Only the reactivity against Forssman glycolipid is presented in this article. As shown in Table 9, the reactivity of NIL cells to anti-Forssman glycolipid reduced significantly at the confluent phase. The anti-Forssman antibody did not react with polyoma-transformed NIL cells; coinciding with that, Forssman glycolipid was completely deleted in the transformed cells (42).

The density-dependent decrease of cell reactivity to the anti-Forssman antibody was not clearly observed when the cells were treated with trypsin. Cells treated and separated from cultures with ethylenediamine tetraacetate (EDTA) showed only a slight difference between sparse-growing and confluency.

As the chemical quantity of Forssman glycolipid and its synthesis increased at higher cell population densities, the decreased Forssman reactivity of cells at the confluent phase indicates that a majority of Forssman glycolipids could have been masked at the confluent phase. Since EDTA-treated cells did not show a decreased reactivity against anti-Forssman antibodies at their confluency, a masking group is dissociable with EDTA. Such grouping might be a peptide furnished on surfaces of counterpart cells.

The results indicate that intercellular linkage might possibly occur at Forssman glycolipids on cell surfaces; thus, confluent NIL cells are conjuncted through Forssman hapten and a complementary peptide chain on surfaces of counterpart cells.

An increased synthesis of ceramide trihexosine, glo-boside or Forssman glycolipid at contact-inhibited, con-fluent cells could be the purposeful mechanism for better intercellular linkages. Malignant cells are lacking in both synthesizing normal level of globoside and Forssman glycolipid and their complementary peptides on cell sur-faces, thus showing decreased intercellular interaction and loss of contact-inhibition. This view is illustrated in Figure 5.

Conclusions and Some Speculations

The glycolipid profile is characteristic for cells.
Some cells showed a simple pattern containing only hematoside and lactosylceramide, as seen in some BHK C13 cell
lines. Some of them showed a complex pattern, as chick
embryonic fibroblasts, containing glucosylceramide, mono-
sialohematoside, disialohematoside, and higher gangliosides. Neutral glycolipids are a major component for some
cells (e.g. NIL cells), while others contain gangliosides
as a major component (e.g. chick fibroblasts). Even cell
lines under the same name showed significant variation.
(Some BHK C13 cells did not contain CTH, some did; some
3T3 cells contained higher gangliosides, others did not).
The transformation-dependent changes of glycolipids
are oriented, however, in principle, in the same direction,
i.e. simplification of carbohydrate chains by deletion of
non-reducing terminals. The deletion often occurs at
1) sialyl residue of sialyl-galactose, the terminal residue
of sialyl-sialyl; 2) α-galactosyl residue of α-galactosyl-
β-galactosyl; and 3) α-N-acetylgalactosaminosyl residue of
N-acetylgalactosaminosyl-β-(N-acetyl)galactosaminosyl.
Change of glycolipid composition also occurs in confluent, contact-inhibited cells; i.e. those glycolipids
containing α-galactosylgalactose, sialylgalactose, and
sialyl-sialyl residues increased in confluent cell sheets
of BHK, NIL, human 8166 cells, and chick embryonic cells.
Thus some membrane glycolipids increased on activation of
UDP-sugar:glycolipid transferase, when cells met and induced contact inhibition. The cell-contact-dependent increase of some glycolipids, therefore, could be a result
of glycosyl extension of the precursor lipid by activation
of glycosyl transferase. Such glycosyl extension is not
observed in transformed cells.
The reactivities of cell surface glycolipids to their
specific antisera decreased significantly when the cells
were crowded and contact-inhibited; transformed cells did
not show this decreased reactivity, nor did EDTA-treated,
nor did trypsin-treated cells. As suggested in Figure 5,
when normal cells became confluent, intercellular linkage
must have been formed between some glycolipids of one cell
surface and proteins (or glycoproteins) of the counterpart
cell surface. The linkages could have occurred through a
bivalent cation, which is sensitive to EDTA treatment. The

receptor protein or glycoprotein should be easily eliminated by protease treatment.

Loss of contact inhibition by a low concentration of
trypsin in the culture medium has been reported (56). A
similar phenomenon was observed in our laboratory by a
protease ficin and a α-galactosidase; NIL cells lost their
orderly orientation, escaped from their shape, and grew as
mimic polyoma-transformed cells. Pure α-galactosidase did
not induce this behavior, probably because the enzyme can
not reach α-galactoside of CTH fixed on cell membranes.
This phenomenon is well-suited to the hypothesis that low
contact-inhibition and intercellular adhesion is induced
by cleavage of the receptor protein. As shown in Figure 5,
transformed cells have glycolipids with shorter carbohydrate chains with new sugar terminals; therefore, the
peptides on the counterpart cell surfaces cannot fit onto
the glycolipids. The intercellular linkages are thus hampered in transformed cells.

It is unknown, however, whether the altered synthesis
of glycolipids in transformed cells is the result of loss
of contact inhibition or whether their changes cause loss
of contact inhibition, but it can obviously be a cause of
altered intercellular interaction between malignant cells.
A hypothetical relation between contact inhibition, intercellular adhesion, and the altered glycolipid synthesis in
membranes is envisaged below:

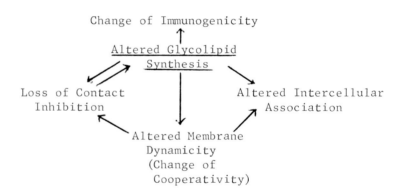

The altered glycolipid (or glycoprotein) synthesis
can induce altered intercellular adhesion as shown above.

It can also alter the membrane dynamicity and cooperativity of membrane particles through their conformational changes. Although it is not known at this time how the membrane dynamicity and conformational cooperativity of membrane particles are related to nuclear DNA replication, a certain mechanism must exist which links membrane dynamicity and regulation of nuclear DNA replication (e.g., a release of de-repressor from inside the membrane by changing conformational cooperativity of membrane particles; the de-repressor can then direct nuclear DNA replication). The altered glycolipid synthesis can induce loss of "contact inhibition" and could be one of the most essential steps in establishing malignancy of cells. Further extensive studies are needed to disprove this optimistic view.

Change of immunogenicity of cell surfaces must be only a secondary phenomenon consequent to the structural and functional changes of cell surface proteins, glycoproteins and glycolipids. Loss of intracellular adhesion must be an important basis for understanding the mechanism of metastasis.

References

1) Sacks, L., _Nature_, 202, 1272 (1965).
2) Abercrombie, M., _Europ J. Cancer_, 6, 7 (1970).
3) Edwards, J.G., and Campbell, J.A., _J. Cell Sci._, 8, 666 (1970); _Nature (New Biol.)_, 231, 147 (1971).
4) Loewenstein, W.R., Develop. Biol. Suppl., 2, 151 (1963).
5) Rennkonen, O., Gahmberg, C.G., Simons, K., and Kaariainen, _Acta Chem. Scand._, 24, 733 (1970).
6) Klenk, H.D., and Choppin, P.W., _Proc. Nat. Acad. Sci. U.S._, 66, 57 (1970).
7) Weinstein, D.B., Marsh, J.B., Glick, M.C., and Warren, L., _J. Biol. Chem._, 245, 3928 (1970).
8) Dod, B.J., and Gray, G.M., _Biochim. Biophys. Acta_, 150, 397 (1968).
9) Rappaport, M.M., and Graf, L., _Cancer Res._, 21, 1225 (1961); _Prog. Allergy_, 13, 273 (1969).
10) Masamume, H., and Kawasaki, H., _Tohoku J. Exp. Med._, 68, 81 (1958).
11) Masamume, H., and Hakomori, S., _Symposia of Cell Chemistry_, 10, 37 (1960).
12) Kay, H.E.H., and Wallace, B.M., _J. Nat. Cancer Inst._, 26, 1349 (1966).

13) Davidsohn, I., Kovanik, S., and Lee, C.L., Arch Pathol., 81, 381 (1966).
14) Kawasaki, H., Tohoku J. Exp. Med., 68, 119 (1958).
15) Iseki, S., Furukawa, K., and Ishikawa, K., Proc. Japan Acad., 38, 556 (1962).
16) Hakomori, S., Koscielak, J., Bloch, K.J., and Jeanloz, R.W., J. Immunol., 98, 31 (1967).
17) Hakomori, S., and Andrews, H.D., Biochim. Biophys. Acta, 202, 1225 (1970).
18) Hakkinen, I., Int. J. Cancer, 3, 582 (1968).
19) Hakkinen, I., J. Nat. Cancer Inst., 44, 1183 (1970).
20) Yang, H.J., and Hakomori, S., J. Biol. Chem., 246, 1192 (1971).
21) Hakomori, S., and Murakami, W.T., Proc. Nat. Acad. Sci., U.S., 59, 254 (1968).
22) Hakomori, S., Teather, C., and Andrews, H.D., Biochim. Biophys. Res. Commun., 33, 563 (1968).
23) Mora, P.T., Brady, R.O., Bradley, R.H., and McFarland, V.W., Proc. Nat. Acad. Sci., U.S., 63, 1290 (1969).
24) Brady, R.O., Borek, C., and Bradley, R.M., J. Biol. Chem., 244, 6552 (1969).
25) Cumar, F.A., Brady, R.O., Kolodny, E.H., McFarland, V.W., and Mora, P.T., Proc. Nat. Acad. Sci., 67, 757 (1970).
26) Siddiqui, B., and Hakomori, S., Cancer Res., 30, 2930 (1970).
27) Cheema, P., Yogeeswaran, G., Morris, M.P., and Murray, R.K., FEBS Letters, 11, 181 (1970).
28) Hakomori, S., Proc. Nat. Acad. Sci., 67, 1741 (1970).
29) Sheinin, R., Onodera, K., Yogeeswaran, G., and Murray, R.K., Second Le Petit Symposium - The Biology of Oncogenic Viruses (in press).
30) Wu, H., Meezan, E., Black, P.H., and Robbins, P.W., Biochemistry, 8, 2518 (1969).
31) Meezan, E., Wu, H.C., Black, P.H., and Robbins, P.W., Biochemistry, 8, 2509 (1969).
32) Buck, C.A., Glick, M.C., and Warren, L., Biochemistry, 9, 4567 (1970).
33) Buck, C.A., Glick, M.C., and Warren, L., Science, 172, 169 (1971).
34) Den, H., Schultz, A.M., Basu, M., and Roseman, S., J. Biol. Chem., 246, 2721 (1971).
35) Burger, M.M., Proc. Nat. Acad. Sci., Wash., 62, 994 (1969).

36) Inbar, M., and Sachs, L., Nature, 223, 710 (1969).
37) Pollack, R.E., and Burger, M.M., Proc. Nat. Acad. Sci., Wash., 62, 1074 (1969).
38) Hayry, P., and Defendi, V., Virology, 41, 22 (1970).
39) Sela, B.A., Lis, H., Sharon, N., and Sachs, L., J. Membrane Biol., 3, 267 (1970).
40) Nicolson, G., Nature (in press).
41) Hakomori, S., Saito, T., and Vogt, P.K., Virology, 44, 609 (1971).
42) Hakomori, S., and Kijimoto, S., unpublished.
43) Shen, L., and Ginsburg, V., "Biological Properties of Mammalian Surface Membranes", edited L.A. Mason (Wiston Institute Symposium Monograph No. 8, 1968, p. 67.
44) Robbins, P.W., and MacPherson, I., Nature, 229, 569 (1971).
45) Siddiqui, B., and Hakomori, S., J. Biol. Chem., 246, 5766, (1971).
46) Kijimoto, S., and Hakomori, S., Biochem. Biophys. Res. Comm., 44, 557 (1971).
47) Hakomori, S., Siddiqui, B., Li, Y-T., Li, S-C., and Hellerqvist, C.A., J. Biol. Chem., 246, 2271 (1971).
48) Wintzer, G., and Uhlenbruck, G., Zs. Immunitatsforsch, Allergie u. Klin. Immunologie, 133, 60-67 (1967).
49) Yamakawa, T., Irie, R., and Iwanaga, M., J. Biochem., 48, 490 (1960).
50) Koscielak, J., Hakomori, S., and Jeanloz, R.W., Immunochem., 5, 441 (1968).
51) Hakomori, S., Vox sanguinis, 16, 478 (1969).
52) Singer, S.J., and Nicolson, G., Science, 175, 720 (1972).
53) Nicolson, G., Nature, 233, 244 (1971).
54) Ozanne, G., and Sawbrook, J., Nature (New Biol.), 232, 156 (1971).
55) Cline, M.J., and Livingston, D.C., Nature, (New Biol.) 232, 155 (1971).
56) Burger, M.M., Nature, 227, 170 (1970).
57) Siddiqui, B., and Hakomori, S., J. Biol. Chem., 246, 5766, (1971).
58) Siddiqui, B., Kawanami, T., Li, Y-T., and Hakomori, S., J. Lipid Res., submitted.
59) Roseman, S., Chemistry and Physics of Lipids, 5, 270 (1970).

Table 1
Blood Group Changes in Some Human Tumors of Glandular
Epithelial Tissue (Adenocarcinoma) As Compared
To Normal Glandular Tissues

1. Deletion of blood group A or B hapten (10-13)
2. Persistence of blood group H hapten (14-16)
3. Increase and co-presence of Le^a and Le^b
   hapten (15-17)
4. Appearance of incompatible blood group
   antigens (16-19)
5. Presence of X-hapten glycolipid that is a
   positional isomer of Le^a glycolipid (20)

Table 2
Structures of Membrane-Bound Heteroglycans and
Their Changes Associated with Malignant Transformation

	Notions	Transformation-Dependent Changes
1. primary structure	a) sequence of sugar b) length of chain c) position of inter-linkages d) anomeric structure	sequence unchanged; length of chain shortened; penultimate sugar becomes exposed to be terminal (21-34)
2. organizational structure	a) degree of interaction of carbohydrates with peptides and lipids b) order of the arrangement of carbohydrate chain	cryptic to be exposed? (21,35-39)
3. topological structure	a) location and distribution pattern	relocation and redistribution (40)

Table 3
Transformation-Dependent Changes of Glycolipids

Cell line and reference	Glycolipids identified in non-transformed cells	Changes in transformed cells (see footnote)*
BHK$_1$(21)	Gal → Glu → Cer	↗↘ py
	NANA → Gal → Glu → Cer	↓
BHK$_2$(28)	Gal → Glu → Cer	↑
	αGal → Gal → Glu → Cer	↓↓ →0 py
	NANA → Gal → Glu → Cer	↓
3T3$_1$(22)	Glu → Cer	↑
	NANA → Gal → Glu → Cer	↓
	NGNA → Gal → Glu → Cer	↓ sv
3T3$_2$(23,25,29)	NANA → Gal → Glu → Cer	↑
	Gal → GalNAC → Gal → Gal → Glu → Cer ↑ NANA	↓↓
	Gal → GalNAC → Gal → Gal → Glu → Cer ↑ ↑ NANA NANA ↑ NANA	↓↓→0
Chick embryonic (c/0 c/B)(41)	Glu → Cer	↑
	NANA → Gal → Glu → Cer	↓
	NANA → NANA → Gal → Glu → Cer	↓↓→0 RS
	Unknown I	↓↓→0
	Unknown II	unchanged
Human diploid (28)	Glu → Cer	unchanged
	Gal → Glu → Cer	unchanged
	NANA → Gal → Glu → Cer	↓
	NANA → NANA → Gal → Glu → Cer	↓↓ sv
	NANA → Gal → Glu → Cer ↑ GalNAC	↑↑
	NANA → Gal → Glu → Cer ↑ Gal → GalNAC	↓↓→0

Table 3 continued next page

Table 3, continued

NIL$_{(42)}$	Glu → Cer	unchanged
	Gal → Glu → Cer	↑
	αGal → Gal → Glu → Cer	⇊→0 py
	GalNAC → Gal → Gal → Glu → Cer	⇊→0
	NANA → Gal → Glu → Cer	unchanged
	GalNAC → GalNAC → Gal → Gal → Glu → Cer ⇊→0	

Rat liver (26,27)

Glu → Cer

Gal → Glu → Cer   Morris

NANA → Gal → Glu → Cer   hepatoma

NANA → Gal → Glu → Cer   Chemical
　　　　　　↑   carcinogen

Gal → GalNAC

　　　NANA → Gal → Glu → Cer
　　　　　↑   ⇈

NANA → Gal → GalNAC

NANA → NANA → Gal → Glu → Cer
　　　　　　　↑

NANA → Gal → GalNAC   ⇊→0

↓ :decreased, ⇊ :strongly decreased, ⇊→0:decreased and almost non-detectable, ↑ :increased, ⇈:greatly increased (several times), py:transformed with polyoma virus, rs: transformed by Rous sarcoma virus, sv:transformed by Simian virus 40.

Table 4

Activity of Ceramide Trihexoside: α-galactosidase in a particulate fraction (P-2 fraction)*

	μm mole CTH hydrolyzed per mg during 16 hours
NIL 2E sparse	13.5
confluenced	12.3
NIL PY high	44.5
BHK confluenced	12.3
BHK PY high	23.0

*See reference (46) for detail.

Table 5
The amount of galactose incorporated
into various glycolipids of NIL cells
(Kijimoto & Hakomori, unpublished data)

Contact inhibitory NIL 2E cells and their py-transformants were grown in Eagle's medium at high and low cell population densities. Cell population densities are indicated below. The medium was removed, and the cell sheets were washed with saline and then incubated with a medium having no glucose, but containing ^{14}C- and ^{12}C-galactose. The specific activity was 2.5 µC/1 mg/ml, and the incubation time was 60 minutes. Cells were collected with rubber and extracted with chloroform-methanol (2:1). The glycolipids were separated by thin-layer chromatography. Spots were counted by a scintillation counter and calculated as shown below.

Cell population density		µµ moles of galactose incorporated into each glycolipid derived from 10^7 cells during 60 minutes	
		NIL	NIL-py
$3.5 \times 10^4/cm^2$	CDH	68	293
	CTH	122	46
	Globoside	122	28
	Forssman	500	20
	Hematoside	550	450
$1.4 \times 10^5/cm^2$	CDH	74	367
	CTH	166	15
	Globoside	160	24
	Forssman	1470	46
	Hematoside	2120	*
$2.5 \times 10^5/cm^2$	CDH	47	139
	CTH	174	19
	Globoside	178	24
	Forssman	800	*
	Hematoside	1770	*

*Failed to determine

Table 6
Change of absorption capability of anti-
globoside, agglutination and hemolysis by
anti-globoside of erythrocytes by proteolysis

	Native	After Proteolysis
Agglutination titer	1:10	1:1500
Hemolysis titer	1:10	1:1500
Absorption	30 µgN*	125 µgN*

*Amount of anti-globoside eluted from 5 ml of a
membrane suspension. The amount of membrane on
lyophilization yielded 40-45 mg of dried residue.

Table 7
Reactivity of fetal, new-born and adult erythrocytes to anti-globoside and anti-lactosylceramide*

		Fetal Erythro-cytes (Mean of 6 Cases)	Cord Erythro-cytes (Mean of 3 Cases)	Adult Erythro-cytes (Mean of 10 Cases)
Reactivities vs. Anti-globoside	Agglutination			
	Non-treated	200	10	10
	Trypsinized	500	320	320
	Hemolysis Titer			
	Non-treated	320	20	10
	Trypsinized	640	320	320
	Absorption			
	Non-treated	10.5 µgN+		1.4 µgN+
	Trypsinized			5.6 µgN+
Reactivities vs. Anti-lactosylceramide	Agglutination			
	Non-treated	5	5	0
	Trypsinized	80	80	80
	Hemolysis			
	Non-treated	20		0
	Trypsinized	80		80

*Reconstructed from Hakomori (51); the value for cord
blood and the value for the absorption value of trypsinized erythrocytes was added later.
+The amount of anti-globoside antibody absorbed by 1 ml
of packed erythrocytes.

Table 8
Reactivities of Anti-Hematoside Antiserum to
"Normal" and "Transformed" Fibroblastic Cells*

	Reactivity Degree of Cr^{51} release: % CPM (Released CPM/total cell-bound CPM)		
	non-treated	trypsinized	%
BHK	42	75	45% increase
3T3	49	70	30% increase
PYBHK	70	70	0
$SV_{40}3T3$	69	65	0

*Constructed from Hakomori, et al (22).

Table 9
Difference in Reactivity of Cells to Some Anti-Forssman
Glycolipid Antisera at Growing and Confluent Phases*

		Reactivity: CPM of anti-glycolipid antibodies per 10^5 cells
Experiment 1		Anti-Forssman
Cells	Pop. density per cm^2	
NIL K** Grow	0.65×10^5	320
NIL K Conf.	1.20×10^5	160
PY Low	1.20×10^5	12
Experiment 2		
NIL K Grow	0.15×10^5	1904
NIL K Semiconf.	0.5×10^5	830
NIL K Conf.	1.1×10^5	750
NIL PY Low	0.5×10^5	25
High	1.2×10^5	50
Experiment 3		
NIL K Grow	0.24×10^5	1500
NIL K Conf.	1.0×10^5	525
PY Low	0.62×10^5	15
PY High	1.0×10^5	12

* Hakomori and Kijimoto, unpublished results
**NIL K is a contact-inhibited clone isolated from NIL2E
originally donated by Dr. Leila Diamond.

Figure 1. Density-dependent changes of enzyme activities
of NIL and BHK cells for synthesis of α-gal →
β-glu → cer (CTH) and β-gal → glu → cer (CDH)

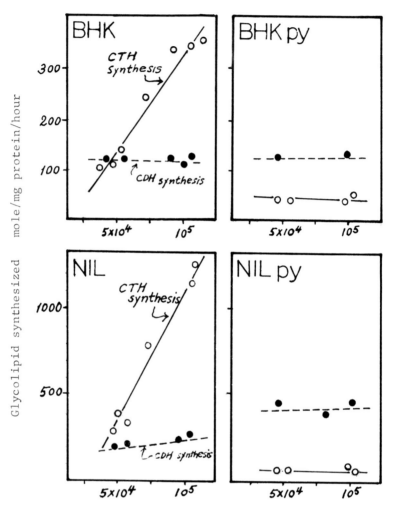

Cell population densities/cm²

Open circles: UDP-gal:lactosylceramide α-galacto-
                syltransferase
Black circles: UDP-gal:glucosylceramide β-galacto-
                syltransferase

Figure 2.
Structure of Some Glycolipid Antigens

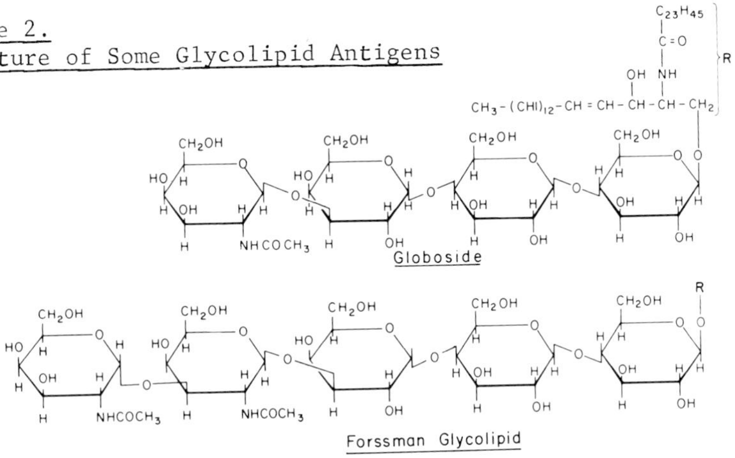

Globoside

Forssman Glycolipid

Globoside and Forssman glycolipid antigen. These structures are identical except that Forssman glycolipid has an extra-molecule of α-N-acetylgalactosamine at the end. (Hakomori, et al (47); and Siddiqui and Hakomori (57).

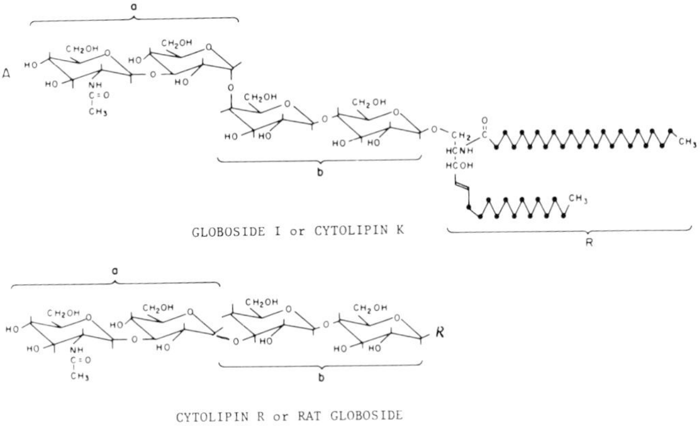

GLOBOSIDE I or CYTOLIPIN K

CYTOLIPIN R or RAT GLOBOSIDE

Human globoside (Cytolipin K) and rat kidney globoside (Cytolipin R). These structures are identical in carbo-hydrate sequence and anomeric linkages, but they are still immunologically distinguishable. They differ in positions of linkage between two oligosaccharides a and b (Siddiqui, et al (58).

273

Figure 3. Radioautogram of Forssman precipitin lines of
NIL cells grown under different conditions
(labelled with ^{14}C-galactose for short periods)

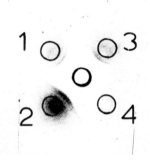

Center: anti-Forssman glycolipid rabbit antisera
1: glycolipid of actively-growing NIL cells
2: glycolipid of confluent NIL cells
3: glycolipid of highly confluent cells
4: glycolipid of polyoma-transformed NIL
cells.

The condition of labelling is the same as in Table 5.
Precipitin lines were radioautographed by exposing
for one week.

Figure 4. A Model of Glycolipid and Glycoproteins on Membranes.

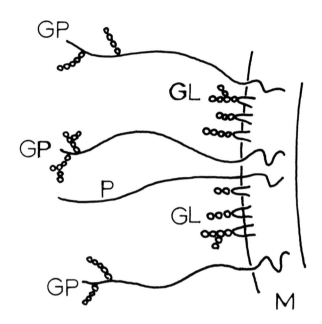

Carbohydrates are shown as circles linked together; glycolipids (GL) are directly imbedded in membrane matrix (M); glycolipids (GP) and some proteins (P) are stretched towards extracellular environment.

Figure 5.   A Model for Intercellular Interaction Through
Some Glycolipids and Stretching Proteins (or
Glycoproteins) on Cell Surfaces.

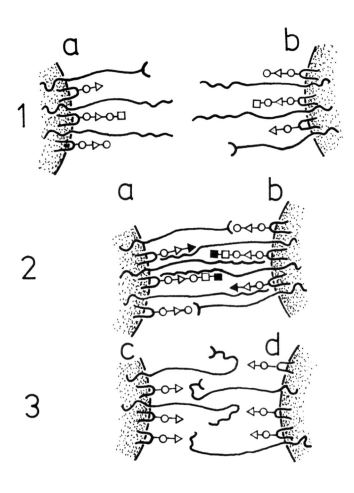

See next page for explanation.

Figure 5, continued.

1: growing cells, a certain proportion of glycolipids
    and proteins (or glycoproteins), arranged in a
    certain order. The structures and the organiza-
    tion (order) of carbohydrates and proteins should
    be complementary.

2: confluent cells, when cell A and B meet, 1) the
    carbohydrates and proteins are linked together
    through complementary structure, 2) some carbo-
    hydrates can extend their chain for the better
    linkages between complementary proteins. These
    linkages should be non-covalent, and possibly
    mediated by bivalent cations as cells are dissoci-
    able by EDTA.

3: transformed malignant cells; carbohydrate chains
    are incomplete, consequently no complementary
    structures were found between glycolipids and
    proteins. Intercellular linkages were therefore
    not formed.

This model is based on 1) the decrease of reactivity
of cells to anti-glycolipids, and 2) glycosyl extension of
some glycolipids when cells contact cells (see the text).
A similar model has been proposed by Roseman (59) based on
the presence of glycosyltransferase on cell surfaces. In
his model, glycosyltransferases are the complementary proteins that can be "sites" for binding surface glycans of
counterpart cells. Temperature-dependency of intercellular
adhesion (3) supports this model.

VI

MEMBRANE-DNA INTERACTIONS

DNA-MEMBRANE ASSOCIATIONS
IN THE DEVELOPMENT
OF A FILAMENTOUS BACTERIOPHAGE, M13

Jack Griffith and Arthur Kornberg
Stanford University School of Medicine

The morphogenesis of the filamentous bacteriophage, M13, includes several steps in which the membranes of the host cell appear to be involved (1, 2) (see Figure 1). We would like to identify those steps in the life cycle which will provide functional assays for specific DNA-membrane interactions. This report focuses on electron microscopic observations consistent with membrane association of certain replication and assembly events.

Structure of M13

The DNA is a 2.3-μ single-stranded circle, surrounded by a capsid constructed essentially of a single protein species (coat protein of molecular weight 4,900). The structure could be that of a cylinder of protein with the DNA contained in the core, or of a thin torus twisted to form a filament.

Chloroform is known to inactivate M13 without releasing any of the major components. Chloroform-treated phage when spread on a water surface open into toroids with half the width and approximately 2.3 times the contour length of untreated phage. This result is consistent with a thin, collapsed torus twisted about itself (Figure 2).

Technique for EM Study of DNA-Membrane Complexes

Most techniques for studying cellular complexes require a preliminary fractionation and concentration of the DNA-containing components. Adventitious molecular associations may occur and valid ones may be lost because of these manipulations; the elapsed time during the procedure is considerable. DNA-surface-spreading techniques which employ basic proteins would, of course, obscure interactions between a specific protein and DNA.

We have developed a procedure in which cells are made osmotically sensitive by brief lysozyme treatment. A few cells in sucrose are placed on a microscope grid and ruptured with a small drop of buffer. The cell and its contents, bound to the grid are dehydrated and visualized by high resolution tungsten evaporation (3). This method has the advantage of being direct and rapid. In addition, the cells are at a very high dilution, and there is minimal opportunity for artifactual associations following lysis.

Uninfected cells spread by this method typically show two bacterial chromosomes bound to opposite ends of the cell envelope, and spread away from the disrupted envelope. No structures such as the M13 intermediates, to be discussed below, were observed in uninfected cells.

Phage Entry and Location of Parental RF

There are several indications that the entering phage is in close association with inner membrane (4) and that uncoating of the capsid protein is dependent on the conversion of the single-stranded, circular, viral DNA to the double-stranded RF (parental RF) (5, 6). There is also recent evidence that the several molecules of adsorption protein in the phage are retained completely in a complex with parental RF in the inner membrane (7).

We have localized the parental RF by infecting cells with an M13 mutant (gene 2) which is unable to further replicate the RF. A DNA circle of M13 length was found in 5 of 30 cells examined, and, under conditions of gentle lysis, always appeared to be attached to the cell envelope (Figure 3). Similar observations were made with both amber mutants and with temperature-sensitive mutants at a restrictive temperature. Were the DNA part of a replication complex and were the attachments to the membrane valid, then we may infer that the gene 2 product is not required for this binding.

Viral Strand Synthesis

Following the production of RF, there is an asymmetric synthesis of circular viral strands which requires the protein product of gene 5. Cells infected with M13 defective in gene 5 accumulate RF. In a cell examined after a 10-min infection with M13 am5 (Figure 4), many copies of the M13 RF are clearly visible. All the circular DNA appear to be located at the envelope, as in an infection aborted by a gene 2 block and examined under these conditions of gentle lysis. Similar results were obtained with cells made fragile by freezing and thawing. However, with a more violent osmotic disruption of the cell, after stronger lysozyme treatment (100 μg/ml, 0°C, 30'), the circles were found separated from the cell envelope.

Phage Morphogenesis

The capsid protein (gene 8 product) is stored exclusively in the inner membrane (8, 9). In addition to this protein, the phage contains a few molecules of adsorption protein (gene 3 product) and possibly the gene 6 product. The products of three other genes (1, 4 and 7) are still unidentified and are required for phage morphogenesis. Mutants defective in any one of these three genes or in gene 8 accumulate single-stranded viral circles, coated with protein, presumably the gene 5 protein product (Figure 5). Here again the circles invariably seem to adhere to the cell envelope.

Phage Release

At about one hour after infection, hundreds of phages are being assembled and many can be seen still associated with the cell envelope (Figure 6). The tiny spheres along the cell wall may be part of the lipopolysaccharide-rich outer membrane which adheres to the phage particles; this appearance of the cell envelope was not seen earlier during infection nor in uninfected cells.

Under conditions of limited capsid protein production (growth of M13 amber 8 in suppressor strains), partially completed phage filaments were isolated attached to what appeared to be fragments of both inner and outer membranes (Figure 7). In such fragments, there was a significant enrichment in phospholipase Al, beyond the levels ordinarily found in the outer membrane to which the location of this enzyme is restricted. These fragments may represent zones of adhesion between inner and outer membranes (10).

Conclusion

The life cycle of the filamentous bacteriophage M13 includes several steps in the replication of the DNA and in the assembly of the phage which are membrane-based. These EM studies help to visualize some of these membrane associations. Such EM studies cannot provide definitive evidence for membrane associations nor can they at present elucidate the chemical nature of the complexes involved. However, combined with genetic and biochemical tools, EM studies should enlarge our insights into the overall process and may even supply sensitive and specific assays for individual stages.

References

1. A.B. Forsheit and D.S. Ray (1971) Virology 43, 647.

2. W.L. Staudenbauer and P.H. Hofschneider (1971) Biochem. Biophys. Res. Comm. 42 #6, 1035.

3. J. Griffith, J. Huberman and A. Kornberg (1971) J. Mol. Biol. 55, 209.

4. M. Jazwinski, personal communication.

5. D. Ray, personal communication.

6. D. Brutlag, R. Schekman and A. Kornberg (1971) PNAS 69, 1826.

7. M. Jazwinski and R. Marco, personal communication

8. H. Smilowitz, H. Lodish and P. Robbins (1971) J. Virology 7 #6, 776.

9. R. Schekman, personal communication.

10. M. Bayer (1968) J. Virology 2, 346.

Abbreviations: RF refers to the double-stranded circular form of M13 DNA.

Filamentous Virus Infection
in *E. coli*

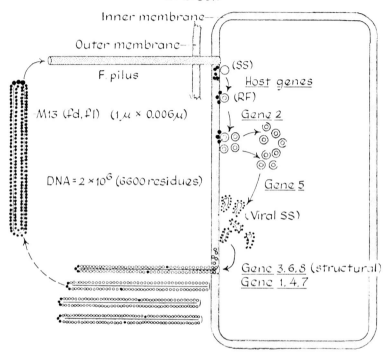

Figure 1 - Hypothetical model of the M13 life cycle.
The phage is shown in its opened conformation (far
left). The small circles along the DNA loop represent
coat protein molecules and the three large circles the
adsorption proteins. Uncoating of the viral DNA may be
dependent on its replication and localized in or on the
inner membrane; adsorption proteins remain associated
with the DNA throughout this process.

Following RF replication, gene 5 protein (small
dots attached to single stranded DNA circles) complexed
with progeny viral strands is replaced by coat protein
stored in the inner membrane. The intracellular localization of the various stages in the M13 life cycle is discussed in the text.

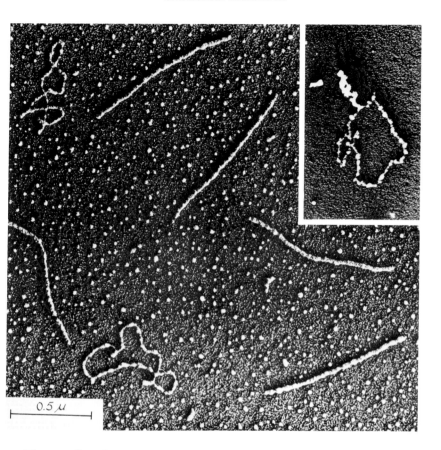

Figure 2 - The structure of M13 is a twisted circle. M13 and M13 treated with chloroform were mixed and spread on a water surface with cytochrome-C. Inset: without cytochrome-C.

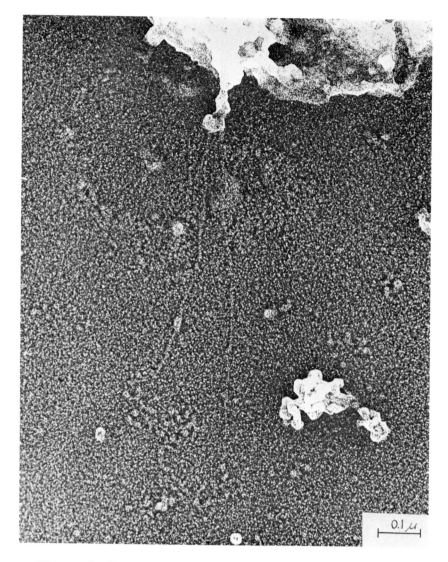

Figure 3 - Parental RF bound to the cell envelope. RF replication was blocked by a defect (amber mutation) in gene 2. A portion of the cell envelope is seen (top) with a 2.3 μ DNA loop attached.

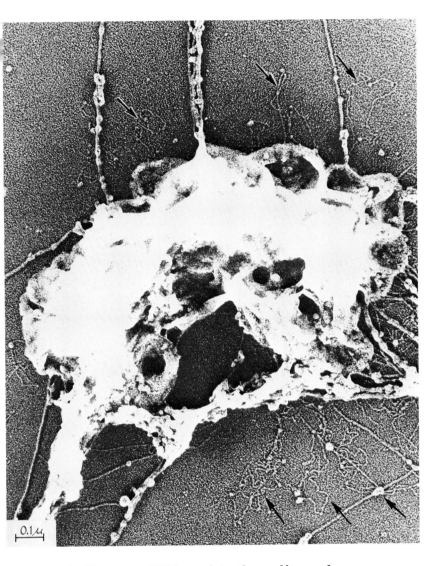

Figure 4 - Progeny RF bound to the cell envelope.
Viral strand production was blocked by a defect
in gene 5. The cellular DNA is condensed (thick fibers)
and the M13 RF circles (arrows) are bound to the disrupted envelope (central mass).

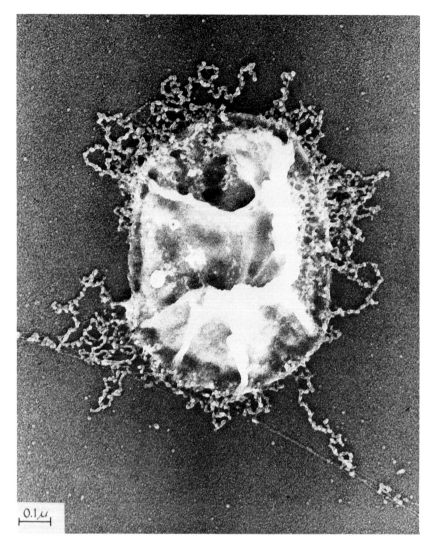

Figure 5 - Viral strand - gene 5 protein complexes bound
to the cell envelope.
    Phage maturation was blocked by a defect in gene 7.
M13 length filaments having the contour width of DNA
gene 5 protein complexes appear to be attached to the cell
envelope.

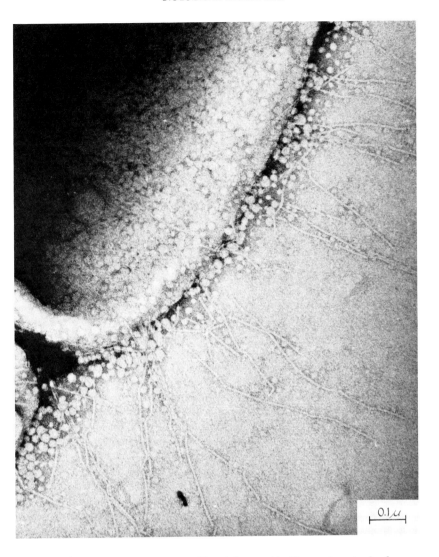

Figure 6 - M13 infected cell with partially extruded phage.
Portion of an infected cell (top left) shows extruded
phage filaments and spheres presumed to be fragments of
the cell envelope.

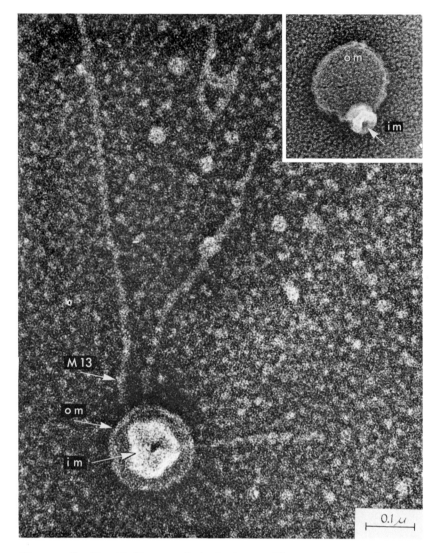

Figure 7 - Partiall completed phage filaments attached to a fragment of the cell envelope.
    Fragments of the outer membrane appear as flat plates and those of the inner membrane as more rounded vesicles. Inset: inner and outer membrane fragments isolated from disrupted uninfected cells by sucrose density sedimentation appear attached.

MEMBRANE INVOLVEMENT IN THE INITIAL EVENTS IN THE
REPLICATION OF BACTERIOPHAGE φX174

Dan S. Ray and Bertold Francke

Molecular Biology Institute and
Department of Zoology, UCLA
Los Angeles, California, 90024

ABSTRACT. One of the earliest events in the replication of
the φX174 double-stranded RF* DNA is the viral gene A-
mediated strand-specific nicking of the parental RF. The
number of such initiation events is limited to approximately four RF's per cell and is in good agreement with
both the number of parental genomes capable of participating in replication and the number of newly-replicated
molecules physically associated with cell membrane
fractions. Membrane association of RF is proposed to be
a prerequisite to the strand-specific nicking event
involved in the initiation of φX174 RF replication. The
cis-limited function of the φX174 gene A may result from
an association of the A protein with a limited area of
the cell membrane in the vicinity of the DNA attachment
site.

INTRODUCTION

Considerable evidence has accumulated in the past
few years to suggest that the chromosomes of bacteria (1-6)
and bacterial viruses (7-17) are physically associated
with the cell membrane. This association has been
proposed as a means of segregating daughter chromosomes
upon completion of a round of replication (18). Both the

_____
*Abbreviations: Cam, chloramphenicol; hcr, host-cell
reactivation; m.o.i., multiplicity of infection; RF, the
duplex replicative form; RFI, the covalently-closed type
of RF; RFII, the class of RF molecules containing one or
more single-strand discontinuities.

origin of replication and the replication fork of the
bacterial chromosome appear to be membrane bound, possibly
at a single site on the membrane (4,6). Contact between
the nuclear body and the cell membrane has been
visualized in electron micrographs of Bacillus subtilis (3).
Enzymes thought to be essential for DNA replication also
appear to be membrane associated (5,19).

The bacteriophage system which probably has been
studied in most detail with respect to membrane involvement in replication is the small coliphage ∅X174. This
minute phage is at one extreme in the spectrum of DNA
viruses. It contains only enough genetic information to
code for approximately ten or fewer proteins. Replication
of the circular single-stranded DNA is outlined in Fig. 1.
The first step in the DNA replication cycle is the conversion of the single-stranded circular DNA into the
parental double-stranded replicative form. The parental
RF molecule replicates to form a pool of progeny RF's
which subsequently serve as material precursors of the
progeny viral strands by an asymmetric mode of synthesis.
The resulting progeny single strands are all of the viral
type.

The second stage of synthesis, the replication of the
parental RF's, might reasonably be expected to reflect
some general features of the replication of circular
double-stranded DNA. Consequently, this aspect of ∅X174
replication has been of considerable interest with respect
to a possible DNA-membrane association. We shall attempt
to summarize here some of the results that have come out
of these investigations and to relate them to our own
studies on the initiation of RF replication (23,24).

BACKGROUND

We now know a number of important details of ∅X174
replication. Perhaps the least expected finding was that
only a very limited number of RF molecules are capable of
semi-conservative replication (7,8,25) Specifically, only
parental RF's replicate; progeny RF's may participate in
single strand synthesis but not in RF replication. However,
at higher multiplicities of infection, not even all of the
parental RF's can replicate. Those molecules that do
replicate are found exclusively on the bacterial membrane.

The limited participation of parental RF's in the
replication process was demonstrated quite elegantly by

the experiments of Yarus and Sinsheimer (26) in which
single bursts of starved cells infected with several genetically distinguishable ϕX174 genotypes were found to
yield most often only a single, randomly chosen, genotype.
Unstarved cells, by contrast, yielded up to four different
genotypes. It was proposed that the bacterial cell
contains a limited number of "sites" at which replication
can occur. This number is reduced to only a single site
as a result of starvation. All of the infecting single
strands can be converted to RF's but only those RF's in
possession of the essential site will be able to contri-
bute genetic markers to the progeny phage. Thus, in the
case of multiply-infected starved cells, only a single
parental RF may be capable of replication. This limitation suggests an explanation for the earlier experiments
of Denhardt and Sinsheimer (27) in which it was observed
that ^{32}P decays in the parental RF formed in starved cells
was lethal even after unlabeled progeny RF's had been
allowed to form. Possibly some necessary function is
performed by the site-associated RF which cannot be
carried out by the other RF's. In such a case, ^{32}P decay
in the parental RF would be lethal even in the presence
of unlabeled progeny RF's.

The limited number of sites for RF replication suggests
that the site might be the bacterial replication apparatus
itself. It is known, for example, that the chromosome of
B. subtilis has a single replication point at slow growth
rates and several at high growth rates (28). The
increased number of ϕX174 capable of replicating in
rapidly growing cells might reflect an increased number
of bacterial replication points relative to that present
in starved cells.

The small number of ϕX174 RF's capable of replication
is in close agreement with the number of newly replicated
RF's found associated with a rapidly-sedimenting fraction
containing the bacterial membranes following a shift of
infected bacteria from a heavy (^{13}C and ^{15}N) to a light
(^{12}C and ^{14}N) medium (7). RF molecules of hybrid density
were found exclusively in the membrane fraction of lysates
of infected cells. Only one to two RF's were replicated
in pre-starved cells and only three parental RF's were
replicated even in cells grown continuously in nutrient
medium. Progeny RF's appeared to be released from the
membrane and were found largely in the fraction of freely-

sedimenting RF molecules. However, it is questionable whether the experimentally defined fast-sedimenting "DNA-membrane complex" reflects a functionally unique entity. Release of progeny RF's from the membrane is not concomitant with the terminal events in a round of RF replication, and some of the newly formed progeny RF's can be found associated with the membrane. Some parental RF's not involved in replication, likewise, can be found in the membrane fraction (7,8). It should be emphasized, though, that the parental molecules involved in replication were found exclusively in the membrane fraction. Thus, with respect to DNA replication, there are both functional and non-functional membrane associations. With high multiplicities of infection, the number of molecules associated in a non-functional way can far exceed that of the molecules functionally-associated with the membrane (8). Whether or not the non-functional associations have any biological significance is still uncertain. Conceivably, they could be formed artifactually at the time of lysis, however, their formation is specific to intracellular RF molecules since addition of RF DNA to infected cells prior to lysis does not generate DNA-membrane associations (13,29).

The experimental limitations to a detailed biochemical analysis of functional ØX174 DNA-membrane complexes can be summarized as follows:

1. The number of replicating RF molecules is limited to from 1 to 4 per cell.

2. Non-replicating parental and newly-formed progeny RF's also can be membrane associated.

3. DNA-membrane preparations usually contain the bulk of both the bacterial DNA and the cellular membranes.

These difficulties are not completely insurmountable but they do pose rather serious obstacles to progress in this area. Attempts to characterize further the ØX174 DNA-membrane complexes have shown that the attached RF can be released from the membrane as a DNA-protein complex containing proteins synthesized both before and after infection (29).

METHODS

All experimental methods described here have been published in references 23, 24 and 31. Bacterial strains used were: E. coli HF4704 (her⁻, thy⁻), a non-permissive host for ϕX174 amber mutants and E. coli Rep₃⁻, a derivative of HF4704 which is unable to replicate ϕX174 RF. Phage strains used were: ϕX174 A⁺, a strain wild-type for gene A and ϕX174 am A, a strain carrying an amber mutation (am 8) in gene A. The ϕX174 A⁺ strain was mutant in the lysis gene (am 3). Bacterial and phage strains were kindly provided by Dr. D. T. Denhardt and Dr. R. L. Sinsheimer.

RESULTS

The experiments which I will describe here allow us to define more precisely some of the initial events in replication and suggest that some of the steps preceeding the replicative events per se are also membrane associated. We have focused our attention on the role of the ϕX174 gene A in RF replication. The parental RF is formed but not replicated in the absence of gene A function. For these studies, we have used the host cell mutant Rep₃⁻ (32), which is unable to support replication of the parental RF even though viral genes can be expressed in the mutant bacterium.

Velocity sedimentation analysis of ϕX174-specific DNA formed under various conditions which do not allow RF replication shows two forms of parental RF, one sedimenting at the rate of the covalently closed RF I and the other at the rate of the nicked RF II. Figure 2 shows five such experiments which use as non-replicating conditions: Rep₃⁻ mutation of the host cell, amber mutation of the ϕX174 gene A (ϕX174 am A), the addition of 150 µg chloramphenicol/ml at 20 minutes prior to infection and combinations thereof. In all cases, the cells are infected with ³²p'-labeled phages in the presence of ³H-thymidine. DNA was isolated from the cells at 20 minutes after infection and analyzed by velocity sedimentation in 5-20% sucrose gradients. Under all of the conditions used, the amount of ³H label contained in the RF's was on the order of one phage DNA equivalent per infecting viral strand.

The RF II peak was always a minor component representing only a few per cent of the total RF's except in the case of infection of Rep$_3^-$ with ϕX174 carrying a wild-type A gene (ϕX174 A$^+$). Similar experiments performed over a wide range of input multiplicities are summarized in Table 1. In infections of Rep$_3^-$ with ϕX174 A$^+$, the proportion of RF II increased with decreasing multiplicities of infection. The absolute number of RF II molecules was limited to about 1 to 4 RF's per cell. In all other cases, the RF II molecules represented a small and fairly constant percentage of the total RF population. These results suggested that a specific RF II might have been formed in the infection of Rep$_3^-$ by ϕX174 A$^+$ while that formed under all of the other conditions examined might have resulted from a small amount of random nicking.

Each RF II species was analyzed by alkaline velocity sedimentation under conditions such that the duplex DNA was denatured and the resulting circular and linear strands separated on the basis of their different sedimentation rates. ^{3}H and ^{32}P radioactivity should have been found in both linear and circular strands if the nicking were entirely random. Figure 3(a) shows the alkaline velocity sedimentation of the RF II species formed under conditions allowing gene A to be expressed. The data show unequivocally that the ^{3}H-labeled complementary strands all sediment at the rate of circular single strands (indicated by the peak of infectious DNA) while the ^{32}P-labeled viral strands all sediment at the rate of unit-length linear single strands. The RF II molecule giving rise to these denaturation products must have contained a single discontinuity in only the viral strand of the duplex DNA. In all cases where gene A could not be expressed, this viral strand specific discontinuity was not observed. Rather, it was observed that parental ^{32}P label and newly-synthesized ^{3}H-DNA were found in both the circular and linear single strands resulting from denaturation of the RF II. In the presence of chloramphenicol, there was even a preference for RF II molecules in which the viral strand was closed and the complementary strand open. We believe this to be the result of inhibition of closure of the newly formed complementary strand in the presence of a high concentration of chloramphenicol.

Further evidence supporting the proposed structure of the RF II made in the presence of a functional gene A was obtained by neutral and alkaline equilibrium sedimentation of the RF II molecules. In the neutral CsCl gradient (Fig. 4(a)), both the 3_H and 32_P labels were found in a unimodal peak at the density of duplex RF. In the alkaline CsCl gradient (Fig. 4(b)) where the RF is denatured, the parental 32_P label was found exclusively at the density of an infectious viral strand marker while the progeny 3_H label was contained in a peak at the density of complementary single strands.

Our finding that gene A mediates a strand-specific cleavage of the $\emptyset$X174 RF led us to investigate the gene A function in mixed infections in which both gene A mutant and wild-type phages were present. This was of particular interest in view of Tessman's finding (33) that gene A mutants could complement mutants in other cistrons but yet could not be complemented themselves, i.e. gene A mutant phages were not obtained in the phage yield from a mixed infection.

To examine the molecular basis of the asymmetric complementation of gene A mutants, we have infected Rep_3^- simultaneously with 3_H-labeled $\emptyset$X174 A^+ and 32_P-labeled $\emptyset$X174 am A. Both phages were present at a multiplicity of five phage per cell so that there was a very low probability of a cell having been infected with only a single genotype. Velocity sedimentation analysis of DNA isolated from the infected bacteria (Fig. 5) showed a large fraction of the 3_H label (A^+) sedimenting at the rate of RF II while only a small amount of the 32_P label (am A) sedimented as RF II. These results indicate that the gene A product acts only upon the wild-type DNA which coded for the gene product and not upon the DNA carrying a mutant A gene. This interpretation is supported by the further observation (Fig. 6) that the viral strand of the wild-type RF II was largely linear while that of mutant-type RF II was found as both linear and circular forms. The mutant-type RF II therefore probably was formed as a result of a small amount of non-specific nicking of the DNA during isolation, whereas the much larger amount of wild-type RF II was formed by a strand-specific nicking of the RF.

DISCUSSION

The strand-specific nicking of $\emptyset$X174 parental RF

under the control of the viral gene A probably represents one of the earliest steps in the replication of the duplex RF (23,24). The limitation of the number of specifically nicked molecules to 1-4 per cell, in agreement with both the number of genomes participating in replication and the number of replicating RF molecules associated with the bacterial membrane, suggests that the specific nicking of the parental RF must also occur on the membrane. We propose that the elements necessary for the introduction of the specific nick include the gene A protein, the RF DNA and a structure localized on the bacterial membrane and present in a very limited amount per cell. How these components interact and what the role of the A protein is in this process remain to be determined. Whether the A protein is directly involved in the nicking of the parental RF or whether it plays an indirect role is also unknown. Any proposal regarding the function of the A protein must account for the cis-limited nicking by the wild-type protein. In some way, the synthesis and action of this protein must be limited in space; an active A protein cannot be freely diffusable. An interesting possibility would be that transcription and translation of the A gene occur in the vicinity of the membrane attachment site and that the A protein binds to the membrane. Such a mechanism would insure localization and subsequent functioning of the A protein in the vicinity of the DNA-membrane attachment site. It is of interest to note that attachment of lambda DNA to membrane depends on the absence of lambda repressor and on active transcription mediated by the N gene-product (10,11). Not only is lambda DNA replication not required for membrane attachment, but also transcription of the lambda genes O and P (genes required for lambda DNA replication) are not required. Similarly, transcription of ϕX174 RF might be necessary for membrane attachment and/or specific nicking of the RF. Replication of RF also might not be required for attachment of ϕX174 RF to the membrane since association of RF with the membrane has been reported for the Rep_3^- mutant. This interpretation is subject to criticism, however, in that the association might have been of the "non-functional" type in this case. Further purification of functional DNA-membrane complexes could provide greater insight into the replication processes occurring on the cell membrane.

ACKNOWLEDGMENTS

This research was supported by a grant from the National Science Foundation (GB18074).

REFERENCES

1. A. T. Ganesan and J. Lederberg, Biochem. Biophys. Res. Commun. 18, 824 (1965).
2. D. W. Smith and P. C. Hanawalt, Biochim. Biophys. Acta 149, 519 (1967).
3. A. Ryter and O. E. Landman, J. Bact. 88, 457 (1964).
4. N. Sueoka and W. G. Quinn, Cold Spring Harbor Symp. Quant. Biol. 33, 695 (1968).
5. R. Knippers and W. Strätling, Nature 226, 713 (1970).
6. P. Fielding and C. F. Fox, Biochim. Biophys. Res. Commun. 41, 157 (1970).
7. R. Knippers and R. L. Sinsheimer, J. Mol. Biol. 34, 17 (1968).
8. W. O. Salivar and R. L. Sinsheimer, J. Mol. Biol. 41, 39 (1969).
9. W. O. Salivar and J. Gardinier, Virol. 41, 38 (1970).
10. L. Hallick, R. Boyce and H. Echols, Nature 223, 1239 (1969).
11. A. R. Kolber and W. S. Sly, Virol. 46, 638 (1971).
12. D. Botstein and M. Levine, Cold Spring Harbor Symp. Quant. Biol. 33, 659 (1969).
13. A. F. Forsheit and D. S. Ray, Virol. 43, 647 (1971).
14. W. L. Staudenbauer and P. H. Hofschneider, Biochim. Biophys. Res. Commun. 42, 1035 (1971).
15. F. R. Frankel, C. Majumdar, S. Weintraub and D. M. Frankel, Cold Spring Harbor Symp. Quant. Biol. 33, 495 (1968).
16. A. W. Kozinski and T. H. Lin, Proc. Nat. Acad. Sci. U. S. 54, 273 (1965).
17. R. C. Miller, Jr. and A. W. Kozinski, J. Virol. 5, 490 (1970).
18. F. Jacob, S. Brenner and F. Cuzin, Cold Spring Harbor Symp. Quant. Biol. 28, 329 (1963).
19. A. T. Ganesan, Proc. Nat. Acad. Sci. U. S. 61, 1058 (1968).
20. D. S. Ray, in Molecular Basis of Virology, ed. by H. Fraenkel-Conrat, pp. 222-254, 1968, Reinhold, New York.

21. D. Pratt, Ann. Rev. Genetics 3, 343 (1969).
22. R. L. Sinsheimer, Progress in Nucleic Acid Res. and Mol. Biol. 8, 115 (1968).
23. B. Francke and D. S. Ray, J. Mol. Biol. 61, 565 (1971).
24. B. Francke and D. S. Ray, Proc. Nat. Acad. Sci. U. S. 69, 475 (1972).
25. A. B. Stone, Biochim. Biophys. Res. Commun. 26, 247 (1967).
26. M. J. Yarus and R. L. Sinsheimer, J. Virol. 1, 135 (1967).
27. D. T. Denhardt and R. L. Sinsheimer, J. Mol. Biol. 12, 663 (1965).
28. M. Oishi, H. Yoshikawa and N. Sueoka, Nature 204, 1069 (1964).
29. L. J. Loos, P. M. Tessler and W. O. Salivar, Virol. 45, 339 (1971).
30. A. J. Burton, J. Virol. 6, 455 (1970).
31. B. Francke and D. S. Ray, Virol. 44, 168 (1971).
32. D. T. Denhardt, D. H. Dressler and A. Hathaway, Proc. Nat. Acad. Sci. U. S. 57, 813 (1967).
33. E. S. Tessman, Virol. 25, 303 (1966).

TABLE 1

Parental viral DNA found in RF II under various non-replicating conditions

Bacterial strain	Phage strain	150 µg Cam/ml.	Input (m.o.i.)	Recovered (m.o.i.)	Label in RFII (%)	RFII molecules/cell
Rep$_3^-$	ϕX174 A$^+$	−	100	40	8.7	3.5
			10	5.5	38.5	2.2
			1	0.85	65.0	0.77
		+	100	85	7.2	6.1
			10	8	8.4	0.67
			1	0.98	9.2	0.088
	ϕX174 am A	−	100	27	7.2	1.92
			10	5	4.0	0.2
			1	0.7	1.4	0.02
		+	100	38	3.5	1.35
			10	7.5	3.6	0.27
			1	0.92	2.0	0.018

Table 1 (continued)

Bacterial strain	Phage strain	150 µg Cam/ml.	Input (m.o.i.)	Recovered (m.o.i.)	Label in RFII (%)	RFII molecules/ cell
HF4704	ØX174 amA	−	100	25	5.7	1.42
			10	4.5	2.85	0.13
			1	0.6	4.85	0.029
		+	100	35	8	2.7
			10	6	5.9	0.35
			1	0.85	2.2	0.03
	ØX174 A$^+$	+	100	30	3.35	1.01
			10	5.5	4.0	0.22
			1	0.96	4.6	0.04

Experimental details were described in the legend to Fig. 2. The recovered multiplicity of infection was calculated from the sum of parental label found in the preparative high-salt sucrose gradient and the specific ^{32}P radioactivity of the phage preparation used for infection. %RFII was calculated from the amount of parental label in the slower sedimenting peak as % of the total recovered parental label. The number of RFII molecules was calculated per total number of cells present at the time of infection. From Francke and Ray (23).

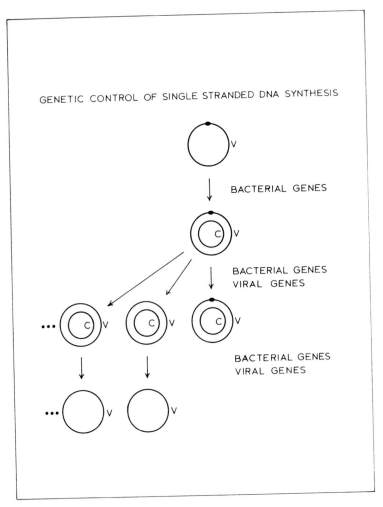

Fig.1. Genetic control of single-stranded DNA synthesis. Viral strands are indicated by V and complementary strands by C. The parental viral strand is distinguished by a single dot.

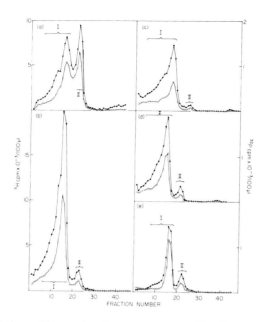

Fig. 2. High-salt neutral sucrose gradients of lysates
from infected cells: ϕX174 A$^+$ in Rep$_3^-$ (a); ϕX174 A$^+$ in
Rep$_3^-$ with 150 µg Cam/ml. (b); ϕX174 am A in Rep$_3^-$ (c);
ϕX174 am A HF4704 (d); and ϕX174 A$^+$ in 4704 with 150 µg
Cam/ml. (e). Cells were grown to 2 x 10^8/ml., and to (b)
and (e) Cam was added for 20 min. Infection was with ^{32}P-
labeled phages at a multiplicity of infection of 10 (the
ϕX174 A$^+$ preparation was 8.4 x 10^{-6} ^{32}P cts/min/plaque-
forming unit, and the ϕX174 am A preparation was
4.0 x 10^{-6} ^{32}P cts/min plaque-forming unit). At the time
of infection, 0.2 mCi [^{3}H]dThd was added. After 20 min at
37°C with aeration, the cultures were chilled in an ice-
bath and washed 3 times with 10 ml. borate-EDTA. After
resuspension in 1 ml. of 0.05 M-borate, they were lysed
with lysozyme (100 µg/ml.) and EDTA (0.003 M). SDS (1%)
and pronase (200 µg/ml.) were added and incubated for 2.5
hr at 37°C. Linear sucrose gradients (5 to 20%) (in
1 M-NaCl, Tris-EDTA, vol. 34 ml.) were formed and kept at
4°C. The lysates were poured directly on top of the
gradients and centrifuged for 15.5 hr at 24,000 rev/min
and 5°C in an SW27 rotor. Fractions of approx. 0.8 ml.
each were collected and 100 µl. of each fraction assayed
for radioactivity. Direction of sedimentation in this
and all other Figures was to the left.——●——●——,
^{32}P cts/min (parental label); ——○——○——, ^{3}H cts/min
(post-infection label). From Francke and Ray (23).

305

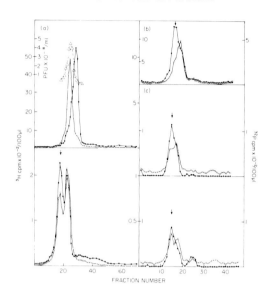

Fig. 3. Long alkaline sucrose sedimentation of RFII from
Rep⁻ infected with ØX174 A⁺ without (a) and with (b) Cam;
Rep⁻ infected with ØX174 am A (c); HF4704 infected with
ØX174 am A (d) and HF4704 infected with ØX174 A⁺ with
Cam (e). The fractions indicated by (II) in Fig. 2 were
pooled, concentrated by ethanol precipitation, mixed with
an unlabeled viral DNA marker (except in (a)), and
sedimented through 5 to 20% alkaline sucrose gradients
(containing 0.25 M-NaOH and 0.025 M-EDTA, 3.5 ml. total
vol.). Centrifugation was for 5.5 hr at 55,000 rev/min
and 5°C in an SW56 rotor. Fractions of 3 drops each
((a) and (e)) or 5 drops each ((b), (c) and (d)) were
collected into 100 μl. Tris-EDTA. 100 μl. of each
fraction were used for the radioactivity assay. Infectivity
assays were done after dilution into 0.05 M-Tris, pH 7.5
——●——●——, ³²P cts/min (parental label); ——O——O——,
³H cts/min (post-infection label); ---△---△---, plaque-
forming units. The arrow indicates the position of the
marker DNA in (b), (c), (d) and (e). From Francke and
Ray (23).

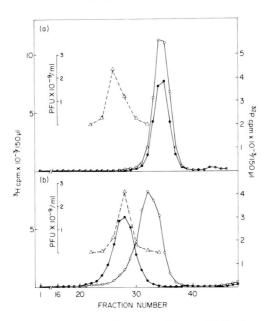

Fig. 4. Neutral and alkaline CsCl equilibrium gradients of RFII from Rep_3^- infected with ϕX174 A^+. The same sample as used for the alkaline sucrose gradient (Fig. 3 (a)) was centrifuged to equilibrium in neutral (a) and alkaline (b) CsCl with an added unlabeled infectivity marker (single-stranded viral DNA). Density increased from right to left. ——●——●——, ^{32}P cts/min (parental label); ——O——O——, ^{3}H cts/min (post-infection label); --△---△--, plaque-forming units (single-stranded viral marker DNA). From Francke and Ray (23).

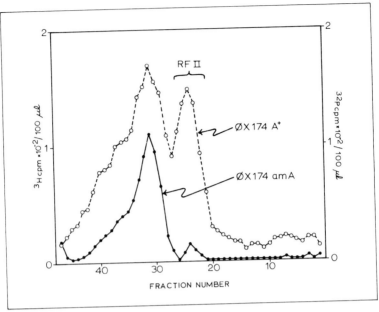

Fig. 5. Parental RF from Rep$_3^-$ cells doubly infected with ØX174 A$^+$ and ØX174 am A. 20 ml of Rep$_3^-$ cells at 2 x 10^8 cells/ml were simultaneously infected with 5 ^{3}H-labeled ØX174 A$^+$ and 5 ^{32}P-labeled ØX174 am A per cell. After 30 min at 37°C, the infected cells were washed, lysed and sedimented through a neutral sucrose gradient as described in the legend to Fig. 2. ●——●, ^{32}P cpm (am A); O---O, ^{3}H cpm (A$^+$). From Francke and Ray (24).

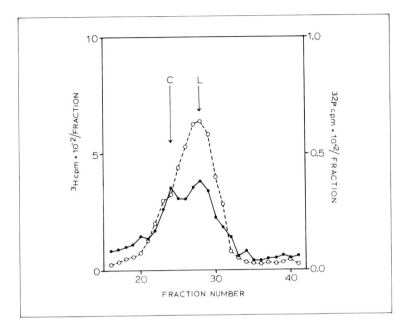

Fig. 6. Alkaline velocity sedimentation of the RFII region from the sucrose gradient shown in Fig. 5. Sedimentation conditions were as described in the legend to Fig. 3. Fractions 16-41 of a total of 75 fractions collected are shown. For clarity, the P^{32} scale has been expanded 10-fold. The sedimentation positions of single-stranded circles (C) and unit-length linear molecules (L) are indicated by arrows. ●———●, ^{32}P cpm; 0----0, ^{3}H cpm. From Francke and Ray (24).

INTERACTIONS BETWEEN CELL SURFACE COMPONENTS AND DNA DURING GENETIC TRANSFORMATION OF BACTERIA

Alexander Tomasz
The Rockefeller University, New York

ABSTRACT

The DNA binding capacity of transformable bacteria (pneumococci) is the result of a complex alteration of the cell surface in which several specific macromolecular factors take part: a novel antigenic determinant; a low molecular weight, cationic "activator" substance; a receptor for the activator; an "agglutinin." Evidence is presented suggesting that DNA binding is restricted to a highly dynamic area of the cell envelope at the cell wall growth zone.

Introduction

Some form of physical association between the bacterial chromosome and the cell surface has been proposed on theoretical grounds and has also been experimentally demonstrated. Attachment to a surface structure was proposed mainly to provide a mechanism for chromosome segregation during cell division and also to explain a hypothetical triggering of DNA synthesis in the male bacteria during contact with cells of opposite sexual polarity (1). Unequivocal evidence to support these propositions is lacking at present. On the other hand physical association of nascent DNA or actually most of the chromosomal DNA with cell surface fragments or cellular debris has been demonstrated repeatedly in a number of laboratories using cytological or biochemical methods (2-6). However, in practically all of these cases, the functional importance of these associations remains obscure in spite of numerous attempts to find such functions. Neither is the chemical nature of the bonds known which hold the DNA attached to particulate cell fractions.

Interaction between cell surface elements and DNA molecules present themselves in somewhat different form in biological phenomena in which DNA molecules have to traverse cellular boundaries, such as in viral invasion, conjugation or genetic transformation. In these processes DNA "packaged" in various ways encounters the cell surface during entry into bacterial hosts.

For studying DNA-cell surface interaction a particularly advantageous situation exists in the case of genetic transformation. Unlike in the cases of virus invasion or conjugation, in genetic transformation the "reactants" are bare DNA molecules and recipient cells and therefore the entire mechanism for DNA binding and transport must reside in the recipient cell. Furthermore, DNA binding to the cell surface in this case has a clear functional significance, it is an essential prerequisite for intracellular transport of the DNA molecules and their eventual genetic integration.

My review will be restricted to the discussion of DNA binding by pneumococci. Detailed description of the methodology used in these types of experiments can be found in the quoted references.

Binding of Polydeoxynucleotides.

Figure 1 demonstrates some of the typical features of DNA binding and uptake in pneumococci: 1) the rate of the process is a function of DNA concentration and shows saturation kinetics; 2) under optimal conditions, using autoradiography (see inserts in fig.) one can demonstrate that a) all or nearly all the bacteria of the population bind DNA; b) both homologous and heterologous DNA are taken up equally well; c) at limiting cell concentrations molecules compete with each other for cellular binding and d) DNA molecules fragmented by pretreatment with endonuclease (DNase II) are not bound by the bacteria (7).

Table 1 shows that DNA binding can be separated into two components: irreversible binding that resists DNase treatment and binding in a form that can be subsequently removed by externally added pancreatic DNase. In cells with blocked energy metabolism most of the cell-associated molecules are present in this reversibly bound form which presumably represents molecules captured on the surface but not yet transported further inside the cells.

Table 1. Reversible and Irreversible DNA binding

Condition	DNA-H^3 Bound per ml Cells (CPM/ml)		
	Reversible	Irreversible	(%)
Full Medium	6659	4981	25.2
Without Energy Source (- Glucose, - Sucrose)	1571	368	76.7
+ Iodo-acetamide (10^{-3}M)	1300	282	76.8

While the importance of cellular DNA binding for genetic transformation is obvious the specificity of the surface structures responsible for the binding is not a priori clear. One may assume for instance that the surface of transformable bacteria has a non-specific "stickiness" to associate with all kinds of polyanions and the actual intake of DNA occurs through some sort of osmotic drag provided by the experimental conditions.

However, on this trivial level at least, it is easy to demonstrate that DNA binding and absorption does not lack specificity: while all kinds of double stranded polydeoxy-nucleotides seem to bind to transformable bacteria, single-stranded molecules do not; furthermore, the binding capacity of the bacteria can be inactivated by relatively gentle treatments, and protoplasts and cell walls prepared from transformable cells do not bind radioactive DNA (Table 2).

Table 2. Polynucleotide binding and the lability of the cellular binding component

Polydeoxynucleotide	Association with cell (CPM bound/ml)	
Pneumococcal DNA-H^3	Yes	(18974 cpm; 0 046 µg)
Adeno Virus II DNA-H^3 (a)	Yes	(913 cpm; 0 038 µg)
pdAdT	Yes*	
pdAT	Yes*	
pdGdC	Yes*	
Phage Fl DNA, Double Stranded, Circular (b)	Yes*	
Phage Fl DNA, Single Stranded (b)	No*	
Pneumococcal DNA-H^3, Single Stranded (Denatured)	No	(210 cpm; 0 0004 µg)

Treatment of Recipient Cells	DNA-H^3 Bound (CPM/ml)
None, Competent Cells	13853
Competent Cells,	
Heat 65 C/10 min	250
Formalin, 1%, 10 min	220
Protoplasts	89
Reaction at 0° C, instead of 30°C	100
Cell Wall preparation (1.8 mg/ml)	89
Genetically incompetent cells	861
Physiologically incompetent cells	620

*Determined by competition assay (8). a) Provided by Dr. Walther Doerfler; b) provided by Dr. Ken Horiuchi.

Raised on a more sophisticated level, the question of specificity of DNA binding and uptake in transformable bacteria is less easy to answer. In this presentation I shall summarize several lines of experimental evidence which suggest that at least in the case of transformable pneumococci the DNA binding capacity is the product of a set of highly specific surface alterations.
Antigenic alteration on the surface of cells with DNA binding capacity.

The capacity pneumococci (and a few other bacterial species as well) to adsorb DNA from the environment is usually referred to as "competence" (9) and I am going to use the term throughout the paper in this sense. Competence is a genetically determined capacity in pneumococci since there are mutant strains known which cannot adsorb DNA under any of the common growth conditions. Interestingly, even in strains which do have this genetic potential, competence is not a constitutive property of the cells but is restricted to a transient "phase" of the culture cycle (Fig. 2). The onset of this competence phase occurs abruptly as cultures of pneumococci enter a critical range of cell concentration. Next, within 10 to 20 minutes, the competent condition spreads to practically all the cells present and then it declines again to undetectable levels. Formalin-killed cells taken from this competent phase can invoke a specific antibody response in rabbits (10): the γ-globulin fraction of such sera can inhibit genetic transformation and DNA binding by pneumococci but have no effect on heterologous transformation systems (11). Rabbits immunized by non-competent cells do not produce such an activity (Table 3).

Table 3. Inhibition of DNA binding by antiserum (from J. Bacteriol. **90**, 1226, 1965).

Preincubation of cells	Transformed	DNA-C^{14} Bound (CPM/ml)
With anti competent serum	4.3×10^5	100
Without antiserum	2.4×10^7	2715
With anti incompetent serum	2.8×10^7	3200

The antigenic alteration which takes place on the surface of competent cells does not seem to be an extensive one. This conclusion is suggested by two findings: the antibodies directed against the DNA binding surface represent only a few percent of all the antibodies present in the sera. In addition, it was found that sera prepared against competent pneumococci could also inhibit DNA binding by transformable cells of streptococcus viridans D. While this streptococcus is taxonomically related to pneumococci, its major surface antigens are completely different

from those of pneumococcus (11,12).
The "activator" substance: a specific inducer of DNA
binding capacity.
    In addition to the appearance of the new antigenic
determinant on the cell surface, bacteria in the competent
phase also produce a protein factor, which is capable of
inducing the capacity for DNA binding (and genetic transformation) in cells as yet incompetent (13). This substance (called "activator" or "competence factor"-CF) can
be extracted from competent bacteria by relatively mild
methods which leave most cellular material cell-bound. The
CF can be further purified by gelfiltration and chromatography on carboxymethyl (CM) cellulose. With the purest
preparations CF containing about 0.1 μg protein can induce
the appearance of DNA binding capacity ("competence") (Fig.
3) in 10^8 cells, at 30°C, in 15-20 minutes:

I. Incompetent cells + CF $\xrightarrow[\text{enzymes}]{\text{proteolytic}}$ competent cells

II. Competent cells + DNA $\xrightarrow[\text{//}]{\text{nucleases}}$ DNA binding cells

    Process I shows saturation kinetics and optimal con-
ditions require mercaptoethanol and a pH around 8. The
elution of CF from CM-cellulose is shown in Fig. 4 and some
properties of the pneumococcal CF are summarized in Table 4.

Table 4. Properties of the Pneumococcal CF

Molecular weight   10000
Charge at pH 7-8   Cationic
Optimum activity requires mercaptoethanol (3mM)
Activity is species specific
CF is sensitive to Pronase, Trypsin, Chymotrypsin,
                Subtilisin

        Insensitive to RNase (pancreatic), Lysozyme,
                        Phospholipase C, Snake Venom

CF lacks the following activities:
        Autolysin, Hemolysin,
        Pneumocin, Exonuclease,
        Protease (casein and pneumo-
        coccal protein substrates)

Experiments with the CF
    Since bacteria prior to treatment with CF do not show
the competence-specific antigenic surface alteration nor do

such cells bind DNA even in a reversible form, the CF seems to be responsible for the induction of the earliest recognizable features of the competent state. For this reason we have performed a number of experiments with the purpose of detecting biochemical changes that may occur in CF-treated bacteria.

A frequently used design of our experiments was as follows: pneumococci were grown in synthetic medium at low pH, (a condition which prevent spontaneous expression of competence). At a convenient cell concentration the pH was adjusted to 8, the culture was divided in two, to one of which highly purified CF was added, while the other culture received "inactive" CF, i.e., pneumococcal extract of identical protein concentration but prepared from incompetent cells. By adjusting the concentration of cells and CF one can obtain this way cultures in which one of the cultures will become highly competent in 10-20 minutes while the other culture remains incompetent. It is reasonable to say that such "twin" cultures only differ from one another in properties related to competence and thus this system has the potentiality of revealing specific biochemical features of the competent state. Indeed, a number of specific changes can be detected in such bacteria (Table 5).

Table 5. Specific Effects of CF on Pneumococci

	Genetic Trans- for- mation	Poly- deoxy- nucleo- tide binding	Compe- tence Anti- gen	Agglu- tinin for- mation	Cell- bound CF
Cells + CF	5-10% (Str marker)	Yes	Yes	Yes	More CF formed
Cells, No CF	$<10^{-6}$	No	No	No	No Activity Detectable

A large number of the general properties of such "twin" cultures were compared in attempts to find some further biochemical correlates of the competent state. Careful measurements were made to compare rates of cellular polymer syntheses (DNA, RNA, protein, teichoic acid), growth rates, rates of K^+ efflux and intake, leakage of intracellular markers in competent (i.e., CF-treated) versus incompetent bacteria (14). No detectable differences were found. These results imply that the activation of cells to the DNA binding capacity does not involve a deep disturbance of the general metabolism of bacteria. This lack of non-specific effects is consistent with and provide further support for the specificity of the DNA binding capacity.

Cellular receptors for CF

Recent experiments provided evidence for the existence of cellular receptors for CF (15). These can be recognized by their ability to a) physically bind and b) reversibly inactivate CF. The receptor substance seems to be located on the protoplast membrane. Brief exposure of purified CF to membrane preparations can cause disappearance of biological activity. After centrifugation followed by brief heating of the pelleted membranes in mercaptoethanol containing salt solution, CF activity can be quantitatively recovered in soluble form (Table 6).

Table 6. CF receptors in pneumococcal membrane preparations (from BBRC 41, 1342, 1970).

	CF (% Activity Recovered)	
	After Incubation with "Receptor"	After incubation with "Receptor" followed by heating at 65° C for 10 mins
Protoplast Membranes:		
From Pneumococcus	0	100
B. Subt. 168	100	–
Streptococcus (Group A)	100	–
Pneumococcal Cell Walls	50-100	–
Pneumococcal Protoplast Membranes, heated 65°C 10 mins	100	–

CF + Protoplast Membranes; 30 C, 10 mins
(100 U)   (5x10^7 cells)

Centrifuge

Pellet   Supernatant
65 C/10 mins   65 C/10 mins

CF activity: +   none
Recovery: 100%   –

Pneumococcal cell walls, protoplast membranes prepared from
B. subtilis cells or Group A streptococci do not have such
activity. Recently Dr. Ziegler in our laboratory succeeded
in solubilizing the receptor activity by sonication or
detergent treatment of washed protoplast membranes. After
extensive dialysis against mercaptoethanol containing salt
solution the preparations were further purified about 500-
fold by gelfiltration and chromatography on DEAE-cellulose
(Fig. 5). Purified, soluble "receptor" preparations cause
rapid inactivation of CF, do not affect stability of competent cells and cause about 50% decrease in the biological
activity of transforming DNA (Fig. 6). The nature of this
latter activity is being further investigated at present.

By experiments of appropriate design it was recently
demonstrated that during treatment of cells with CF, the
bulk of the CF becomes associated with the bacterial protoplast membrane and can be recovered in active form by brief
heating. This finding is consistent with attachment of CF
molecules to the membrane-bound receptor substance as a
first step in the induction of DNA binding capacity of the
cells.

Formation of new protein needed for DNA binding.

Even with saturating concentrations of CF added to the
bacteria, competence to bind DNA seems to require at least
two additional and subsequent metabolic events. These can
be recognized by the use of selective metabolic inhibitors.
The earlier one of these two events seems to require protein synthesis, the latter one requires the incorporation
of nascent cell wall material into the growth zone of the
cell envelope. First I shall briefly discuss the requirement for protein synthesis. Addition of inhibitors of protein and/or RNA synthesis to bacteria saturated with CF
prevents the appearance of DNA binding capacity (Table 7)
(16).

Table 7. Effect of inhibitors on the binding of CF and DNA

Inhibitor	Cellular Binding of CF	Transformability and DNA Binding by the CF-Treated Cells
Chloramphenicol (100μg/ml)	Normal	Inhibited (>90%)
Tryptazan (10μg/ml)	–	Inhibited (>90%)
Leucine Starvation	Normal	Inhibited (99%)
Rifamycin (0.1μg/ml)	–	Inhibited (99%)

Yet, one can show that under these conditions binding of CF occurs to the "correct" cellular sites, since upon removal of the protein synthesis block such bacteria can rapidly develop the competent condition even in the presence of a proteolytic enzyme in the medium, i.e., even under conditions that typically would inhibit "activation" of cells by yet unadsorbed, extracellular CF (Fig. 7). This finding suggests that cells can correctly adsorb CF in the presence of e.g. CAP and that upon removal of the drug a second, consecutive process essential for DNA binding is completed in the very cells that had a head start in adsorbing CF. The sensitivity of this process to tryptazan implies that the requirement is for the synthesis of a specific, qualitatively correct protein (or class of protein molecules) rather than for a gross increase in cellular mass.

Recent experiments in our laboratory indicate that at least one of these proteins can be recognized as an "agglutinin." In 1968 we noted the tendency of competent pneumococci to agglutinate (7). In 1970 Pakula reported the dramatic agglutination of competent streptococci in dilute acid (17). We repeated Pakula's experiment with pneumococci. Pneumococci activated to competence showed a striking and immediate agglutination upon resuspension in acid while the same bacteria treated with inactive CF showed no agglutination. These experiments fully confirm Pakula's findings and extend their validity to pneumococci also (Fig. 8) (18).

The agglutination is caused by some trypsin-sensitive material--presumably protein, since pretreatment of competent bacteria with trypsin prior to agglutination yielded negative tests. The agglutinins seem to be located on the plasma membrane since competent bacteria converted to

spheroplasts retained their agglutination; spheroplasts prepared from incompetent cells gave negative agglutination
tests. Agglutination is not caused directly by CF molecules
attached to plasma membrane, since
    1) membrane-preparations "loaded" with CF did not
agglutinate.
    2) Competent bacteria which were forced to lose their
transformability and cell-bound CF by chloramphenicol
treatment still retained their original agglutinin titer.
    3) A genetically incompetent strain of pneumococcus
(RA7) which binds normal quantities of CF gave no agglutination. The precise role of this protein in DNA binding
is not known at present.
The cell envelope growing zone and DNA binding.

    If pneumococci are treated with CF in a medium allow-
ing protein synthesis but lacking choline (a nutritionally
required component of the cell wall teichoic acids (19) of
this bacterium), the cells will bind CF, deposit agglutinin
and yet will not be capable of binding DNA molecules (Table
8). By observing DNA binding in bacteria that are fed by
brief "pulses" of limiting concentrations of choline, one
can notice a dramatically fast fluctuation in cellular competence paralleling the cessation and resumption of cell
wall biosynthesis (Fig. 9). The extreme rapidity with
which this property responds to the incorporation of
choline molecules into macromolecular material directs
one's attention to the topographic area of the bacterial
surface at which nascent cell wall material incorporates
into the cell envelope (20). Clearly, one may imagine
several different ways by which continuous functioning of
the cell wall growth zone might be needed for DNA binding.
It may be, for instance, that DNA binding requires a lateral
movement of the cell envelope, driven by a growing zone.
An alternative interpretation is that the growing zone(s)
of the cell envelope represents the actual site of entry of
DNA molecules. The experiment illustrated in Fig. 9 indicates that the DNA binding capacity is an extremely dynamic
property of the bacterial surface. One may as well emphasize this, since this experiment quite clearly shows how
erroneous the often encountered view is that considers the
surface of bacteria a rather static and inert structure.
Summary and Discussion

    The observations I described briefly allow one to iden-
tify several specific macromolecular factors as well as

metabolic requirements which participate in the surface
alterations that enable pneumococci to capture and absorb
double stranded polydeoxynucleotides from the environment.
These surface alterations probably involve a relatively
minor area of the cell surface only. The specific macromolecular components identified as participants in DNA
binding are located both in the outer and inner layers of
the cell envelope. This fact together with the transience
and lability of the competent condition suggests that the
DNA-binding surface is part of, or is in some way linked
to, a highly mobile, dynamic portion of the cell envelope,
which contains both cell wall and membrane material. The
experiment illustrated in Fig. 9 suggests that this area
may be the cell envelope growth zone, which was recently
localized at the cellular equator, i.e., at the zone of
cell division (20).

The acquisition of "competent state" in pneumococci
seems to occur in several stages and the model in Fig. 10
is an attempt to group these stages into a temporal sequence. The nature of the protein synthesis requiring step
needs some comment. Formation of new protein molecules may
be interpreted as the de novo synthesis of molecules specifically involved with DNA binding or transport. Alterna-
tively, protein synthesis may represent the production of
new autolysin molecules, possibly at the cell wall growth
zone, which in turn may perform the job of unmasking some
DNA binding and transporting mechanism which may be located
in the plasma membrane and which is continuously present in
the cell. In this latter version of our model the surface
alteration in competent bacteria would be an unmasking
and/or reorganizing process, somewhat analogous to the
present view of certain types of surface changes in
mammalian cells (21-23).

Epilogue

In the phenomena I described DNA molecules become
attached to the outer surface of bacteria and therefore
this type of binding need not have anything to do with the
much looked for association between resident chromosomal
material and the inner surface of the cell. Yet, transformation provides a unique and perhaps the only case in
which a direct interaction between DNA and cell surface
components does occur and also has an obvious physiological
function. Studies on this system may provide generally
useful information about the chemical nature of DNA-cell

surface interactions.

Acknowledgements

Investigations reported in this paper were supported by grants from the U.S. Atomic Energy Commission and NIH. I want to thank the editors of J. Bacteriology, Proc. National Academy and Academic Press for permissions to reproduce published illustrations.

References

1. Cuzin F., Buttin, G., and Jacob, F.   J. Cell Physiol. (Suppl. 1) 70 77- (1967).
2. Ryter, A.   Bact. Rev. 32 39- (1968).
3. Goldstein, A. and Brown, B.J.   Bioch. Bioph. Acta 53 19- (1961).
4. Ganesan, A.T. and Lederberg, J.   Bioch. Bioph. Res. Comm. 18 824- (1965).
5. Smith, D.W. and Hanawalt, P.C.   Bioch. Bioph. Acta 149 519- (1967).
6. Tremblay, G.Y., Daniels, M.J., and Schaechter, M. J. Mol. Biol. 40 65- (1969).
7. Javor, G. and Tomasz, A.   Proc. Natl. Acad. Sci. U.S. 60 1216- (1968).
8. Hotchkiss, R.D.   In Chemical Basis of Heredity, p. 321- (The Johns Hopkins Press, 1957).
9. Tomasz, A.   Ann. Rev. Genetics 3 217- (1969).
10. Nava, G., Galis, A., and Beiser, S.   Nature 197 903- (1963).
11. Tomasz, A. and Beiser, S. J. Bacteriol. 90 1226- (1967).
12. Bracco, R.M., Krauss, M.R., Roe, A.S., and MacLeod, C.M. J. exp. Med. 106 247- (1957).
13. Tomasz, A. and Mosser, J.L.   Proc. Natl. Acad. Sci. U.S. 55 58- (1966).
14. Tomasz, A.   In Symposium on the "Cellular Uptake of Informational Molecules" Mol, Belgium, North Holland Publishers, 1971.
15. Ziegler, R. and Tomasz, A.   Bioch. Bioph. Res. Comm. 41 1342- (1970).
16. Tomasz, A.   J. Bacteriol. 101 860- (1970).
17. Pakula, R., Ray, P., and Spencer, L.R.   Can. J. Microbiol. 16 345- (1970).
18. Tomasz, A. and Zanati, E.   J. Bacteriol. 105 1213- (1971).
19. Tomasz, A.   Science 157 694- (1967).
20. Tomasz, A., Zanati, E., and Ziegler, R.   Proc. Natl. Acad. Sci. U.S. 68 1848- (1971).

ALEXANDER TOMASZ

21. Burger, M.M. and Goldberg, A.R. Proc. Natl. Acad. Sci.
    U.S. <u>68</u> 1848- (1971).
22. Nicolson, G.L. Nature <u>233</u> 244- (1971).
23. Taylor, R.B., Duffus, W.P.H., Raff, M.C., and dePetris,
    S. Nature <u>233</u> 225- (1971).

TABLE 8. EFFECT OF CHOLINE DEPRIVATION ON VARIOUS PHASES OF CELLULAR COMPETENCE

Reprinted from the Proc.Natl.Acad.Sci.US <u>68</u>: 1848 (1971)

	Amino-alcohol in the medium	Cell-bound CF (units per ml)	Agglutinin	H^3-DNA bound (cpm/ml) Total	Nuclease-resistant	Transformants per ml
Cells in choline phase	choline	0	0	7	8	$< 10^2$
Cells in choline phase + CF	choline	2-400	++++	10795	7300	$2,3 \times 10^5$
Choline-phase cells, 20 minutes after shift to aminoalcohol free medium + CF	none	50-70	+++	47	16	$< 10^2$
Choline-phase cells, in limiting concentration of choline (experimental design like in Fig. 3) plus CF, at 0 minute (A)[*]	limiting	-	-	10827	7331	$2,3 \times 10^5$
at 10 minutes (B)[*]	"	-	-	1509	1038	$3,4 \times 10^4$
at 20 minutes (C)[*]	"	-	-	79	48	$5,0 \times 10^1$
at 40 minutes (D)[*]	"	-	-	53	48	< 30

In all experiments physiologically incompetent bacteria (i.e. cells grown at low pH) were used at cell concentrations of $1 - 2 \times 10^7$ per ml. CF was used at 100 units per ml concentration.

[*]Letters A,B,C and D refer to time points, as indicated in Figure   (-): no determination made.

324

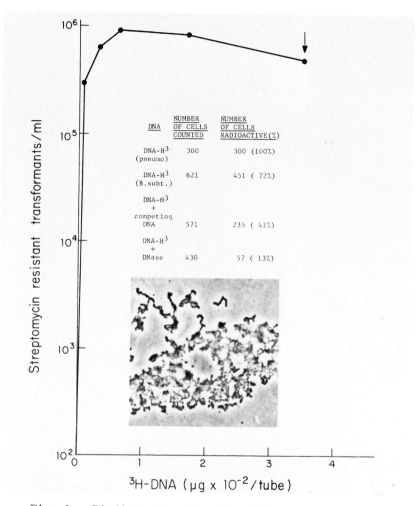

Fig. 1. Binding of radioactive DNA by <u>Pneumococci</u>. The autoradiographic procedure used has been described in reference 7.

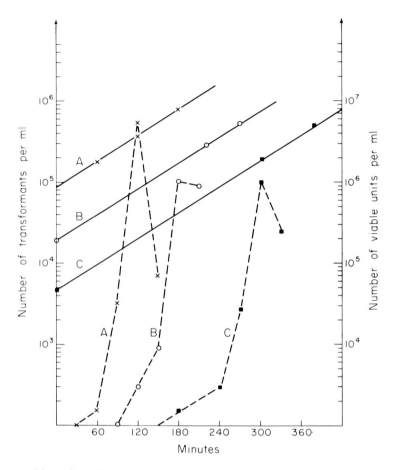

Fig. 2. The competent phase of pneumococcal culture cycle. Reprinted from J. Bacteriol., 90: 1050 (1966).

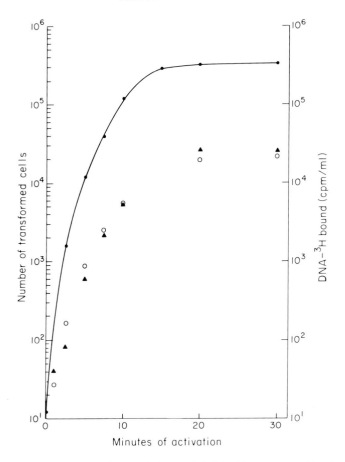

Fig. 3. Induction of the DNA binding capacity by the competence factor. Solid line: number of transformed cells; triangles: reversible binding; empty circles: irreversible binding (cpm/ml).

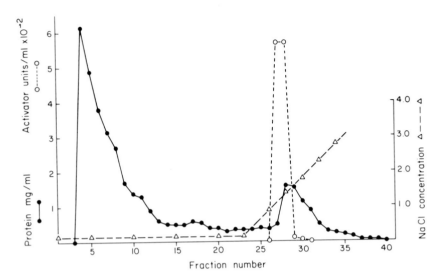

Fig. 4.   Fractionation of the competence factor on CM-cellulose.   Unpublished experiment of Dr. J.L. Mosser.

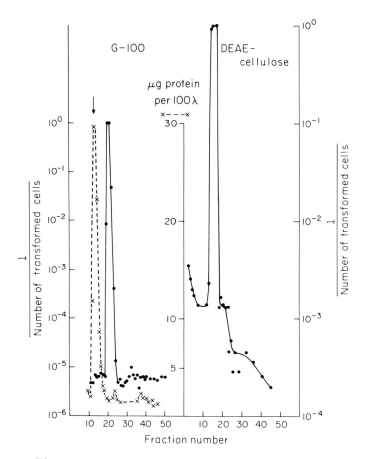

Fig. 5. Fractionation of the CF-receptor.

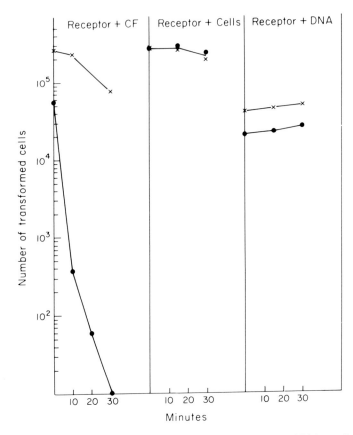

Fig. 6. Effect of receptor on the stability of CF, competent cells and transforming DNA.

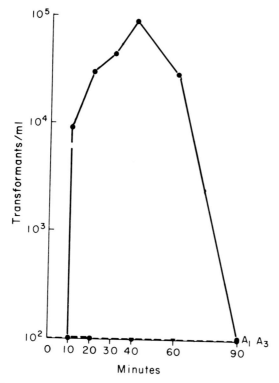

Fig. 7. Adsorption of CF to cells in the absence of protein synthesis. Reprinted in modified form from J. Bacteriol., 101: 860 (1970). Curves indicate the appearance of competence in two cultures: A1 culture was exposed to CF in the presence of CAP (chloramphenicol, 100 μg/ml), culture A3 was not. Both cultures were resuspended subsequently in medium free of CAP but containing subtilisin (0.1 μg/ml) and the appearance of competent cells was monitored by the usual method.

Fig. 8. Agglutination of competent bacteria in acid.
Reprinted from J. Bacteriol., 105: 1213 (1971). Test tube
on left contains incompetent bacteria, on the right: competent cells.

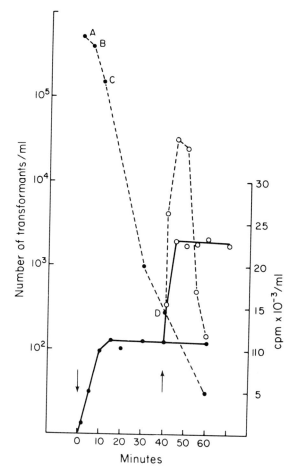

Fig. 9. Incorporation of choline into the cell walls and reactivity of the bacteria to CF. Reprinted from the Proc. Natl. Acad. Sci. U.S. 68: 1848 (1971). Solid lines indicate the incorporation of choline into macromolecular material; dashed lines show response of the cells to short pulses of "activation" by CF (5 minutes treatment). Arrows show addition of choline-H^3 to the medium (0.1 µg and 1 µC). Letters A, B, C, D indicate times at which binding of radioactive DNA was also determined.

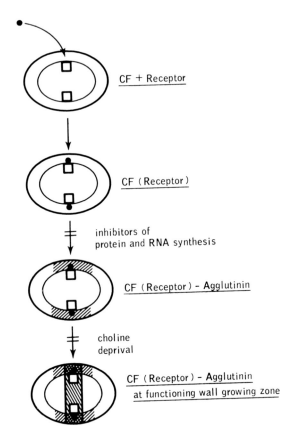

Fig. 10. Model for the stages in the development of DNA binding capacity. Reprinted from paper quoted in Fig. 9 legend.

VII

CELL-CELL INTERACTIONS

ISOLATION AND CHARACTERIZATION OF GAP JUNCTIONS FROM MOUSE LIVER

Daniel A. Goodenough

Department of Anatomy
Harvard Medical School
Boston, Mass. 02115

ABSTRACT. A method is reported for the isolation of mouse hepatic gap junctions. The method involves enzyme digestion, treatment with detergent, ultrasonication, and continuous sucrose gradient ultracentrifugation. Electron microscopy reveals that the gap junctions are isolated morphologically intact, and that there is little non-junctional contamination. The junctions are analyzed with SDS-polyacrylamide gel electrophoresis, thin-layer chromatography, and x-ray diffraction. These methods reveal that the gap junction is a homogeneous structure composed of one major protein, one major phospholipid, and some neutral lipid, with several quantitatively minor components. The junction is characterized by an 86 A center-to-center hexagonal lattice, visible with the electron microscope and with x-ray diffraction.

INTRODUCTION

Since the development of cell fractionation techniques, numerous preparations of intracellular and plasma membranes have been isolated for study. With the exceptions of the purple membrane fraction isolated from a halophilic bacterium (1, 2), many of these membrane systems contain a complex mosaic of many functional sites. Biochemical and physical measurements will represent an average of these many sites, making difficult any correlation between structure and function.

337

The gap junction, or nexus, is a differentiated portion of two interacting adjacent plasma membranes, which may be distinguished morphologically from other membranes. A complex polygonal lattice of substructures may be seen within the junction with lanthanum (3, 4), freeze-cleave (4), and negative stain (4, 5); this lattice serves as a morphological marker useful during isolation and final purity determinations.

A steadily increasing body of literature indicates that the gap junction is the site of electrical transmission in excitable tissues (6, 7), and the site of electrotonic coupling in other tissues (8). Recently (9), the gap junction has been reported to be involved in electrotonic coupling and "metabolic cooperation" (10) between cells in culture. The coexistance of gap and septate junctions in some invertebrates (11, 12, 13) seriously questions earlier speculations that electrotonic coupling is mediated by the septate attachments (14, 15), and implicates the gap junction for this role.

The gap junction is thus a differentiated domain in the plasma membranes of two adjacent cells which is morphologically distinguishable from other areas of the cell surface, and which probably represents the site of low intercellular resistance to the passage of ions and larger molecules.

A method is reported in this paper for the isolation of gap junctions from mouse liver. The isolated junctions do not appear altered in their complex morphology, and the preparation is suitable for direct chemical characterization, antibody production, and x-ray diffraction.

METHODS

Detailed isolation procedures will be published elsewhere (16); only the major steps of the isolation will be outlined here.

A crude membrane pellet from the livers of 36 mice was prepared as described earlier (4). This preparation was washed twice in 0.9% NaCl buffered at pH 7.4 with 1 mM $NaHCO_3$, then digested one hour in 0.1% collagenase (Sigma, type I) in the bicarbonate saline at room temperature (23°C). The membranes were then washed in 1 mM $NaHCO_3$ (pH 8.2) and dissolved in 0.5% Sarkosyl NL-97 (Geigy Industrial Corporation) in the bicar-

bonate buffer. The detergent-membrane solution was ultrason-
icated at medium power for six seconds (23°C), allowed to stand
for 10 minutes (23°C) and then centrifuged to a pellet (48,000 g,
15 minutes, 3°C). The residual pellet was resuspended in 1 ml
of bicarbonate buffer, placed on a continuous sucrose gradient
(d= 1.20 to d= 1.127) and centrifuged at 283,000 g for 40 hours.
    Gradients were collected in 12 drop aliquots and read
at 280 mμ. The one band at d= 1.16 was pooled and pelleted in
bicarbonate buffer at 283,000 g for 1 hour. This final pellet
was used for all subsequent manipulations.
    Thin-layer plates were run as previously described (4).
Electrophoresis gels were 7.5% acrylamide with 1% sodium
dodecyl sulfate (SDS). Samples were soaked 1 hour at 23°C
prior to electrophoresis in 1% SDS; this treatment appeared to
disrupt the junctions by morphological criteria.
    X-ray diffraction patterns were taken with a Franks
camera (17) using nickel-filtered CuKα radiation. Pellets of
junctions were examined wet and following drying at 4°C in 0.5
mm glass capillaries.

RESULTS

    The final pellet from each isolation run contains a very
pure preparation of gap junctions as seen in figures 1 and 2.
The profiles of the gap junctions in thin section, as in figure 1,
retain the 20 A "gap" (fine arrows) chacteristic of the junctions
in whole tissue. The morphology of the junction is not affected
by prolonged exposure (24 hours) to Sarkosyl NL-97, indicating
a unique insolubility of the junction in this detergent relative to
other membranes. Figure 1 also demonstrates an amorphous
contamination (C) present in the preparation. This contamination is frequently seen associated with the ends of the junctions
(large arrow) and therefore may represent incompletely solubilized non-junctional plasma membrane. The total thickness of
the sectioned junctions following isolation is 150 A, similar to
values obtained from whole tissue (4).
    Negative staining of the isolated preparation with sodium
phosphotungstate (pH 7.0) reveals that the gap junctions have retained their characteristic hexagonal lattice of subunits (figure 2).

Most of the junctions appear to have completely separated from the adjacent plasma membrane, although the amorphous contamination may still be seen (arrow). Optical diffraction of such images reveal a center-to-center spacing between the subunits of 84 A.

The gap junctions are disrupted by 1% sodium docecyl sulfate (SDS) as judged by morphological methods. Polyacrylamide gel electrophoresis with SDS of the dissolved junctions, followed by staining with Coomassie Blue (0.25%), reveal one major protein band, migrating between trypsin (MW= 23,000) and RNase (MW= 13,000) standards, flanked by two minor bands of heavier and lighter molecular weight, respectively. It is not known if these minor bands are junctional proteins, or if they are related to the amorphous contamination. The carbohydrate portion of these proteins must be determined before an accurate molecular weight can be assigned to the junctional protein (18).

Thin-layer chromatography of chloroform-methanol extracts of the isolated junctional pellet reveal that there is one major and one minor phospholipid, migrating with R_f's similar to phosphatidylcholine and phosphatidyl ethanolamine, respectively. There are also undetermined neutral lipids migrating at the solvent front.

X-ray diffraction patterns from wet junctional pellets are characterized by two rings at 73 A and 43.5 A. Spacings were calculated from the films as Bragg reflections. These two rings index closely to the first two maxima expected from a hexagonal array of subunits with a center-to-center distance of 86 A. Dried specimens show these same reflections on the equator with an additional broad maximum at 25-28 A. This broad maximum spans the expected location of the fourth and fifth hexagonal reflections. The third order, expected at 39 A, has not been seen. With drying, the junctions stack parallel to each other; a strong profile diffraction is seen on the meridian of films from dried specimens. Maxima are seen at 155, 75, 50, and 39.5 A which index close to the first four diffraction orders from a lamellar structure 150 A in thickness. On the films from dried specimens, two higher angle rings are seen at 10.5 and 4.35 A, probably arising from protein and lipid, respectively (19).

DISCUSSION

The isolation method described in this paper yields a
relatively pure preparation of gap junctions from mouse liver.
The yield is small, ranging from 0.1-0.5 mg protein from 36
animals as measured spectrophotometrically. The isolated
material contains varying amounts of amorphous contamination,
visualized in thin-sectioned and negatively stained preparations.

Chemical analysis performed thus far have revealed that
the gap junction contains one major protein, with a molecular
weight between 13,000 and 23,000, and two minor proteins. Although these proteins must account for the hexagonal lattice seen
in the electron microscope, it is possible that junction-associa-
ted proteins, essential for the physiological function, have been
lost during the process of isolation. Chemical analyses have
also revealed phospholipids and neutral lipids in the gap junction.
This result is consistent with earlier observations that phospholipids were an integral part of gap junction structure (4).

Preliminary studies have revealed that the hexagonal
lattice of the gap junction may be seen with x-ray diffraction,
although studies thus far have been limited to measuring and
indexing of the maxima. Diffraction from the dried junctions
are characterized by a strong profile diffraction on the meridian, with the hexagonal maxima on the equator. These films
lack a third order hexagonal reflection. The absence of inter-
vening reflections has been found in myelin (20) and provided
pertinent structural information.

Diffraction patterns from wet junctions show two rings
which index close to the 86 A hexagonal lattice. One would also
expect to find on these films the continuous diffraction from the
junctional profiles (21). Estimates may be computed assuming
an electron density profile of two parallel bilayers for the junction; Fourier analysis reveals that the maxima expected from
such an electron density distribution would fall at the origin
and close to the two observed rings. Thus it can be concluded
that the two rings seen in the wet diffraction patterns may
contain contributions from both the junctional hexagonal array
and the continuous profile diffraction.

This preparation of gap junctions allows one to study the molecular architecture of a differentiated segment of the plasma membrane. The regular array of structure in the junction presents the possibility of solving the detailed molecular architecture using x-ray diffraction. While certain features of the arrangement of the molecules in the gap junction will be highly specialized, there will certainly be fundamental properties of hydrophobic protein-protein and protein-lipid interaction in the junction which are common to other cell membranes. Additionally knowledge of the detailed structure of the gap junction will allow a better understanding of the molecular basis of intercellular communication.

ACKNOWLEDGEMENTS

This work was supported in part by Postdoctoral Fellowship 40017 from the NSF and grant HE 06285 from the NHLI (to W. Stoeckenius). The author is indebted to Walther Stoeckenius for his friendship and collaboration. The helpful discussions and criticisms of Drs. D. L. D. Caspar, W. Wober, V. K. Miyamoto, G. T. King, and N. B. Gilula are gratefully appreciated.

REFERENCES

1. Osterhelt, D. and Stoeckenius, W., Nature New Biology 233:149 (1971).
2. Blaurock, A. E. and Stoeckenius, W., Nature New Biology 233:152 (1971).
3. Revel, J. P. and Karnovsky, M. J., J. Cell Biol. 33:C7 (1967).
4. Goodenough, D. A. and Revel, J. P., J. Cell Biol. 45:272 (1970).
5. Benedetti, E. L. and Emmelot, P., J. Cell Biol. 38:15 (1968).
6. Asada, Y. and Bennett, M. V. L., J. Cell Biol. 49:159 (1971).
7. Pappas, G. D., Asada, Y., and Bennett, M. V. L., J. Cell Biol. 49:173 (1971)

8. Furshpan, E. J. and Potter, D. D., in Current Topics in Developmental Biology, v. III, Academic Press, N. Y., p. p. 95-127 (1968).

9. Gilula, N. B., Reeves, O. R. and Steinbach, A., Nature 235:262 (1972).

10. Subak-Sharpe, H., Burk, R. R. and Pitts, J. D., J. Cell Sci. 4:353 (1969).

11. Hudspeth, A. J. and Revel, J. P., J. Cell Biol. 50:92 (1971).

12. Gilula, N. B. and Satir, P., J. Cell Biol. 51:869 (1971).

13. Rose, B., J. Membrane Biol. 5:1 (1971).

14. Bullivant, H. S. and Loewenstein, W. R., J. Cell Biol. 37:621 (1968).

15. Gilula, N. B., Branton, D. and Satir, P., Proc. Nat. Acad. Sci. (U. S.) 67:213 (1970).

16. Goodenough, D. A. and Stoeckenius, W. 1972 (in preparation).

17. Franks, A. Proc. phys. Soc. B 68:1054 (1955).

18. Segrest, J. P., Jackson, R. L., Andrews, E. P. and Marchesi, V. T., Biochem. Biophys. Res. Comm. 44:390 (1971).

19. Engleman, D. M., J. Mol. Biol. 47:115 (1970).

20. Caspar, D. L. D. and Kirschner, D. A., Nature New Biology 231:46 (1971).

21. Wilkins, M. F. H., Blaurock, A. E. and Engleman, D. M., Nature New Biology 230:72 (1971).

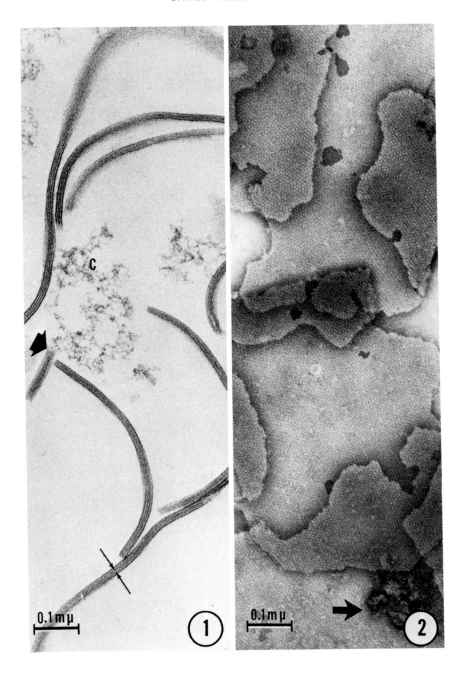

Figure 1. A thin section through the isolated pellet of gap junctions. The junctions retain their 20A gap (fine arrows) and their 150A profile thickness characteristic of the junctions in whole liver. An amorphous contamination (C) is seen in the preparation, frequently associated with the ends of the junctions (large arrow). Magnification: 123,000x

Figure 2. A negative stained preparation of the junctional fraction. The stain was sodium phosphotungstate at pH 7.0. The junctions are characterized by an 84A center-to-center hexagonal lattice of subunits, visible following the isolation procedure. An amorphous contamination (arrow) may be seen. Magnification: 123,000x

A POSSIBLE ROLE FOR CARBOHYDRATES
IN CELL-CELL ADHESION

Edward John McGuire*

Johns Hopkins University
Baltimore, Maryland

Abstract

A mouse embryoma has been shown to require the
metabolism of L-glutamine for the cells to form aggregates. The available evidence is best interpreted that
L-glutamine is converted to hexosamine containing macromolecules of the cell surface which are involved in cell-cell adhesion. A second system has shown, using an assay
that measures specific intercellular adhesion, that a β-galactosyl residue is involved in the adhesive recognition of chicken embryonic neural retina cells. It has
further been shown that intact chicken embryo neural
retina cells catalyze glycosyl transferase reactions to
endogenous acceptors as well as exogenous acceptors.
Evidence has been presented that these enzymes are located on the outer cell membrane and may play a role in
cellular recognition or adhesion.

Introduction

Previous work from our laboratory and others has
demonstrated the anabolic and catabolic pathways involved in the utilization and synthesis of carbohydrates
(1). More recent work has elucidated the mode of synthesis of more complex carbohydrate containing structures, such as glycoproteins (2), glycolipids (3), mucopolysaccharides, etc. Our attention has now been shifted
to an attempt to discover a functional role of carbohydrates.

When examining the literature for clues, the outstanding feature to emerge is the fact that most of the
complex carbohydrate containing macromolecules, with the
exception of storage polysaccharides, occur extracellu-

*Assistant Professor, Department of Biology

larly. That is they are found in the extracellular
fluids (4) or on the external surface of the plasma membrane (5). These facts have been interpreted to suggest
that sugars are added to proteins or lipids during the
secretion process (4). Thus, the role of sugars is, in
some unknown way, to participate in the secretion of certain glycoproteins and glycolipids from cells. There is
a great deal of evidence, radioautographic (6) and sub-
cellular isolations (7), that sugars are added to glycoproteins in the golgi region of the cell and hence that
glycosylation takes place during the secretion of glycoproteins.

There is also a vast body of literature showing
that cells are covered with carbohydrate containing
molecules (8). Currently there is much interest in the
analysis of the role that "cell surface" sugars play in
the well known phenomenon of contact inhibition (9), by
the use of plant lectins (10), chemical techniques (11),
and electronmicroscopic visualization (12).

The fact that cells are covered with carbohydrate
containing molecules and that proteins that are secreted
are almost always glycoproteins is consistent with the
idea that, indeed, sugars are added to proteins and
lipids to get them out of the cell. However, why is
there such a diversity of carbohydrate structures?
Wouldn't a few simple sugars serve to code or participate
in the secretion process? The almost infinite array of
carbohydrate structures suggested to us that this variability could be used to code information. Coupled with
the fact that carbohydrate containing molecules are
found primarily on the exterior of cells, it seems the
type of informational role that sugars could play would
be to communicate from one cell surface to another or
from one extracellular carbohydrate containing macro-
molecule to a "receptor" molecule. The system we chose
to investigate was cell-cell adhesion with an emphasis
on the possibility of carbohydrates being involved in
this process.

Moscona (13) made a fundamental contribution to
this field when he found that embryonic tissues could be
dissociated into single cell preparations by gentle
treatment with the proteolytic enzyme, trypsin. Upon
gentle shaking, these cells then readhered to form small
aggregates in the flasks. Subsequently, Moscona and

Steinberg (14) using histological techniques found (Fig. 1) that the initial aggregates are composed of a random collection of cells. These cells over a period of time rearrange, "sort out" and arrive at a final disposition where "like" cells find "like"cells (isotypic adhesion). This process can be used as an _in vitro_ model of morphogenesis and can be conveniently divided into two stages, one an aggregation (non-specific adhesion) phase and one a recognition (specific adhesion) phase.

Many efforts have been made to quantitate the phases of this process. These attempts fall into two categories. Following the dissociation of the tissues either the appearance of aggregates is followed and measured visually, or the disappearance of single cells, suggesting aggregation is occurring, is followed using a hemocytometer (15), Coulter electronic cell counter (16), or light scattering techniques (17). These techniques have some shortcomings. First, they can only measure the aggregation phase of this process which may be measuring aggregation and sorting out at the same time; and secondly, it has not been possible to do mixed cell types (heterotypic adhesion) and get information about the rates or strengths of adhesion between different cell types. However, they can provide very accurate rates of adhesion for one cell type (isotypic adhesion).

Results

The first system we used to study cell-cell adhesion involved the use of the Coulter counter. The cells used were a mouse ascitic testicular teratoma, 402 AX obtained from Dr. Leroy Stevens at the Jackson Laboratories, Bar Harbor, Maine. This tumor grows in the ascitic cavity as small "embryoid" bodies which resemble mouse embryos at the blastula or morulla stage of development. Fig. 2 shows the way in which the "embryoid bodies" are formed during growth in the ascitic cavity. Stage IB is a single cell stage which is apparently competent to divide and aggregate. As the cells divide and aggregate, there is a tendency for the cells to create a large vacuole. Cells in this stage are Stage II cells. The cells divide around the vacuole to form the blastula-like embryoid body (Stage III). In the normal growth of this tumor, division appears to be limited after the cells have divided completely around the vacuole. DNA synthesis ceases. Many small vacuoles form along the plasma membranes where cell-cell contact is made. These vacuoles appear to be lysosome-like particles since they stain for lysosomal enzymes using histochemical techniques and by the fact that there is a marked rise in the degradative enzymes associated with lysosomes as the cells move into Stage IV. The embryoid bodies then fall apart into single cells (Stage IA) which presumably cannot aggregate and must now resynthesize surface material in order for the next cycle to ensue. This cycle summarizes and is a natural counterpart to much of what I have talked about today. The cells going from Stage IB to Stage III clearly are aggregated and thus must involve cell-cell adhesion. From Stage III to IA involves a natural dissociation, nature's trypsinization procedure. Finally in going from Stage IA to IB may be a healing or lag period during which surface components are being resynthesized.

Dr. Oppenheimer showed that trypsin treatment of these "embryoid" bodies results in single cell preparations. When single cell preparations are placed in complete tissue culture medium and shaken the cells aggregate (18). When aggregation experiments are carried out in simple salts plus glucose medium, no cell adhesion occurs. Further studies indicated that of the 51 components of the tissue culture medium, only one was necessary for aggregation, that compound being L-glutamine. Fig. 3 shows the concentration requirement for L-glutamine to stimulate cell aggregation. Aggregation rates are optimal at about 7 mM. Two amino sugars, D-glucosamine and D-mannosamine, can substitute for L-glutamine and were equally effective in promoting aggregation. Fig. 4 shows the effect of L-glutamine on the adhesion of teratoma cells. Complete tissue culture medium minus glutamine fails to promote aggregation while glucose + L-glutamine show a marked aggregation.

These results are consistent with the known metabolic scheme for synthesis of complex carbohydrates shown in Fig. 5. The key reaction is the transfer of the amide nitrogen of L-glutamine to fructose-6-phosphate, giving glucosamine-6-phosphate. Glucosamine-6-phosphate can be utilized in the synthesis of all other nitrogen contain-

ing sugars of glycolipids and glycoproteins. The teratoma is completely devoid of the enzyme glutamine synthe-
tase and is thereby dependent on exogenous sources of
glutamine. In the absence of glutamine it can be seen
how glucosamine or mannosamine can substitute in these
pathways leading to complex carbohydrate synthesis.

Diazo-oxonorleucine (DON) is a potent inhibitor of
the glutamine transfer reaction thereby preventing the
synthesis of glucosamine-6-phosphate. The addition of
DON should prevent L-glutamine from stimulating aggregation but should have no effect on the D-mannosamine
stimulated effect. Fig. 6 shows that these predicted results are found. The results are shown as the number of
single cells per milliliter and a decrease in number with
time indicates aggregation is occurring. Clearly, DON
prevents glutamine from stimulating aggregation but is
not inhibitory to the D-mannosamine stimulation.

Fig. 7 shows a summary of the evidence suggesting
that the adhesion of the teratoma cells is a consequence
of the metabolism of L-glutamine. All of these data suggest that L-glutamine is a precursor of hexosamines which
are further utilized for complex carbohydrate containing
macromolecules. Thus, in this system, hexosamine containing molecules seem to play a role in the cell-cell
adhesion mechanism. Using this aggregation system
coupled with the use of isotopically labelled D-hexosa-
mines, which are required for aggregation, we hope to be
able to isolate and characterize the complex carbohydrates involved in intercellular adhesion of this embryoid tumor.

Another assay has been used in our laboratory to
measure cell-cell adhesion (19). This assay is based on
modification of the collecting fragment assay of Roth and
Weston (20). The method is diagramatically shown in Fig.
8. Embryonic chick tissues are labelled with $^{32}PO_4$ and
subsequently dissociated by trypsin treatment into
$^{32}PO_4$ labelled single cell suspensions. Unlabelled cell
aggregates of a like cell type (isotypic) or unlike cell
type (heterotypic) are then placed in the labelled cell
suspension and shaken. Periodically the aggregates are
withdrawn, washed and counted. If tissue specific cell
adhesion has occurred, more cells should adhere to iso-
rather than heterotypic aggregates. Specificity has been
demonstrated using the collecting aggregate assay with all

(nine) embryonic tissue types tested from two different
species (mouse and chicken). In all cases isotypic collection is favored over heterotypic.

The effect of varying aggregate size, shaker speed,
cell number, etc. on specific cell adhesion have been
studied. In Fig. 9 the results of placing 8 day embryonic
chick neural retina aggregates and liver aggregates in a
labelled liver cell preparation are shown. At 5 hours
the isotypic liver-liver labelling is 50 times the liver-
neural retina heterotypic labelling. Also there is
clearly a lag time for liver-liver cell pick-up. This 2
hour time period is thought to be the time necessary to
resynthesize components necessary for specific cellular
recognition which were probably destroyed during the
trypsinization procedure.

Fig. 10 shows the mirror-image experiment where unlabelled liver and neural retina aggregates are placed in
a labelled neural retina cell suspension. In this case
at 5 hours, neural retina aggregates have collected 30
times as many neural retina cells as liver aggregates.
Neural retina cells show little lag time before collection. When the temperature of collection is dropped from
37° to 28° C, Fig. 11, a reproducible 25-30 minute lag
time is seen before isotypic collection ensues. This lag
time is again considered the time to resynthesize specific adhesion components (healing time). No such healing
time or lag is observed following trypsinization, if non-
specific aggregation is observed using the electronic
Coulter counter to follow the disappearance of single
cells going to form aggregates suggesting these two
assays are measuring different aspects of cell adhesion
(16).

Isotypic adhesion of neural retina cells is inhibited by low temperature and by inhibitors of ATP production, such as cyanide, dinitrophenol, and sodium azide.
Surprisingly, the protein synthesis inhibitors, cycloheximide and puromycin do not inhibit isotypic collection
or the apparent healing process observed during the lag
time.

Attempts were made to alter the specificity of
pick-up of cells by treatment of the aggregates with proteases. This treatment had no substantial effect on the
collection of iso- or heterotypic cells. In the absence
of an effect by treatment of the aggregate surface with

proteases the neural retina aggregates were treated with
the crude supernatent of Diploccus pneumoniae Type I (21),
a rich source of glycosidases. The experimentally
treated aggregates collected 5 times as many isotypic
cells as the untreated control aggregate in 25 minutes at
20°. At this time the cells have not had sufficient time
to heal and do not adhere to untreated aggregates. This
effect was duplicated by purifying β-galactosidase from
this extract and treating the aggregates. An increase in
collection of neural retina cells by β-galactosidase
treated neural retina aggregates at 25 minutes ranged
from 2.5 to 6 fold depending on the extent of treatment.
β-galactosidase presumably acts by removing a terminal
β-galactosyl residue from the aggregate surface. The
effect of β-galactosidase on the aggregates was completely destroyed if the enzyme was first heat-denatured
or if D-galactal (22), a potent inhibitor of β-galactosidase, was added during enzymatic treatment. β-galactosidase treatment of neural retina aggregates also increased
heterotypic collection of liver cells. A similar treatment of liver aggregates had no effect on iso- or heterotypic collection.

The results of β-galactosidase treatment of neural
retina collecting aggregates suggested that a terminal
β-galactopyranosyl residue plays a role in adhesive recognition of this cell type. Attempts were made to re-es-
tablish specificity of β-galactosidase treated cells by
transferring ^{14}C-galactose from UDP-^{14}C-galactose to these
intact cells using a crude neural retina cell homogenate
as an external source of a glycosyl transferase enzyme.
This type of reaction is shown in Fig. 12. The enzyme in
the presence of Mn^{++}catalyzes the transfer of a sugar
moiety from the activated nucleotide sugar to an appropriate acceptor. This type of enzyme is specific for the
donor nucleotide sugar and for the acceptor molecule.
The result of the experiment with cells and added enzyme
was that indeed you could add ^{14}C-galactose to the cells,
but the added enzyme was not necessary. Apparently the
cells have a glycosyltransferase type enzyme on their
cell surface capable of transferring galactose from UDP-
galactose to some unknown acceptor molecule also on the
cell surface. An examination of neural retina cells for
surface glycosyl transferases was undertaken. The results suggest that there are surface transferases and

further that these enzymes may play a role in intercellular adhesive recognition (23).

Conditions were optimized for incorporation of
galactose so that incorporation into the cell surface
(endogenous) was proportional with time and cell number
and that optimum level of sugar nucleotide and metal
were used. Although nucleotide sugars are purported not
to penetrate the cell surface this possibility was tested
by addition of unlabelled galactose, galactose 1-phos-
phate, and UDP-glucose, all of which would metabolically
equilibrate intercellularly with the labelled UDP-galactose if it penetrated the cell. None of the compounds
reduced the incorporation of galactose, suggesting that
UDP-galactose does not enter the cell.

Preliminary characterization of the endogenous product to which ^{14}C-galactose had been transferred suggested it was a membrane bound glycoprotein. Four glycosyl-
transferases that transfer galactose to glycoproteins are
shown in Fig. 13. Two of these enzymes were found: 1)
"Serum glycoprotein type" UDP-Gal + β-N-acetylglucosa-
mine-R; 2) "Mucin type" UDP-Gal + α-N-acetylgalactosa-
mine-R. Fig. 14 shows the results of adding N-acetylglucosamine acceptor to the intact cells. ^{14}C-galactose is
transferred to the exogenous acceptor N-acetylglucosamine
and further the cells are required for this transfer.
Preincubation of cells for 2 hours and then removing the
cells and testing the supernatant for enzyme shows very
low levels of enzyme. This indicates that galactose
transfer is probably not due to cell breakage and the
subsequent leakage of enzymes to the medium. Further
preincubation of cells for 2 hours before addition of
UDP-^{14}C-galactose shows a marked stimulation over cells
incubated immediately following trypsinization.

Results shown in Fig. 15 show a similar situation
when cells are incubated with the exogenous mucin type
acceptor; in this case sialidase treated ovine submaxillary mucin which has exposed α-N-acetylgalactosaminyl
residues. The acceptor molecule has a molecular weight
of 10^6, which reduces the chance of penetration into the
cell. Again it can be seen that preincubation of the
cells for 1, 2 or 4 hours does not release a significant
amount of enzyme to the medium. It can also be seen with
this acceptor that preincubation of cells, in this case
for 3 hours, markedly stimulates incorporation compared

to freshly trypsinized cells. It is not known whether
during preincubation more enzyme is made or if existing
enzyme is made more accessible to the external substrates
during this time.

A summary of the evidence for the presence of external cell surface galactosyl transferases on neural
retina cells is shown in Fig. 16. The product of the endogenous transfer of galactose is a particulate (membrane) glycoprotein. Exogenous acceptors, some of which
are large enough not to penetrate the cell, act as acceptors. Metabolites which would equilibrate with internal
UDP-galactose do not compete. Preincubation of the cells
for various times and then checking the supernatants for
glycosyl transferase enzymes showed negligible activity.
A radioautographic examination of T^3-galactose labelled
cells showed that most of the cells were labelled and
cells in section showed ring type labelling, indicating
that the majority of incorporation was at the cell surface. Finally, some acceptors which are active as galactosyl transferase acceptors inhibited neural retina cell
pick up when tested using the collecting aggregate assay.
This result suggests that not only is the galactosyl
transferase on the cell surface, but it may also play a
role in intercellular adhesion.

A number of other cells have been tested for the
presence of glycosyl transferases on the outer cell membrane. All cells thus far tested have shown these activities with the exception of red blood cells. It would
seem that there is a ubiquitous distribution of these
enzymes in other cell types and the findings of these
enzymes in the embryonic neural retina is not an isolated
oddity.

Discussion

In this paper we have shown that carbohydrates are
involved in the intercellular adhesion process in two
systems (embryonic chick neural retina and the mouse
ascitic embryoma). We have also suggested that some
glycosyl transferase enzymes are located on the external
surface of cells. Recently, a theory for the potential
function of cell surface glycosyl transferases and complex carbohydrates was proposed (24). Fig. 17 shows a
brief schematic summary of this theory. Briefly the

glycosyl transferase substrate complex is involved in the initial adhesive recognition process between cells. After the adhesive recognition and adherence occurs, if the cell-cell union is to be made permanent, more complex and stronger structures are made such as desmosomes, tight junctions, etc.

A simple extension of this theory is that cells can modify one another by completing the reaction that the individual glycosyl transferase can catalyze. Sugar nucleotide must be supplied for this to occur. UDP-X is supplied in Fig. 17, presumably from within the cells and X is transferred to each cell by the opposing cell transferase. Each cell has thereby modified its opposing cell. This process would convert the acceptor molecule to a product which would either bind poorly or not at all to the enzyme. This could result in dissociation where each cell would have modified the other or if other glycosyl transferases were on the cell surface, these enzymes could then bind to the new substrate, resulting in increased adhesiveness. The major point being that intercellular adhesive strengths could be altered in this fashion. This theory is not meant to show how permanent adhesion occurs but rather is a possible mechanism of recognition.

Recently, two papers (25, 26) appeared suggesting the possibility that blood platelets bind to a wound site by sticking to the carbohydrate moieties of collagen. The plasma membrane of the platelets contain the appropriate glycosyl transferases to form a substrate-enzyme complex. Both authors suggest this mechanism of platelet-wound site recognition. This is the first data which tends to support the hypothesis of cell surface glycosyl transferases acting in a cell-adhesion mechanism.

Roth (27) has suggested that the cell surface glycosyl transferases may play a role in growth regulation and that there are fundamental differences in the ability of these enzymes to function when comparing normal (contact inhibited) and transformed fibroblasts (non-contact inhibited).

All the data of this paper and the papers mentioned in the discussion are consistent with the notion that extracellular carbohydrate provides information. The information provided must be "read" or transduced. The cell surface glycosyl transferases could provide the mechanism for both "reading" and "altering" this information.

References

1. Caputto, R., Barra, H. S., and Cumar F. A., in P. Boyer (Editor), Annual Reviews of Biochemistry, Annual Reviews, Inc., Palo Alto, 1967, p. 211.
2. Spiro, R. G., in E. Snell (Editor), Annual Review of Biochemistry, Annual Reviews, Inc., Palo Alto, 1970, p. 599.
3. Roseman, S., in E. Rossi and S. Stoll (Editors), Biochemistry of Glycoproteins and Related Substances, Proceedings of the Fourth International Conference on Cystic Fibrosis of the Pancreas, S. Karger, Basel, 1968, p. 244.
4. Eylar, E. H., J. Theoret. Biol., 10, 89 (1965).
5. Revel, J. P., and Ito, S., in B. Davis and L. Warren (Editors), The Specificity of the Cell Surface, Prentice-Hall, Inc., Englewood Cliffs, N.J., 1967, p. 211.
6. Peterson, M., and LeBlond, C. P., J. Cell Biol., 21, 143 (1964).
7. Schachter, H., Jabbal, I., Hudgin, R. L., Pinteric, L., McGuire, E. J., and Roseman S., J. Biol. Chem., 245, 1090 (1970).
8. Curtis, A. S. G., The Cell Surface, Its Molecular Role in Morphogenesis, Academic Press, Inc., New York (1967).
9. Abercrombie, M., Heaysman, J. E. M. and Karthauser, H. M., Exptl. Cell Res., 13, 276 (1957).
10. Burger, M. M., Proc. Nat. Acad. Sci., 62, 994 (1969).
11. Hakomori, S. and Murakami, W. T., Proc. Nat. Acad. Sci., 59, 254 (1968).
12. Nicolson, G. L., and Singer, S. J., Proc. Nat. Acad. Sci., 68, 942 (1971).
13. Moscona, A. A., J. Cell Comp. Phys., 60, Suppl. 1, 65 (1962).
14. Steinberg, M. S., Science, 141, 401 (1963).
15. Curtis, A. S. G., J. Embryol. Exp. Morph., 22, 305 (1969).
16. Orr, C. W., and Roseman, S., J. Memb. Biol., 1, 109 (1969).

17. Kemp, R. B., Jones, B. M., Cunningham, I., and Jones, M. C. M., J. Cell Sci., 2, 323 (1967).
18. Oppenheimer, S. B., Edidin, M., Orr, C. W., and Roseman, S., Proc. Nat. Acad. Sci., 63, 1395 (1969).
19. Roth, S., McGuire, E. J., and Roseman, S., J. Cell Biol., 51, 525 (1971).
20. Roth, S. A., and Weston, J. A., Proc. Nat. Acad. Sci., 58, 974 (1967).
21. Hughes, R. C., and Jeanloz, R., Biochemistry, 3, 1535 (1964).
22. Lee, Y. C., Biochem. Biophys. Res. Commun., 35, 161 (1969).
23. Roth, S., McGuire, E. J., and Roseman, S., J. Cell Biol., 51, 536 (1971).
24. Roseman, S., Chemistry and Physics of Lipids, 5, 270 (1970).
25. Bosmann, H. B., Biochem. Biophys. Res. Commun., 43, 1118 (1971).
26. Jamieson, G. A., Urban, C. L., and Barber, A. J., Nature New Biology, 234, 5 (1971).
27. Roth, S., and White, D., Proc. Nat. Acad. Sci., 69, 485 (1972).

Tissue Dissociation and Stages of Reformation

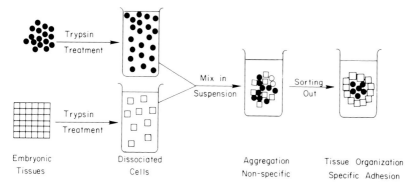

Fig. 1 A schematic representation of trypsin dissociation of tissues to give single cell suspensions. The cell suspensions when mixed and shaken, yield random, mixed aggregates (non-specific adhesion) which subsequently sort out (specific adhesion).

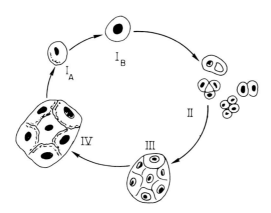

Fig. 2 The cycle of growth of mouse ascitic teratoma (402 AX) in 129 J mice.

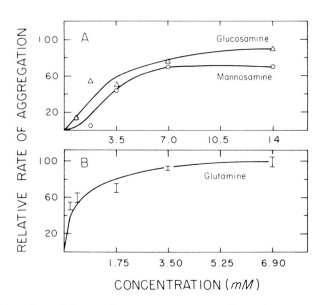

Fig. 3 Effect of concentration of L-glutamine, D-glucosamine, and D-mannosamine upon rate of aggregation of teratoma cells.

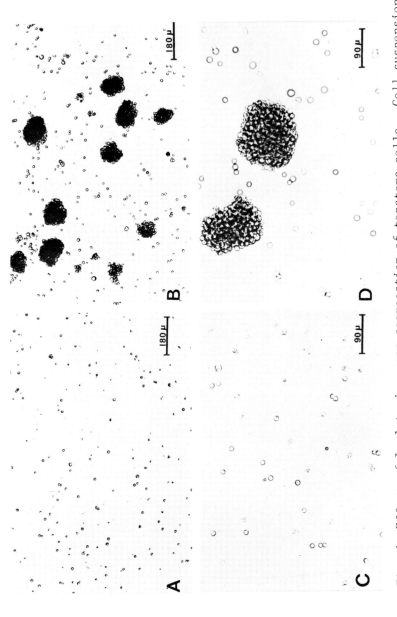

Fig. 4 Effect of L-glutamine upon aggregation of teratoma cells. Cell suspensions were rotated at 76 rpm for 200 minutes in medium 199 minus L-glutamine (left), and Hanks medium plus 3.5 mM L-glutamine (right).

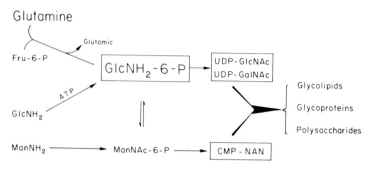

Fig. 5   Metabolic pathway for synthesis of amino sugars from L-glutamine and complex polysaccharide, glycoprotein, and glycolipid synthesis.   Transaminidase synthesizes GlcNH$_2$-6-P from L-glutamine and fructose-6-P; this step is inhibited by (DON) Diazo-oxonorleucine and azaserine.   DON has no effect on D-glucosamine (GlcNH$_2$) or D-mannosamine (ManNH$_2$) metabolism.

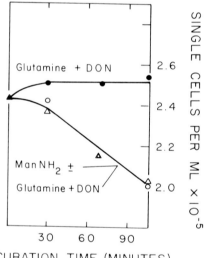

INCUBATION TIME (MINUTES)

Fig. 6   Effect of 6-Diazo-5-oxonorleucine (DON) on aggregation of teratoma cells.   Single cells were counted at indicated times using a Coulter electric cell counter. Disappearance of single cells indicates aggregate formation.   (●), 3.5 mM L-glutamine + 1.5 mM DON; (△), 6.9 mM D-mannosamine; (o) 3.5 mM L-glutamine + 1.5 mM DON + 6.9 mM D-mannosamine.

1   L Glutamine cannot be replaced either by other

   amino acids or by structural analogues ie

   D glutamine

2   L Glutamine-mediated adhesion inhibited by

   NaCN, NaN$_3$, NaF

3   D Glucosamine or D mannosamine but not D or

   L galactosamine can replace L glutamine

4   L Glutamine-mediated adhesion inhibited by DON

5   Inhibition by DON reversed by either D glucosamine

   or D mannosamine

Fig. 7 Summary of evidence for necessity of metabolism of L-glutamine.

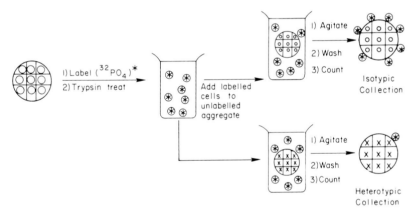

Fig. 8 Schematic representation of specific cell
adhesion assay. Tissues are labelled with ^{32}P-inorganic
phosphate and then dissociated to form ^{32}P-labelled single
cell suspensions. Unlabelled cell aggregates of the same
(iso-) or different (hetero-) are added to the labelled
single cell suspensions and shaken, washed and the number
of single cells adhering to the collecting aggregate is
determined by counting the aggregates. As shown, isotypic
collection is favored.

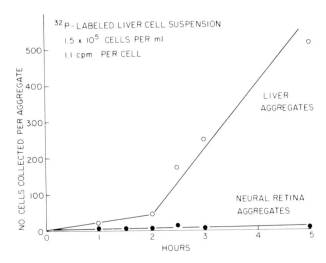

Fig. 9   Collection of ^{32}P-labelled liver cells by liver and neural retina cell aggregates as a function of time.   Cell concentration was 1.5 x 10^5 cells per ml (1.1 cpm per cell) with three aggregates per flask in nutrient medium, and the temperature was 37° C.

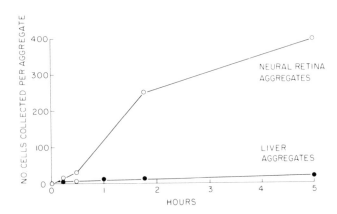

Fig. 10   Collection of ^{32}P-labelled neural retina cells by neural retina and liver aggregates as a function of time.   Cell concentration was 1 x 10^5 cells per ml (1.0 cpm per cell) with three aggregates per flask in nutrient medium, and the temperature was 37° C.

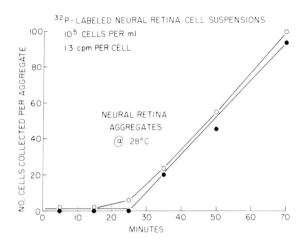

Fig. 11 Collection of ^{32}P-labelled neural retina cells by neural retina aggregates at 28° C. The curve represents the results from an experiment using 1 x 10^5 cells per ml (1.3 cpm per cell). Each bar includes the range of four points, each of which represents one flask (five aggregates per flask).

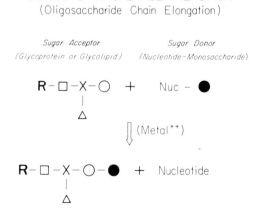

Fig. 12 Glycosyl transferase reaction. Enzymes are specific for both donor nucleotide sugar (nuc-•) and acceptor molecule (R-□-△-o).

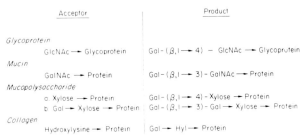

ANIMAL GALACTOSYL — TRANSFERASES

(GLYCOPROTEIN)

UDP – Gal + ACCEPTOR ⇌ Gal – ACCEPTOR + UDP

Acceptor	Product
Glycoprotein	
GlcNAc ⟶ Glycoprotein	Gal-(β,1 ⟶ 4) – GlcNAc ⟶ Glycoprotein
Mucin	
GalNAc ⟶ Protein	Gal-(β,1 ⟶ 3) – GalNAc ⟶ Protein
Mucopolysaccharide	
a. Xylose ⟶ Protein	Gal-(β,1 ⟶ 4) - Xylose ⟶ Protein
b Gal ⟶ Xylose ⟶ Protein	Gal-(β,1 ⟶ 3) - Gal ⟶ Xylose ⟶ Protein
Collagen	
Hydroxylysine ⟶ Protein	Gal ⟶ Hyl ⟶ Protein

Fig. 13 Four known galactosyl:protein glycosyl transferases.

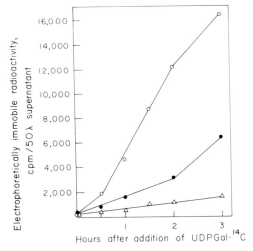

Fig. 14 UDP-galactose: N-acetylglucosamine galacto-syl transferase activity in cell suspensions and in supernatant cell free fluids. Incubations contained 8 day chick embryonic neural retina cells 4 x 10^8 cells per ml or cell free supernatant, Hanks' medium, UDP-C^{14}-galac-tose (25 μM) 10^8 cpm per μmole and N-acetyl-D-glucosamine (5 mM). Incorporation was determined by electrophoresing the reaction mixtures in 1% sodium borate and counting the product region. (●) cell suspension incubated immedi-ately following trypsin dissociation; (o) cell suspension preincubated 2 hours in Hanks' medium before incubation with components of incorporation experiment; (△) cell free supernatant of 2 hour preincubation incubated to check for cell breakage.

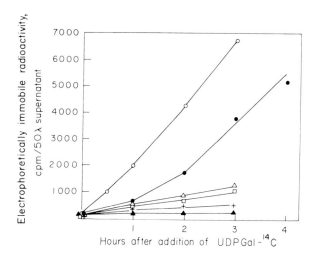

Fig. 15 UDP-galactose:mucin galactosyltransferase activity in cell suspensions and in supernatant fluids. Incubation mixtures were prepared and assayed as described above (Fig. 14) with 25 μM UDP-galactose (10^8 cpm per μmole), and 25 mgs/ml of sialidase-treated ovine submaxillary mucin, and where present, intact cells at 1.3×10^8 cells per ml. The enzyme sources for each incubation were the following: (o) whole cell suspensions which had been preincubated in Hanks' medium at 37° for 3 hr; (•) whole cell suspensions incubated directly with UDP-galactose and mucin; (△,□, +) supernatants obtained by centrifugation at 200 g for 6 min of cells preincubated for 4, 2, and 1 hr, respectively, in Medium H; (▲) supernatant obtained after 2 hr preincubation above, and incubated with UDP-galactose-^{14}C but without mucins.

EVIDENCE FOR SURFACE GALACTOSYLTRANSFERASES
(CHICKEN EMBRYONIC NEURAL RETINA CELLS)

1. INCUBATION OF WHOLE CELLS WITH UDP-GAL*
   a. ENDOGENOUS PRODUCTS ARE PARTICULATE GLYCOPROTEINS
      AND CONTAIN ONLY ONE LABELED SUGAR, GALACTOSE.

   b. EXOGENOUS ACCEPTORS ARE ACTIVE (GLYCOPROTEINS,
      MUCINS, MONOSACCHARIDES), AND SHOW EXPECTED
      SPECIFICITIES.

   c. NO COMPETITION OF INCORPORATION BY UDP-GLUCOSE,
      GAL-1-P, GALACTOSE.

   d. CELL SUPERNATANT INACTIVE.

   e. OVER 90% OF CELLS ARE LABELED AND RADIOACTIVITY
      APPARENTLY LOCATED ON CELL SURFACES
      (RADIOAUTOGRAPHY).

2. SOME EXOGENOUS ACCEPTORS INHIBIT SPECIFIC
   ADHESION WHEREAS STRUCTURAL ANALOGS DO NOT.

Fig. 16 A summary of the evidence for cell surface galactosyltransferases.

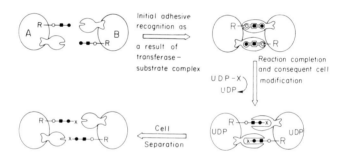

Fig. 17 A possible role for cell surface glycosyl-transferases in cell-cell recognition and cell contact modification (o-■-●) complex carbohydrates of cell surface; (⊃) cell surface glycosyl transferase.

VIII

CELL FUSION

FUSION OF CELLS BY HVJ(SENDAI VIRUS)

Yoshio Okada

Research Institute for Microbial Diseases,
Osaka University, Osaka, Japan

ABSTRACT. Our studies of cell fusion by native HVJ virions
indicate that 1) a kind of damage is produced in cell membranes by a function of the virus, comparable with its hemolytic activity; 2) a function or functions for repairing
the damaged structure are induced in the cells, and the fun-
ction(s) proceed utilizing ATP as a Ca-dependent action; 3)
fusion of cells and repair them proceed in parallel. The
degree of fusion depends on a balance between the extent of
cellular damage and the ability of the cells to repair it.
If the extent of cellular damage exceeds the repair potential, initial fusion products may be formed but do not proceed to stable syncytia.

INTRODUCTION

Fifteen years has elapsed since the first report of
cell fusion by HVJ was published(Okada,et al.,1957).In recent few years, this phenomenon has been utilized as a standard method for the formation of heterokaryons(Harris & Watkins,1965; Okada,1961; Okada & Murayama,1965) and for somatic hybrid formation(Yerganian & Nell,1966). During the initial period of development of somatic cell genetics by
Barski et al.(1961) and the subsequent extension of this
work by Ephrussi et al., hybrids were formed without the
aid of exogenously added virus. The early work has shown
that 1)cells in culture naturally maintain a potentiality
for fusion, 2)fusion between cells of different species can
occur in culture and added virus is not necessary for the
reaction.
The effectiveness of using the virus is mainly due to
the surprisingly higher efficiency of fusion achieved

and due to the minimal influence of the participating virus
on the fused cells. As a result, the efficiency of hybrid
formation increased about 1000-fold when fusion was promoted
by virus(Coon & Weiss,1969; Murayama & Okada,1970). The use
of virus has extended the range of cell types which can be
used for hybrid formation. Of course, cells less sensitive
to the fusion promoting activity of virus exist. Even in
these cases, they could form heterokaryons with highly sensitive cells as indicated by Harris(1965). This finding was
followed by the developement of a new technique to insert
macro-materials artificially into cells, such as virus par-
ticles(Enders et al.1967; Hanafusa et al.1970), spermatozoa
(Sawicki & Koprowski,1971) and chromosomes.

The mechanism of cell fusion by the virus has not been
clear. However, it may be considered that the phenomenon is
valuable for understanding animal cell membranes. It may
include several kinds of reactions; such as of the virus
envelope with polysaccharide chains as the receptor, fusion
of virus envelope with cell membranes, fusion of two cell
membranes, and rearrangement of the fused cell membranes.

METHODS

Viruses. HVJ propagated in embryonated eggs or in primary chick embryo fibroblasts was washed and suspended in
BSS (Okada & Murayama,1966). The former was used for cell
fusion and the latter(HVJ$_{CF}$) was a control showing no fusion.

Cells. Ehrlich ascites tumor cells(ETC), KB and FL
cells were used in suspension form(Okada & Murayama, 1966).
For the electrical work, FL cells were suspended in a growth medium (MEM + 10% calf serum + Hepes 20 mM).

Cell fusion. For cell fusion BSS containing 1 mM CaCl$_2$
was used as the medium. The degree of cell fusion was indicated as Fusion Index(FI).

$$FI = \frac{\text{Cell number in control sample without virus}}{\text{Cell number in test sample}} - 1.0$$

Electrical measurement. Glass microelectrodes filled
with 3 M KCl were used, which had resistance of 10 to 35 M
ohm. FL cells in suspension form were introduced in a small
chamber and the chamber was placed on a microscope stage
with temperature control. The cells were inserted with ele-
ctrodes, while being visualized microscopically.

RESULTS

Observation of the cell fusion reaction by HVJ.
When Ehrlich ascites tumor cells(ETC) were exposed to
HVJ in ice cold, the cells promptly agglutinated together.
The primary reaction is adsorption of the virus to the receptors on cell surfaces. The cell aggregates grew largely
with the concentration of the virus and if the cells were
pretreated with receptor destrying enzyme no aggregation
appeared.
When the cell-virus mixture was incubated at $37^{\circ}C$, fusion of cells in the aggregates occurred within a few minutes. What kinds of reactions proceeded at $37^{\circ}C$? Energy supply and the presence of Ca ions are essential for cell fusion reaction at $37^{\circ}C$ (Okada $1962_{a,b}$; Okada et al.1966; Okada
& Murayama,1966). If the conditions were not satisfied, the
cell aggregates disaggregated and the redispersed cells
were lysed.
I. Energy supply and cell fusion. When the air in the
reaction tube was replaced by N_2 gas, cell fusion was inhibited. The addition of 2,4-dinitrophenol also inhibited the
reaction. If the cell-virus mixture was suspended in a medium containing glucose, a strong drop of pH of the medium
was seen during the course of cell fusion. During these experiments we observed an interesting phenomenon. When ETC
harvested from a mouse was preincubated aerobically at $37^{\circ}C$,
the efficiency of fusion of the cells increased sharply.Pre-
incubation under anaerobic conditions decreased the fusion
capacity of the cells. This preincubation effect was great
but completely reversible. The change in the capacity seems
to be correlated with the level of ATP in the cells but not
to the synthesis of proteins. For the cell fusion reaction,
the endogenous ATP might be necessary but the ATP added into medium was no effect.
II. Ca ions and cell fusion. If Ca ions were minimum,
cell aggregation with HVJ at low temperature was not affected but fusion of cells at $37^{\circ}C$ was inhibited and cells were
lysed even when cells keeping a high level of ATP were used
(Fig. 1). The amount of Ca ions required for the reaction is
a function of the virus concentration used. In Fig. 2, the
correlation is shown. Various concentrations of HVJ were
added to a fixed concentration of ETC in medium with or
without Ca ions. After the completion of the cell fusion
reaction for 60 min at $37^{\circ}C$, the total number of cells and

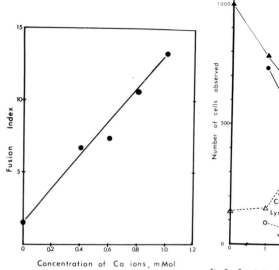

Fig. 1. Requirement of Ca ions for ETC fusion

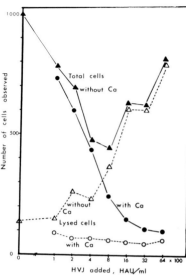

Fig. 2. Correlation between fusion of ETC and lysis of ETC by HVJ under the conditions with or without Ca ions

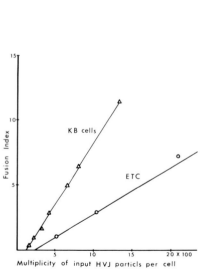

Fig. 3. Dose response curves of cell fusion to HV

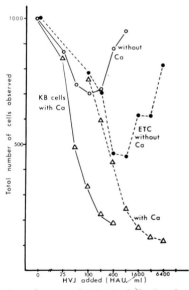

Fig. 4. Comparison of sensitivities of KB cells and ETC to HVJ; fusion with and lysis without Ca ions

374

the number of lysed clls were counted. In all samples, no
aggregation of cells appeared after the completion of the
reaction. In samples with lower concentrations of the virus,
the total number of cells decreased as the concentration of
HVJ increased, even in medium without Ca. In these samples,
almost all cells were intact and small fused cells appeared
in both media. In the medium without Ca ions, further increase in the concentration of virus increased the total number of cells(over 800 HAU in Fig. 2). In these samples almost all cells were lysed. On the other hand, in the medium containing Ca ions, the total number of cells decreased
with increasing virus concentration, and no degenerating
cells appeared. A possible explanation of this result is
that i) cell membranes are damaged by HVJ at $37^{\circ}C$, ii) the
damaged sites could be repaired in the presence of Ca ions,
iii) when the number of the damaged sites on each cell membrane is high, there is a better chance for fusion of cell
membrane with those of neighboring cells in medium containing Ca ions.

III. Cell fusion and cell lysis. In Fig. 3, dose response curves of fusion of KB cells and ETC are shown. A
straight line is given in each case. The two lines are different in the slope and in the multiplicity of the virus per
cell required for minimum fusion. Both cells are readily
fused by HVJ, but the sensitivity of cell membrane to cell
fusion reaction induced by one virus particle is clearly
higher in KB cells. This pattern was reflected in the higher
sensitivity of KB cells to lysis by HVJ in medium without Ca
ions, as shown in Fig. 4.

Several kinds of cells showing minimal fusion capacity
exist. Of these, there are two types. One is the cells showing less sensitivity to lysis in medium without Ca ions,
such as neural retina and chondrocytes dissociated from
chick embryos, and lymphocytes. The other is those cells
which are sensitive to lysis, such as chicken red blood
cells. In this case, the initial effects of virus may take
place but the cells may be deficient in their ability to
carry out the subsequent repair reactions.

Presence of a state of cell membranes loosened by HVJ and
the reversion during the course of cell fusion.

From the above findings, a loosened state of the cell
membranes must be detected at the early stage of the cell
fusion reaction under conditions with Ca ions. For this pur-

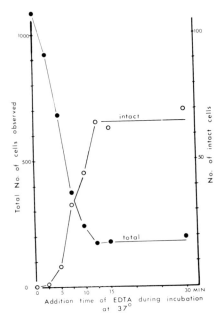

Fig. 5. Kinetics of fusion
reaction of ETC by HVJ.
ETC and HVJ were mixed in
BSS containing Ca ions.
After keeping the mixture
at $4°C$ for 10 min, they
were incubated at $37°C$ with
shaking. During the course
of the incubation, EDTA was
added to the mixture at intervals of 2.5 min and then
they were incubated for a
further 60 min at $37°C$.

Table 1. Changes in Potential and Membrane resistance of
FL cells during the course of fusion by HVJ at
$37°C$.

Treatment of FL cells	Potential (-mV) determined at		Membrane Resistance (ohm.cm^2) determined at	
	$10°$	$35°$	$10°$	$35°$
non-treated	18.2 ± 2.1	22.0 ± 6.5	491 ± 149	437 ± 71
with HVJ, $37°$ 0 min	6.1 ± 2.5	26.0 ± 8.7	162 ± 84	
with HVJ, $37°$ 5 min	3.5 ± 1.6	32.2 ± 7.0	92 ± 46	513 ± 192
with HVJ, $37°$10 min	10.5 ± 6.7	30.0 ± 9.1		
with HVJ, $37°$15 min	20.5 ± 7.3	34.8 ± 5.3		
with HVJ, $37°$30 min	23.2 ± 6.3	29.3 ± 7.4		

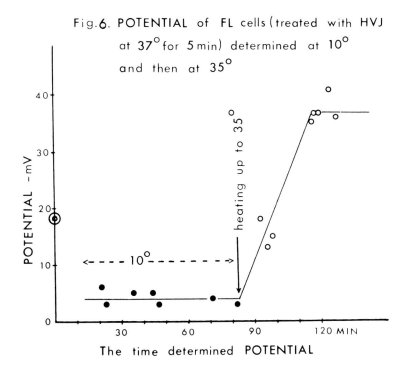

Fig.6. POTENTIAL of FL cells (treated with HVJ at 37° for 5 min) determined at 10° and then at 35°

The time determined POTENTIAL

Table 2. Comparison of HVJ_{egg} and HVJ_{CF} in the effects to Potential of FL cells.

Treatment of FL cells	Potential (-mV)
non-treated	22.1 ± 10.4
with HVJ_{CF}, 37° 0 min	20.9 ± 5.9
with HJV_{CF}, 37° 5 min	23.7 ± 4.8
with HVJ_{egg}, $37^\circ C$ 5 min	7.9 ± 3.9

377

pose, electrical measurements on the cells were done. For the experiments, FL cells were chosen. They showed a high fusion capacity (Okada & Tadokoro, 1963), and insertion of microelectrodes into FL cells was rather easier than into ETC. In Fig. 5, kinetics of ETC fusion is indicated. Based on the kinetic curves, the experiment was planned. FL cells and HVJ(1500 HAU) were mixed in the presence of Ca ions and kept in ice cold for 5 min. The mixture was incubated at 37° for 0, 5, 10, 15 and 30 min, respectively. They were chilled in an ice bath for 5 to 10 min to stop the reaction, then introduced into a chamber for electrical measurement. Firstly, the measurement was done at low temperature (10° to 15°). After the completion (about 60 to 80 min passed), the temperature of the microscope stage was heated up to $35^{o}C$, the resting potential of the cells in the same chamber were measured again. At the low temperature, the cell fusion reaction did not proceed microscopically, but proceeded at the high temperature. As shown in Table 1, the potential across membrane of FL cells determined at the low temperature was reduced greatly from native one (-18.2 mV as mean value) when the cells were treated at $37^{o}C$ for 5 min (-3.5 mV).The potential, then increased with increasing the time of incubation at $37^{o}C$ with HVJ; -10.5 mV in 10 min incubation, -20.5 mV in 15 min incubation and -23.2 mV in 30 min incubation. Membrane resistance also decreased by 5 min incubation.

In the sample of 5 min incubation, fusion was still unclear and the contours of the individual cells in aggregates were discernible. During the measurements at the low temperature, the fusion reaction did not proceed further and the low potential was maintained, but when heated to $35^{o}C$ the potential increased rappidly and reached a level higher than that of non-treated FL cells, as shown in Fig. 6. Membrane resistance was also reversed (Table 1). This indicates that the FL cell membrane is loosened by HVJ and then reverses during incubation at $37^{o}C$.

When HVJ_{CF} was used instead of HVJ, the loosened state was not observed as shown in Table 2. HVJ_{CF} grown in chick embryo fibroblasts has no hemolytic nor cell fusion activities.

DISCUSSION

When viewing viruses, the viruses showing syncytium

formation are distributed in the group of enveloped viruses. Recently, evidence has been reported showing that the entry of enveloped viruses into cells is preceeded by fusion of their envelope with the cell membrane. Now, it must be considered whether the cell fusion phenomenon correlates directly with this event or not.

The viruses forming syncytium are all enveloped viruses but not all enveloped viruses show the syncytium forming activity. The envelope-cell membrane fusion has been observed in Influenza virus(Morgan & Rose,1968), Sendai virus(Morgan & Howe,1968), Herpes simplex virus(Morgan et al.1968), and Vesicular stomatitis virus(Heine & Schnaitman,1971) infection. In those viruses, Herpes virus and Sendai virus have cell fusion activity but the other two have not.

HVJ and Influenza virus, both belonging to Myxoviruses, have hemagglutinating (HA) and neuraminidase activities. However, HVJ has both hemolytic(HL) activity and fusion activity, while Influenza virus is deficient in both of these activities. Recently, Hosaka and Shimizu(1972) of our laboratory have observed an interesting phenomenon. They solubilized HVJ envelopes and separated the protein and the lipid components by centrifugation in CsCl gradients. The protein fraction showed HA and neuraminidase activities but not HL activity. However, the envelope reconstituted from the mixture of the solubilized protein fraction and the lipid fraction showed HL activity as well as the former activities. The viral protein fraction was essential for the reactivation of HL activity but the lipid fraction from Influenza virus was also effective as well as from HVJ. They determined that phosphatidyl ethanolamine from a non-viral source was equally effective, phosphatidyl choline and sphingomyelin were only effective at high concentrations and phosphatidyle serine was ineffective. The biological character of the HVJ envelope is clearly different from the character of that of Influenza virus.

A peculiarity of host controled variation of HVJ has attracted the attention of some Japanese virologists, since the virus was first isolated(Ishida & Homma,1960; Mastumoto & Maeno,1962). The standard host of the virus is embryonated eggs, and the virus propagated in entodermal cells of chorioallantoic membrane is used as the standard virus. A kind of variant was isolated by one step growth of HVJ in L cells or mouse lung cells (HVJ_L and HVJ_M). After a single cycle of growth in eggs, they returned to the standard HVJ. HVJ_L was

not infectious for L cells but infectious for eggs. Furthermore, HVJ_L lost HL activity. HVJ_M was similar to HVJ_L. From
subsequent observations, it has been known that the HVJ_L
type viruses keep HA and neuraminidase activities but loss
HL and cell fusion activities. Recently, Homma(1971) reported that the infectivity for L cells and HL activity were
regained by the treatment of HVJ_L with trypsin.

Holmes and Choppin(1966) reported the presence of cont-
rasting effect of parainfluenza SV5 in two cell types. After
the infection with SV5, newly propagated viruses are released in great quantity from primary cultured monkey cells(MK
cells), but the release was apparently less from BHK21 cells
an established line from hamster kidney. They suggested that
the difference was due to the difference in the efficiency
of budding process, rather than the difference in growth of
viral antigens in cells.Furthermore, a contrast in syncytium
formation was reported. The formation was significant in BHK
cells but rare in MK cells.

Recently, we found that an established line of cells
from Potoroo kidney (Potoroo cells) could produce and release virus as well as embryonated eggs(unpublished). The
virus(HVJ_P) had the same characters of the standard HVJ.
HVJ_{CF} from chick embryo fibroblasts showed the character of
HVJ_L and the virus yield from the cells was small. Syncytium
formation of Potoroo cells was rare in both early and late
stages of infection.

As described above, some interesting findings about
parainfluenza viruses have been uncovered, and HL activity
of the native virions seems to correlate closely with cell
fusion. The inactivation rate of each biological activity of
HVJ by UV-ray also showed that there is a close correlation
of cell fusion activity with HL activity, but not with infectivity, HA or neuraminidase activities(Okada & Tadokoro,19
62). In spite of the UV results, the HL activity can be dissociable from fusion activity; when HVJ is sonicated, HL
activity increases but cell fusion activity decreases (Okada
& Murayama,1968; Okada,1969). This problem has not yet been
dissolved.

Our studies of cell fusion by native HVJ virions indicate that 1) a kind of damage is produced in cell membranes
by a function of the virus, comparable with its hemolytic
activity; 2) a function or functions for repairing the damaged structure are induced in the cells, and the function(s)
proceed utilizing ATP as a Ca-dependent action; 3) fusion of
cells and repair them proceed in parallel. The degree of

fusion depends on a balance between the extent of cellular damage and the ability of the cells to repair it. If the extent of cellular damage exceeds the repair potential,initial fusion products may be formed but do not proceed to stable syncytia. Howe and Morgan(1969) showed the fusion of Sendai virus envelope with the human red blood cell membrane The finding may correspond to the irreversible binding of NDV with human RBC reported by Burnet & Anderson(1946). They cited the correlation of the binding with hemolysis. Ca ions are essential for cell fusion and an inhibitor of hemolysis by parainfluenza viruses. The step of virus envelope- cell membrane fusion may be included in the course of the cell-cell fusion, but it is possible that Ca ions are not necessary for the step of envelope-cell membrane fusion.

Results of the electrical measurement show the presence of a loosened state of cell membranes in the course of cell fusion. It has not yet been clear whether this correlates with the step of rearrangement of membranes fused together as indicated by Frye & Edidin(1970).

ACKNOWLEDGMENTS

The author is greatly indebted to Dr. Y.Kanno and Mr. Y.Mastui who kindly undertake the electrical measurement.

REFERENCES

Barski,G.,Sorieul,S.,& Cornfert,Fr., J.Nat.Cancer Inst., 26, 1269 (1961).
Burnet,F.M. & Anderson,S.G., Brit.J.exptl. Path., 27, 236 (1946).
Coon,H.G. & Weiss,M.C., Proc.Nat.Acad.Sci., 62, 852 (1969).
Enders,J.F., Proc.Nat.Acad.Sci., 57, 637 (1967).
Frye,L.D. & Edidin,M., J.Cell Sci., 7, 319 (1970).
Hanafusa,H., Miyamoto,T. & Hanafusa,T., Proc.Nat.Acad.Sci., 66, 314 (1970).
Harris,H. & Watkins,J.F., Nature, 205, 640 (1965).
Harris,H., Nature, 206, 583 (1965).
Heine,J.W. & Schnaitman,C.A., J.Virol., 8, 786 (1971).
Holmes,K.V. & Choppin,P.W., J.exptl.Med., 124, 501 (1966).
Hosaka,Y. & Shimizu,Y., Virology, in press
Howe,C. & Morgan,C., J.Virol., 3, 70 (1969).
Ishida,N. & Homma,M., Tohoku J.exptl.Med., 73, 56 (1960).
Mastumoto,T. & Maeno,K., Virology, 17, 563 (1962).

Morgan,C.,Rose,H.M. & Mednis,B.,J.Virol., 2, 507 (1968).
Morgan,C. & Rose,H.M., J.Virol., 2, 925 (1968).
Morgan,C. & Howe,C., J.Virol., 2, 1122 (1968).
Murayama,F. & Okada,Y., Biken J., 13, 1 (1970).
Okada,Y., Suzuki,T. & Hosaka,Y., Med.J.Osaka Univ., 7, 709 (1957).
Okada,Y., Biken J., 4, 145 (1961).
Okada,Y., Exptl. Cell Res., 26, 98 (1962).
Okada,Y., Exptl. Cell Res., 26, 119 (1962).
Okada,Y. & Tadokoro,J., Exptl. Cell Res., 26, 108 (1962).
Okada,Y. & Tadokoro,J., Exptl. Cell Res., 32, 417 (1963).
Okada,Y. & Murayama,F., Biken J., 8, 7 (1965).
Okada,Y., Murayama,F. & Yamada,Y., Virology, 27, 115 (1966).
Okada,Y. & Murayama,F., Exptl. Cell Res., 44, 527 (1966).
Okada,Y. & Murayama,F., Exptl. Cell Res., 52, 34 (1968).
Okada,Y., Current Topics in Microbiol. Immunol., 48, 102 (1969).
Sawicki,W. & Koprowski,H.,Exptl. Cell Res., 66, 145 (1971).
Yerganian,G. & Nell,M.B., Proc.Nat.Acad.Sci., 55, 1066 (1966).

BIOLOGICAL PARAMETERS OF FUSION FROM WITHIN
AND FUSION FROM WITHOUT

Michael A. Bratt and William R. Gallaher[*]

Department of Microbiology and Molecular Genetics
Harvard Medical School
Boston, Massachusetts   02115

ABSTRACT. Fusion from without (FFWO) and fusion from within (FFWI) are distinct, though not necessarily mechanis-
tically different, phenomena. The former is a direct response of cells to the virus particles in the infecting inoculum and the latter results only from alterations in
membrane biogenesis associated with productive infection.
Newcastle disease virus strains differ in their ability to
induce these two types of fusion, and for some strains the
two types are mutually exclusive. While no explanation is
available for this mutual exclusion there is suggestive
evidence that the ability of a virus to induce FFWO is
associated with a higher affinity for, or tightness of
binding to, cell membranes. The strain specificity of FFWO
is identical in chick embryo, BHK-21 and MDBK cells. The
strain specificity of FFWI is identical in chick embryo
and BHK-21 cells, but NDV appears to be unable to induce
FFWI in MDBK cells. Inability to induce FFWI in MDBK cells
is the result of their failure to undergo the necessary
changes in membrane biogenesis even though they are productively infected.

INTRODUCTION

Recent years have seen a burst of interest in cell-cell
interactions, and in the cellular membranes which mediate
these all important interactions. Our laboratory has been
interested in what must certainly be the most intimate of
all cell-cell interactions--namely, the cell fusion reaction, which involves the fusion of the plasma membranes of

---

[*]Present address: Division of Biochemical Virology,
Scheie Eye Institute, Philadelphia, Pa.   19104

two or more cells. Fusion of biological membranes is actually a widespread phenomenon which plays an important
role in many processes essential for the normal functioning
of animal cells. These include not only the systematic
intercellular fusion of myogenic cells during myogenesis,
but also the extremely commonplace intracellular fusion of
membranes during cell division and pinocytosis, during the
coalescence of vacuoles with each other or with lysozymes,
and in secretory processes. Cell fusion also occurs as
either a cytotoxic or cytopathic response to infection by a
number of common viruses including members of the paramyxovirus group with which we have concerned ourselves. Al -
though it represents a question of considerable interest,
our knowledge of these fusion reactions has not yet allowed
us to determine whether the molecular events involved in
normal cell membrane fusion and in virus-induced cell
fusion are similar.

Paramyxovirus-induced fusion was first observed in
humans infected with measles virus by Finkeldey (1) and
also by Warthin (2) in 1931. Subsequently it was found
in measles virus (3) and mumps virus (4) infection of
tissue cultures. Interest in paramyxovirus-induced cell
fusion took its greatest leap forward in 1958 when Okada
(5) demonstrated that infection by very high multiplicities
of Sendai virus can result in rapid cell fusion. This discovery took on added significance with the demonstration
that this rapid cell fusion can be induced by virus rendered non-infectious by exposure to ultraviolet irradiation.
This fact and the finding that cells fused by inactivated
Sendai virus can remain viable, has subsequently been ex-
ploited in numerous studies where heterologous nuclei and
cytoplasms have been placed within the same plasma membrane either in the form of heterokaryons or cell hybrids.

Our interest in paramyxovirus-induced cell fusion lies
not in its important use as a biological tool, but rather
in the mechanism or mechanisms of cell fusion, and the
possible relationship of this reaction to viral pathogene-
sis. To date, our achievements in this area consist of an
analysis of the biological parameters which affect cellular
fusion induced by Newcastle disease virus (NDV). We hope
that the information we have gained, and our current studies which take advantage of these findings, will ultimately lead us to an understanding of the molecular basis of

fusion.

Several years ago, on the basis of findings with NDV we suggested (6,7) that a clear-cut distinction should be made between the type of rapid cell fusion which Okada (5) had shown is induced soon after Sendai virus infection at high multiplicities, and the fusion which occurs late in productive infection. We created the term fusion from without (FFWO) to describe the early fusion which is a direct result of the interaction between cell membranes and the virus particles in the infecting inoculum. We borrowed the term fusion from within (FFWI) from Kohn (8) to describe the late fusion which results from productive infection, and appears to be cell-mediated rather than virus particle-mediated. These concepts of FFWO and FFWI are directly analogous to those of "lysis from without" and "lysis from within" which Doermann (9) used to distinguish between the rapid lysis of E. coli by high multiplicities of a T-even bacteriophage (lysis from without) and the lysis which occurs later in productive infection as a result of the accumulation of viral lysozyme (lysis from within). The intention of our terminology was not to suggest that different mechanisms are involved in these two types of fusion--a possibility which has neither been proven nor ruled out--but rather to emphasize that one type (FFWI) involves viral replication and the processes of membrane biogenesis, while the other (FFWO) does not.

In what follows we will first summarize the distinctions between FFWO and FFWI. Then using the chick embryo cell system as a model we will elaborate on the strain specificity of, and the mutual exclusion which in certain cases exists between, these two types of fusion. Next we will present evidence which suggests that the ability of a virus to induce FFWO may be a function of its affinity for cell membranes. Then we will discuss the general applicability of our conclusions to fusion of other cell types. Finally we will suggest that in at least one case, that of NDV infected MDBK cells, failure to undergo FFWI is not due to inate infusibility, but rather to failure to undergo those alterations in membrane biogenesis which are required in order for FFWI to occur.

METHODS

The methods we have used are essentially those we have described before. Purified stocks (10) were prepared from allantoic fluid-grown virus (11) derived from cloned seed stocks (6) of 14 naturally occuring strains of NDV. These strains include: NDV-AV (Australia-Victoria-1932); NDV-B1 (B1-Hitchner-1948); NDV-EH (England-Herts-1933); NDV-F (N.J.-Roakin-1946); NDV-GB (Texas-GB-1948); NDV-HP (Israel-HP-1953); NDV-I (Iowa-125-1947); NDV-IM (Italy-Milano-1945); NDV-IS (Iowa-Salisbury-1949); NDV-L (L-Kansas-1948); NDV-M (Mass.-MK-1945); NDV-N (N.J.-LaSota-1946); NDV-RO (Calif.-RO-1944); NDV-W (Wis.-Appleton-1950).

Virus infectivity titrations in secondary cultures of chick embryo cells (6,11), and hemagglutination titrations (10), as well as neuraminidase treatments and antibody neutralization procedures (12) and the hemadsorption assay (13,14,15) have been described elsewhere.

Fusion experiments were carried out in 60 mm monolayer cultures of secondary chick embryo cells, BHK-21* cells, or MDBK* cells, in supplemented Eagle's Minimal Essential medium. At various times after appropriate infection, the cultures were fixed with 95% methanol, and stained with Giemsa stain. Fusion was then quantitated by calculating a "Fusion Index" (6,7,8)--which is actually the average number of nuclei per cell--and then subtracting the fusion index of uninfected cells from the fusion index of the test cultures in order to obtain a measure of the FUSION EVENTS PER CELL (14,15).

RESULTS AND DISCUSSION

Table 1 summarizes the major distinctions between FFWO and FFWI. Viral infectivity is required only for FFWI. FFWO occurs prior to the end of the viral latent period, while FFWI is detected well after viral replication has begun. FFWO requires high multiplicities, while FFWI is

---

*The BHK-21F line of baby hamster kidney cells and MDBK (Madin Darby Bovine Kidney) cells were generously provided by Dr. Purnell W. Choppin of the Rockefeller University.

induced optimally at low multiplicities of infection. FFWO shows only a slight pH dependence in the range of pH 7.0-8.4 while FFWI shows a sharp optimum at pH 8.1-8.2. There is no requirement for DNA-dependent RNA synthesis for either type of fusion, and protein synthesis is only required for FFWI. Antiviral antibody is only capable of preventing FFWO if added very soon after viral attachment, but is capable of almost completely suppressing FFWI if added as late as 5 hours after infection. As will be seen below 5 out of the 14 strains tested are good inducers of FFWO and poor inducers of FFWI, while 3 strains are good inducers of FFWI and never induce any FFWO. Finally, we will show that while the strain specificities for FFWO are identical in chick embryo, BHK-21, and MDBK cells, for FFWI the strain specificities are similar only in chick embryo and BHK-21 F cells; little or no FFWI is induced in MDBK cells.

Figure 1 shows the relative ability of chick embryo cells treated with 10,000 HAU/ml of 14 different NDV strains to undergo FFWO. The strains have been arranged in decreasing order with the best inducer being NDV-N on the left. Strains NDV-L, NDV-IS, NDV-F, and NDV-AV have never been observed to induce any FFWO in these cells.

In Table 2, which contains a comparison of the strain specificity of FFWO and FFWI in chick embryo cells, we have divided the strains into 4 groups. Group I contains strains which induce little or no FFWO, but considerable amounts of FFWI. The strains in group II induce the greatest amounts of FFWO, and also induce FFWI, but to a lesser extent than those in group I. Group III contains strains which induce little of either type of fusion. Group IV is made up of 4 strains which are relatively avirulent in culture as well as in vivo. The strains in this group induce little or no FFWI, and with the exception of NDV-N which is our best inducer of FFWO, they induce little FFWO. Several important points are contained in this table. First, the best inducers of FFWI (Group I) do not induce FFWO. This mutual exclusion does not hold for all strains, however, since some of the best inducers of FFWO (Group II) also induce some FFWI. Other strains induce little or no fusion of any kind. In addition, as suggested by Brandt (16) and recently by   Reeve and Poste (17) ability to

387

induce FFWI is limited to virulent strains (Groups I and II). However, the inverse of this correlation is not true, since ability to induce FFWI is not associated with all virulent strains (Group III). Also contained in this table is the information that, for all but the avirulent strains of Group IV, infection results in those virus-specific alterations which are necessary to allow hemad-sorption. For this reason it is possible to conclude, at least for the strains in the first 3 groups, that failure to induce FFWI is not due to inability to induce productive infection nor to inability to cause virus-specific changes in membrane biogenesis, per se.

While the mutual exclusion is not absolute for all strains, where it does exist it poses a puzzling problem. We are faced with viruses which possess the genetic information to induce fusion but which will do so only under very different conditions. As we have pointed out, FFWO is a direct response to the virus particles in the in-oculum, while FFWI is a result of virus-specific altera-tions of the cell membrane. We know that the virion which induces FFWO and the altered membrane responsible for FFWI both consist of virus-specific proteins, as well as lipids and carbohydrates selected from the host cell. We don't yet know whether one or a combination of these con-stitute a specific "fusion factor" which causes fusion in a particular instance. It is even more difficult to assign responsibility for failure to induce fusion in a specific case, for such failure may be due either to the absence of a hypothetical fusion factor or to the pre-sence of a factor which acts as an antagonist of fusion.

Although we cannot yet account for the mutual exclu-sion, where it exists, nor describe in molecular terms a mechanism or mechanisms for fusion, we can present sug-gestive evidence to account for FFWO's occurence or failure to occur in a particular instance. This evidence suggests that the ability to induce FFWO may be determined by the affinity of the virus particle for, or its tightness of binding to, the cell membrane. The first piece of evidence concerns the fate of virus particles which have attached to the cell. It has been suggested that the neuraminidase contained in the virus particle, through its promotion of viral elution, might affect the ability of

the virus particle to cause FFWO. We therefore examined
the elution potential of some of our NDV strains. Figure
2 shows an example of such an experiment. Here we have
allowed two concentrations of NDV-IM and NDV-L to adsorb
to monolayer cultures of chick embryo cells at 4°. We then
washed away unadsorbed virus, incubated the cultures at 4,
23, and 38°, and at various times assayed the medium for
eluted infectious virus. The values plotted represent the
ratio of virus in the medium at a given time, to that initially in the medium at 4° at 0 time. At the lower input
concentration NDV-IM barely elutes. At the high input
concentration elution occurs in a temperature dependent
manner. (A similar temperature and concentration dependence has also been reported for the elution of NDV from
erythrocytes to which it has attached (18,19).) The elution of NDV-L which is also concentration and temperature
dependent is much more rapid and dramatic than that of
NDV-IM. The important point here is that NDV-IM as well
as NDV-EH and NDV-HP (not shown), which are good inducers
of FFWO, tend to elute slowly and form more stable complexes with cells. In contrast, NDV-L as well as NDV-IS,
NDV-RO, and NDV-F (which are not shown), which are unable
to induce FFWO, have a high potential for elution. There
is, therefore, the possibility that the virus's elution
potential may act as an antagonist of any tendency to induce FFWO.

A second piece of evidence comes from studies on the
susceptibility of neuraminidase treated cells to infection
by these same strains. It has been known for many years
that treatment of tissue culture cells with neuraminidase
reduces their susceptibility to NDV attachment or infection (12,20,21). We tested the ability of several NDV
strains to attach to several cell types including the chick
embryo, MDBK and BHK-21 cells used in our present study.
In each case, we found that neuraminidase treatment had
little effect on subsequent attachment of strains NDV-IM,
NDV-EH, and NDV-HP. Such treatment radically inhibited
attachment by NDV-IS and NDV-RO and to a somewhat lesser
extent NDV-L and NDV-AV. Thus, strains which induce FFWO
appear to have less stringent requirements for intact
neuraminidase-sensitive receptors, than do strains which do
not induce FFWO. However, the low requirement for intact
receptors shared by the strains which can induce FFWO is

limited to the processes of attachment and infection of these cells, because as Kohn (8) has previously shown, neuraminidase treatment of cells completely reduces their susceptibility to FFWO by these viruses. This discrepancy is probably a reflection of FFWO's requirement for simultaneous interaction of cell surfaces with large numbers of virus particles (22,23,24).

A third piece of evidence is to be found in the analysis of neutralization kinetics by antiviral antibody. An example of this type of experiment is shown in figure 3. Strains NDV-IM, NDV-EH, and NDV-HP are much less susceptible to neutralizationthan the four other strains tested. These differences in neutralization kinetics by a single antiserum do not appear to be an indication of antigenic differences between the strains, because similar patterns with similar relative rates of inactivation are obtained no matter what strain the antibody was prepared against. Rather, these differences in neutralization kinetics appear to reflect differences in the affinity of the virus and antibody, or of the way in which the virus-antibody complexes interact with the cell. Other experiments show that the strains which are capable of inducing FFWO also bind less antibody than do strains which do not induce FFWO.

A final similarity among strains which induce FFWO is their high hemolytic activity measured during interaction with erythrocytes. In Figure 4 the strains have been placed in the same order of decreasing ability to induce FFWO as used in figure 1. It can be seen that, with the possible exception of NDV-F, poor inducers of FFWO are relatively poorly hemolytic, while good inducers of FFWO are highly hemolytic. Evidence presented elsewhere (19, 25) suggests that high hemolytic activity is also associated with greater affinity for, or tightness of binding to, erythrocyte membranes.

In summary, the best inducers of FFWO: 1) have a low elution potential; 2) have low requirements for intact neuraminidase sensitive receptors; 3) have a low affinity for neutralizing antibody. Furthermore, the best inducers of FFWO share a high potential for hemolyzing erythrocytes.

These correlations suggest the possibility that the determining factor in the ability to induce FFWO may be differences in the affinity or tightness of binding of the virus particles to the cell surface. A similar concept of differing affinities of virus particles for cells has been suggested by Choppin and Tamm (26) to explain patterns in virus-cell interactions among different populations of influenza virus particles (which are myxoviruses rather than paramyxoviruses, and as such do not induce fusion).

Now we want to shift to the question of the general applicability of our findings with chick embryo cells to NDV-induced fusion of other types of cells. Our original interest in the distinction between FFWO and FFWI was forced on us by the observations that for chick embryo cells different strains of NDV possessed different cell fusing abilities. The apparent mutual exclusion was surprising, since for three other paramyxoviruses--measles virus (27), mumps virus (4), and SV5 (28)--infection by the same strain resulted in considerable fusion both early and late in infection. For this reason, and because chick embryo cells, while derived from the natural host of this virus, have previously been reported to be poorly susceptible to fusion (8), we felt that it would be of value to compare this fusion with that in other cells. A natural choice for comparison were the BHK-21-F cells which Holmes and Choppin (28) had reported to be highly sensitive to both FFWO and FFWI by another paramyxovirus--SV5--and MDBK cells which were reported to be relatively insensitive to SV5-induced fusion by Klenk and Choppin (29).

Figure 5 shows the strain specificity of the induction of FFWO in BHK-21 and MDBK cells. These values were obtained in the same experiment as that shown in Figure 1, and the strains were placed in the same decreasing order of ability to induce FFWO in chick embryo cells. Panel A shows that the BHK-21 cells are slightly more sensitive to the induction of FFWO than are the chick embryo cells, but the strain specificity is essentially the same in the two cell types. Several strains including NDV-RO, NDV-B1, NDV-I, NDV-M, and NDV-W which are unable to induce detectable FFWO in chick embryo cells, induce low but significant levels of FFWO in BHK-21 cells. However, strains NDV-AV, NDV-F, NDV-IS, and NDV-L do not induce FFWO even in this more sensitive cell type.

Panel B shows that MDBK cells are as sensitive to NDV-induced fusion as chick embryo cells, and thus only slightly less sensitive than the BHK-21 cells. In these cells, too, the order of the strains is similar to that observed in chick embryo cells.

Figure 6 shows examples of the fusion induced by NDV-N in the experiments shown in Figures 1 and 5. Frames a, b, and c illustrate uninfected control cultures of chick embryo, BHK-21. and MDBK cells, respectively. Frames d, e, and f illustrate sister cultures in which the average cell has undergone approximately 1.0 fusion event. At this level of fusion 5-8 nuclei are not uncommon. These results make it clear that the strain specificity of NDV induced FFWO which we had observed in chick embryo cells is a general characteristic of NDV and not just a fortuitous result of the use of the chick embryo cells.

It was next of interest to determine the extent to which cell type is a determinant of the strain specificity of FFWI. In order for this study to be meaningful, however, it was necessary to determine whether the viruses multiplied equally well in these different cell types, since, as already mentioned, the induction of FFWI is associated with productive infection. Preliminary results indicated that both BHK-21 and MDBK cells are capable of supporting the synthesis of NDV-specific RNA. In addition these cells release infectious progeny virus at rates, and in final amounts, similar to those found for chick embryo cells.

Table 3 shows the results of an experiment in which cultures of BHK-21 cells infected by four strains of NDV were incubated for 11 hours at either pH 7.2 or 8.3. As in the case of the induction of FFWI in chick embryo cells, the degree of fusion in these BHK-21 cells was also considerably greater at the higher pH. The strain specificities were also similar to those found in chick embryo cells in that NDV-AV induced massive fusion while NDV-L and NDV-EH induced considerably less; NDV-HP induced practically no fusion at all. Although the amount of FFWI induced by NDV-L was relatively low in this experiment, it is clear that the strain specificity of the induction of FFWI which we had previously seen in the chick embryo cell system is not unique to that system.

MDBK cells were also tested for their ability to undergo FFWI as a result of NDV infection. A typical experiment is shown in Table 4. There is little evidence of FFWI by either NDV-AV, which is the best inducer in the other cells, or by NDV-IM which is a poor inducer of this type of fusion. The failure of these cells to undergo FFWI after infection by either of these strains is obviously not due to inability of these cells to fuse because within the same experiment at high concentration of NDV-N induced massive FFWO.

In order to determine whether the failure of NDV-AV to induce FFWI in MDBK cells is due to their inability to make a hypothetical "fusion factor", or alternately, to respond to one which is made, the experiment shown in Table 5 was done. Cultures of MDBK or chick embryo cells were infected by NDV-AV and incubated for 6 hours. At this point, they were trypsinized, washed, and transferred to uninfected cultures of either chick embryo or MDBK cells. The induction of fusion by transferred chick embryo cells is most successful, resulting in 0.5-0.8 fusion events per cell. Since this induction of fusion after transfer occurs both in the presence and absence of the inhibitor of protein synthesis, cycloheximide, the induction of FFWI does not require further protein synthesis after transfer. The induction of FFWI by infected chick embryo cells in MDBK cell monolayers is somewhat less successful, but reproducibly significant levels of FFWI are achieved in this type of heterologous transfer. In contrast, the transfer of infected MDBK cells to either uninfected chick embryo cells, which are sensitive to FFWI, or to uninfected MDBK cells, does not produce significant levels of FFWI.

The causes of the restriction of FFWI in MDBK cells are not yet known. The sensitivity of these cells to FFWO, and to FFWI when the hypothetical fusion factor is made in chick embryo cells, has been clearly demonstrated. The failure of these cells, when productively infected by NDV-AV, to fuse even to sensitive chick embryo cells, suggests that virus-specific membrane biogenesis in the

MDBK cell may be different from that which occurs in either chick embryo or BHK-21 cells. Whether these differences are due to differences in events occuring prior to the insertion of virus-specific proteins in the plasma membrane, or--as Klenk and Choppin (29) have suggested in the case of SV5 infection--to difference in the membrane composition of these cells, remains to be determined.

Again we would like to stress that in any given instance where fusion--either FFWO or FFWI--does not occur, it is difficult to conclude whether such failure is due to the absence of the hypothetical fusion factor, or to the presence of other factors which are antagonistic to fusion. This must be considered in analyzing the failure of infected MDBK cells to undergo FFWI, in analyzing the correlations of ability to induce FFWO with the affinity of virions for cell membranes, and also in trying to understand the mutual exclusion of FFWO and FFWI, where it exists.

Our observations on the interactions of different strains of NDV with several cell types, and studies in other laboratories on the interactions of other paramyxoviruses with a variety of cell lines--of which the study by Holmes and Choppin (28) on SV5 infection of BHK-21 F and monkey kidney cells is one of the best examples--can only engender a healthy respect for the variability of paramyxovirus infection. These results amply demonstrate that paramyxovirus infection is far more complex than generally believed. In light of this variability it is clear that generalizations about "what NDV does" or what another paramyxovirus does, can be meaningless. For this reason, it is clear that limiting one's study to a single strain of virus or a single cell type can be very misleading.

Finally we might point out that whether or not FFWO and FFWI involve the same mechanism, the study of each is valuable in its own right. Analysis of FFWO makes it possible to distinguish between those cultural conditions essential for the fusion process itself, and those essential for both productive infection and fusion. Furthermore, since the paramyxovirus particles themselves are directly involved in the induction of FFWO, analysis of

structural components and biological properties of virions
may lead to an understanding of the mechanism of FFWO.
FFWI on the other hand is intrinsically interesting because of its intimate involvement with the processes of
membrane biogenesis. Taking advantage of the kind of
biological findings we have described here, we are currently working on the effects of infection on the synthesis of the protein and lipid components which make up
these virus-specific biological membranes.

ACKNOWLEDGEMENTS

Portions of this study were taken from the Ph.D. thesis
of W.R.G., Harvard University, May, 1971.

We are grateful to: Dr. L.A. Clavell for his contribution of the hemolysis experiments; Miss P.A. Dore for
her contribution of the neuraminidase studies; Mrs. H.B.
Keary and Mrs. J.A. Rice for excellent technical assistance; Miss J.R. Larmon and Miss M.T. Denman for assistance in the preparation of this manuscript; and the
National Institutes of Health for financial support (AI-
08367).

REFERENCES

1. Finkeldey, W., Virchow's Arch. 281, 323 (1931).
2. Warthin, A.S., Arch Pathol. 11, 864 (1931).
3. Enders, J.F.and Peebles, T.C., Proc. Soc. Exptl. Biol.
   Med. 86, 277 (1954).
4. Henle, G., Deinhardt, F., and Girardi, A., Proc. Soc.
   Exptl. Biol. Med. 87, 386 (1954).
5. Okada, Y., Biken's J. 1, 103 (1958).

6. Bratt, M.A. and Gallaher, W.R., Proc. Natl. Acad. Sci. U.S. 64, 536 (1969).
7. Bratt, M.A. and Gallaher, W.R., In Vitro 6, 3 (1970).
8. Kohn, A., Virology 26, 228 (1965).
9. Doermann, A.H., J. Bacteriol. 55, 257 (1948).
10. Clavell, L.A. and Bratt, M.A., Appl. Microbiol. 23, in press.
11. Bratt, M.A. and Rubin, H., Virology 33, 598 (1967).
12. Bratt, M.A. and Rubin, H., Virology 35, 395 (1968).
13. Marcus, P.I., Cold Spring Harbor Symp. Quant. Biol. 26, 351 (1962).
14. Gallaher, W.R., Ph.D. Thesis, Harvard Medical School (1971).
15. Gallaher, W.R. and Bratt, M.A., in preparation.
16. Brandt, C.D., Virology 14, 1 (1961).
17. Reeve, P. and Poste, G., J. Gen. Virol. 11, 17 (1971).
18. Sagik, B.P. and Levine, S., Virology 3, 401 (1957).
19. Bratt, M.A. and Clavell, L.A., in preparation.
20. Rubin, H. and Franklin, R., Virology 3, 84 (1957).
21. Marcus, P.I., Salb, J.M., and Schwartz, V.G., Nature 208, 1122 (1965).
22. Compans, R.W., Holmes, K.V., Dales, S., and Choppin, P.W., Virology 30, 411 (1966).
23. Meiselman, N., Kohn, A., and Danon, D., J. Cell Sci. 2, 71 (1967).
24. Hosaka, Y. and Koshi, Y., Virology 34, 419 (1968).
25. Clavell, L.A. and Bratt, M.A., submitted for publication.
26. Choppin, P.W. and Tamm, I., in Cellular Biology of Myxovirus Infection, ed. G.E.W. Wolstenholme, Brown, Boston, pp. 218-245.
27. Cascardo, M.R. and Karzon, D.T., Virology 26, 311 (1965).
28. Holmes, K.V. and Choppin, P.W., J. Exptl. Med. 124, 501 (1966).
29. Klenk, H.D. and Choppin, P.W., Proc. Natl. Acad. Sci. U.S. 66, 57 (1970).

TABLE 1

Comparison of Fusion From Without and Fusion From Within

	Fusion From Without	Fusion From Within
State of Virus Infectivity	Infectious or non-infectious (UV-inactivated)	Infectious only
Time of Detection	3 hours	6 hours
Optimal Virus Concentration	High multiplicities of infection	Low multiplicities of infection (Inhibition at high multiplicities)
pH Dependence in Range of 7.0-8.4	Slight optimum at pH 8.0	Sharp optimum at pH 8.1-8.2
Requirement for DNA-Dependent RNA Synthesis (Actinomycin D)	None	Probably none
Requirement for Protein Synthesis (Cycloheximide)	None	Absolute requirement
Effect of Antiviral Antibody	Prevention if added within 20 minutes of viral adsorption	Prevention if added within 5 hours of viral adsorption
Strain Specificity	NDV-N,IM,HP,GB,EH	NDV-AV,IS,L
Cell Responses	Identical in chick embryo, BHK-21, and MDBK cells	Identical in chick embryo and BHK-21 cells-none for MDBK cells

TABLE 2

Strain Specificity of Fusion in Chick Embryo Cells

Group	FFWO	NDV Strain	FFWI	Hemadsorption at 11 Hours
I (Virulent)	0.1	AV,IS,RO,L	1.0-4.0	$(\overset{+}{-})$---(++)
II (Virulent)	0.5-1.0	IM,HP,EH	0.2-1.0	(+)---(++)
III (Virulent)	0.5	GB	0.1	(++)
	0.1	F,I	0.1	$(\overset{+}{-})$---(++)
IV (Avirulent)	1.0	N	0.1	(-)
	0.3	B	0.1	(-)
	0.2	M,W	0.1	(-)

Typical degrees of fusion expressed in fusion events per cell. Hemadsorption measured by removal of medium at 11 hours and adsorption of erythrocytes at 4°.

TABLE 3

Strain Specificity of Fusion From Within in BHK-21 Cells

NDV Strain	Fusion Events Per Cell at 11 Hours	
	pH 7.2	pH 8.3
AV	0.03	>5.0
L	0.01	0.41
EH	0.18	0.31
HP	0.00	0.02

Monolayers were infected at multiplicities of 10 PFU/cell and incubated at 42° at the indicated pH.

TABLE 4

Attempts to Induce Fusion From Within in MDBK Cells

NDV Strain	Time of Fixation (Hours)	Fusion Events Per Cell
AV	12	0.25
AV	20	0.12
IM	12	0.20
IM	20	0.05
N	2.5 (FFWO)	3.0

Cells were infected with NDV-AV or NDV-IM at multiplicities of 2 PFU/cell or NDV-N at a high multiplicity, and fixed after incubation at pH 8.2 and 38°. At 20 hours considerable cell loss and pycnosis but little fusion was detected.

TABLE 5

Fusion From Within After
Homologous and Heterologous Cell Transfers

Cell Type		Fusion Events Per Cell 90 Minutes After Transfer	
Originally Infected Cells	Recipient Cells	No Cycloheximide	10 ug/ml Cycloheximide
Chick	Chick	0.58	0.78
Chick	MDBK	0.21	0.26
MDBK	MDBK	0.08	0.15
MDBK	Chick	0.05	0.09

Cells infected at 6 PFU/cell and incubated at 42° at pH 8.2 for 6 hours were trypsinized and transferred at a concentration of 10^5 cells to fresh monolayers in the presence or absence of cycloheximide, and fixed after 90 minutes further incubation.

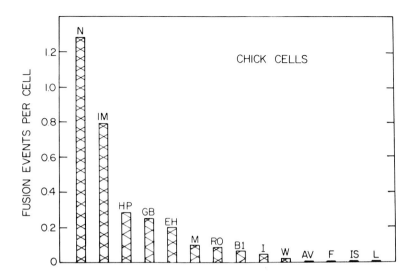

Figure 1. Strain specificity of FFWO in chick embryo
cells. Cells treated with 10,000 HAU/ml at 5° were incubated at 38° for 2.5 hours and then fixed.

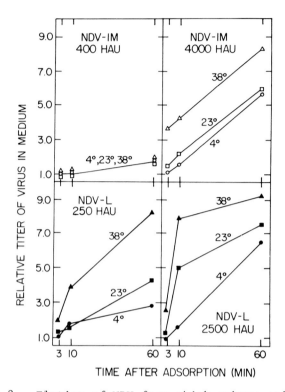

Figure 2. Elution of NDV from chick embryo cells. After adsorption of virus for 1 hour at 4°, monolayers were washed 5 times and then incubated at the indicated temperatures following the addition of 5 ml of preequilibrated medium. Input virus concentrations were 400 and 4,000 HAU/ml (10 and 100 PFU/cell, respectively) for NDV-IM; and 250 and 2,500 HAU/ml (7.5 and 75 PFU/cell, respectively) for NDV-L. The values plotted represent the ratio of virus at each time to that in the medium at 4° at 0 time; the initial level of residual virus was 0.2% of the inoculum except in the case of the higher concentration of NDV-L where it was 0.8% of the inoculum.

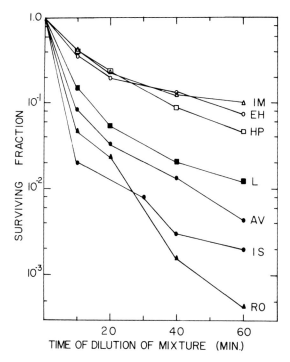

Figure 3. Neutralization of NDV strains by a single antiserum. Anti NDV-B1 at a final dilution of 1/500 and virus at a final concentration of 10^6 PFU/ml were mixed and incubated at 23°. Aliquots were diluted and assayed for residual infectivity at the indicated times.

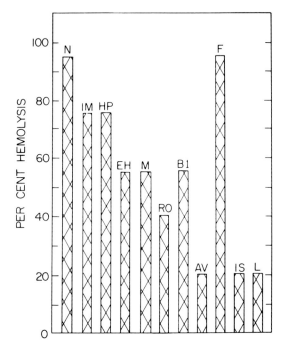

Figure 4. Strain specificity of hemolysis. Hemolysis was measured by mixing virus at a final concentration of 1,500 HAU/ml with erythrocytes at a final concentration of 10^8/ml and incubating at 37° for one hour (10). The strains have been placed in the same order as in Figure 1.

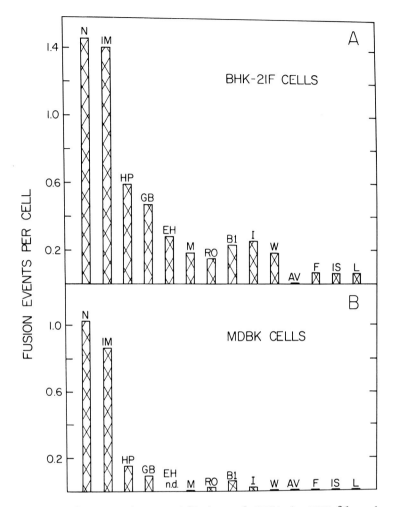

Figure 5. Strain specificity of FFWO in BHK-21 and MDBK cells. The experiment and the ordering of strains within the table were the same as in Figure 1.

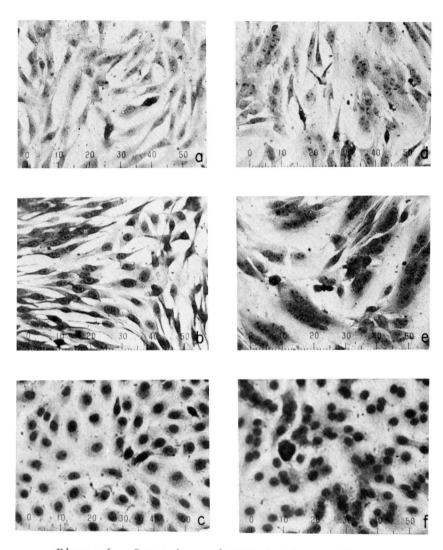

Figure 6. Comparison of FFWO in three cell types. Examples of uninfected control cultures of: a) chick embryo; b) BHK-21; c) MDBK cells. Frames d, e, and f show the same cell types treated with NDV-N.

IX

ENERGY COUPLING AND TRANSPORT

COMPONENTS OF HISTIDINE TRANSPORT

Giovanna Ferro-Luzzi Ames

Biochemistry Department
University of California
Berkeley, California 94720

ABSTRACT. The transport of histidine in *Salmonella typhimurium* occurs through at least five permeases of varying specificity and affinity. The high-affinity histidine permease has been shown to be constituted of at least two systems: the J-P and the K-P systems. Two separate proteins have been found to be essential components of the J-P system: the histidine-binding protein J and the P protein. The histidine-binding protein J is a periplasmic protein which binds histidine with high affinity. The P protein (possibly a membrane component) has not been isolated: its participation in the J-P system has been demonstrated genetically and physiologically. The P protein appears to be a multifunctional component because it is involved in the transport of other substrates besides histidine. It is postulated that the multiplicity of transport systems, which is by no means unique to histidine, might be due to the involvement of each system in different areas of cell metabolism. Some evidence in this respect is presented.

INTRODUCTION

I have been studying the uptake of histidine in *Salmonella typhimurium* for a number of years. A rather complex picture has emerged. Below I will review the knowledge accumulated and I will discuss in more detail the following features:
   a) The concept of *limit concentration*.
   b) Multiple permeases, with different overlapping specificities, transport histidine. I will briefly speculate on the significance of this multiplicity.

c) Two separate proteins have been identified as essential components of the permease with the highest affinity for histidine (J-P system): the histidine-binding protein J and the P protein.

d) The possibility of a direct interaction between these two protein components as part of their function will be discussed.

e) One of these components, the P protein, is necessary for the function of more than one permease: i.e., the P protein is multifunctional.

THE *LIMIT CONCENTRATION*

Two methods were developed for the assay of histidine transport: the starved-cells assay and the growing-cells assay (1). Both have been extensively described recently (2) and here I will only comment briefly on the growing-cells assay, because out of its use has emerged the concept of *limit concentration*. This assay has been developed especially for the measurement of transport under physiological conditions. Uptake of labeled amino acid is rendered limiting for the incorporation of the externally added amino acid into protein, without being limiting for cell growth and without disturbing it in any way. Thus the values obtained for the affinity of permeases by this method are probably representative of their true function *in vivo*. It should be borne in mind however that these "Km" values are a measure of the limiting step in what is probably a complex series of reactions. An interesting parameter which becomes obvious when uptake is assayed by this method is the *limit concentration* (1). This is the minimal external concentration required by the permease in order to supply the cell with enough amino acid to carry on protein synthesis without utilizing any of the biosynthetic amino acid. This is possible because the increase in the internal amino acid pool through the action of the permease (1) results in feedback inhibition of the biosynthetic pathway (3). The limit concentration for histidine is 1.5×10^{-7} M (Figure 1).

The limit concentration for a permease is a biological parameter of more significance than the Km, because it expresses the actual efficiency of the uptake with respect to growth. A low limit concentration indicates a high efficiency. This is desirable from the point of view of the cellular economy, since the energy involved in the uptake of one molecule of amino acid is probably much smaller than that spent in the biosynthesis of the same molecule from glucose. In fact, it has been calculated that for the synthesis of one molecule of histidine, the potential energy corresponding to 41 ATP molecules is sacrificed (3).

EVIDENCE FOR MULTIPLE TRANSPORT SYSTEMS FOR HISTIDINE

Figure 2 is a schematic representation of the multiple means of uptake of histidine by *S. typhimurium*. The five permeases known to transport histidine (in order of decreasing affinity) are named: J-P, K-P (these two systems together form the high-affinity histidine permease), general aromatic permease, and the X and the Y permeases. It is quite possible that other systems exist which transport histidine besides these known five, but more extensive kinetics and, more importantly, more genetic work would be necessary to define them clearly.

The permease with the highest affinity is the J-P permease. Its affinity for L-histidine is best measured in a *dhuA* mutant (which has elevated transport levels) and is approximately 6×10^{-9} M (4). The J-P permease transports also, with much poorer affinity, HIPA, D-histidine, arginine, lysine, citrulline, ornithine and azaserine (5). This permease will be discussed in more detail in the following sections.

The K-P permease has a Km of 2×10^{-7} M for L-histidine (4). Uptake through this permease is measured in a mutant defective in the J-P system because of the loss of the J component of that system. The K-P permease transports HIPA and possibly also other amino acids besides histidine.

The J-P and K-P systems share at least one component with each other: the P protein. Mutants (*hisP*) defective in the P protein do not transport through either the J-P or the K-P system. A schematic representation of transport through these systems is shown in Figure 3.

The general aromatic permease (*aroP*) has been characterized with respect to its transport of histidine (1,6). Its affinity for histidine is comparatively low (Km: 10^{-4} M when measured with the starved-cells assay); the aromatic amino acids are transported with affinities between 10^{-7} M and 10^{-6} M. Histidine uptake through the aromatic permease is clearly detected in the wild type as a sharply deflected portion of the curve in the Lineweaver-Burk plot for histidine uptake (Figure 4, Km = 10^{-4} M), when the starved-cells assay is used. Each of the aromatic amino acids competes efficiently with the uptake of histidine through the aromatic permease, with Ki values between 10^{-6} M and 10^{-7} M, thus justifying the name of general aromatic permease for this transport system. When the growing-cells assay is used, transport of histidine through the aromatic permease is only detected in mutant strains lacking transport through the high-affinity histidine permease (*hisP* mutants). In the wild type uptake through the aromatic permease does not occur significantly at histidine concentrations below the limit concentration; therefore it can't be assayed (1).

In double mutants *hisP aroP*, there is no uptake through either the J-P or K-P systems because of the *hisP* mutation, nor through the general aromatic permease because of the *aroP* mutation. However, histidine is still taken up by such double mutants. Preliminary kinetic experiments indicate that this residual uptake is due to at least two more permeases, temporarily named X and Y. The X permease transports also the aromatic amino acids, arginine and possibly other amino acids. Therefore it also is a system of broad specificity. The specificity of the Y permease has not been studied.

EVIDENCE FOR MULTIPLE TRANSPORT SYSTEMS FOR OTHER AMINO ACIDS. POSSIBLE SIGNIFICANCE OF THE MULTIPLICITY.

It is interesting to speculate about this multiplicity of systems for histidine transport. The first question that comes to mind is whether this is a special feature of histidine transport. The answer is definitely no. Since my first results on histidine and aromatic amino acid transport appeared (1), several cases of multiplicity of transport systems for other amino acids have been reported in the literature. As part of the study on histidine transport I also presented evidence of multiplicity of the aromatic amino acids transport systems (1). Each aromatic amino acid is transported by at least one more system, besides the general aromatic permease, with a more stringent substrate specificity. Figure 5 shows the detection of a specific phenylalanine permease upon inhibition of the

general aromatic permease by tryptophan. Similar results
are obtained also with either tyrosine or tryptophan as
substrates, and any of the other aromatic amino acids as
inhibitor. More recently, Brown has obtained similar results in a study of aromatic amino acids transport in *E.*
coli (7). In *E. coli* the following permeases have been detected: Two transport systems for aspartate have been
demonstrated unequivocally (8); two transport systems for
glutamine can be assayed in an appropriate mutant derivative of *E. coli* (9); two transport systems for lysine, two
for ornithine and probably two or more systems for arginine
have been found (10); at least two systems for the transport of leucine are known (11,12).

If one wishes to extend the observations to transport
of compounds other than amino acids, the example of galactose will suffice: *E. coli* has six transport systems for
galactose (13). In conclusion, transport of one substrate
by more than one system is a very common occurrence indeed
in these bacteria.

I previously speculated (1) that the purpose of the
general aromatic permease might be that of supplying large
amounts of aromatic amino acids or histidine for utilization as carbon or nitrogen sources. I could not test this
hypothesis because none of these amino acids were utilized
by the available strain of *S. typhimurium* as either nitrogen or carbon source. However, recent evidence from this
laboratory (M. Alper, personal communication) indicates
that that hypothesis might be true. The activity of the
general aromatic permease is significantly decreased in a
cya mutant (i.e., unable to produce cyclic AMP), thus indicating that this permease might be under catabolite repres-
sion control, as are other pathways involved in the dissimi-
lation of carbon sources (14). This type of regulation for
the aromatic permease supports the hypothesis that one of
its functions is that of supplying the aromatic and other
amino acids as carbon sources.

No change in the level of the histidine high-affinity
permease has been observed under a variety of conditions.
It is reasonable to speculate that this permease is constitutive and that its primary function is to supply amino
acid to the protein synthesizing system. This hypothesis
is consistent with the low value obtained for the *limit*
concentration (1.5 x 10^{-7} M), which is an indication of the
efficiency of this permease (1). The high-affinity histi-

dine permease is able to remove from the medium any histidine present, even when present at very low concentrations,
thus making it available intracellularly for protein synthesis. In this respect it is also of interest that $hisP$ mutants excrete very small amounts of histidine in the medium, while the wild type parent does not (15). This might
indicate that the high-affinity histidine permease has the
function of recapturing very efficiently any of the biosynthetic histidine which is lost into the medium. The recap-
ture of histidine is a very important event, because of the
expense involved in the biosynthesis of each histidine
molecule (3).

No information is available on the means of regulation
of the multiple permeases for other amino acids. However,
the possibility that they are under different regulatory
controls, and that therefore they perform different functions in the cell is a likely one. More complete characterization of the pattern of regulation and of the specificity of these systems is necessary before drawing final
conclusions.

EVIDENCE FOR MULTIPLE COMPONENTS IN ONE TRANSPORT
SYSTEM: THE J-P PERMEASE

Recent studies (4,5) demonstrated that the J-P permease is made up of at least two proteins: the histidine-
binding protein J, which has been isolated and purified,
and the P protein. The evidence relative to this finding
will be summarized below.

The J Protein. The J protein is a periplasmic protein
which is released by the mild osmotic shock procedure devised by Heppel and his collaborators (16), and which binds
histidine. By selecting and studying appropriate mutants
altered in histidine transport, it has been unequivocally
shown that the J protein is a component of the J-P permease. Mutants have been obtained which have an increased
level of histidine transport and an increased level of the
J protein, and mutants have been obtained which have defective histidine transport and lack the J protein.

Table 1 summarizes the properties of a variety of
mutants with respect to the levels of transport and of the
J protein. The first line refers to the wild type strain,
which transports histidine with a rate of 0.25. The level
of J protein in this strain is taken as 1. It is possible

(4,17) to obtain mutants (*dhuA*) which have elevated histidine transport (rate: 0.43; second line) by selecting for
growth on D-histidine as source of L-histidine. D-Histidine is a poor substrate for the J-P permease and transport
through this permease apparently is the limiting step for D-
histidine utilization (17); therefore one of the mechanisms
leading to D-histidine utilization is the raising of the
level of D-histidine transport through the J-P permease.
In fact, mutants in *dhuA* (possibly a regulatory site for
genes involved in histidine transport) have elevated levels
of both D- and L-histidine transport (4,17). These strains
also have elevated levels of the J protein (5- to 10-fold
the wild type level). Extensive analysis of J protein
isolated and purified from a *dhuA*-containing strain yielded properties identical to those of pure J protein obtained
from the wild type (18). Therefore, the increase in J protein level is not due to a change in the affinity of the J
protein for histidine, but rather to a true increase in the
amount of J protein.

From *dhuA*-containing strains, mutants have been isolated which have a lowered level of histidine transport
(rate: 0.02), by selecting for inability to grow on D-histidine (4). The third line in Table 1 describes the properties of one of such mutants of genotype *dhuA hisJ*: no
J protein is detectable.

This indicates that histidine transport under these
conditions is dependent upon the presence of the J protein,
i.e., that this protein is an indispensable component of
histidine transport through the J-P permease. The possibility existed that the *hisJ* mutation was in a regulatory
gene, rather than in the structural gene for the J protein.
This seemed quite unlikely, but if it were true, it would
have invalidated the conclusion that the J protein is an
indispensable component of transport, because its higher
and/or lower levels might just be coincidental to a raising and lowering of transport as part of a regulation
mechanism.

To rule out this possibility a revertant of a *hisJ*
mutant has been isolated which produces an altered J protein and has a concomitantly altered transport (5, fourth
line in Table 1). The J protein from this revertant differs from wild type J protein in chromatographic and
electrophoretic mobility, heat stability and affinity for
histidine. The histidine transport in this revertant is

temperature-sensitive and its D̲-histidine growth capability is temperature-sensitive. These results unequivocally identify the *hisJ* gene as the structural gene for
the J protein and the J protein as an indispensable component of the J-P permease. In agreement with this, the correlation between the pattern of specificity of the J protein
in vitro and of transport *in vivo* has been demonstrated to
be excellent (5).

A histidine-binding protein from *S. typhimurium* has
been described recently (19), which has different properties
than the J protein. Because the evidence concerning its
involvement in transport is not complete, it can't be ascer-
tained what is its relationship to the J protein. It is
possible that it is one of the other three histidine-binding proteins present in the shock fluid besides the J protein (18).

The P Protein. It has become clear that the J protein
is not the only component of the J-P transport system: in
fact mutants are available (*hisP*) which have greatly lowered levels of histidine transport, even though they have
the same (or higher) level of J protein as the wild type.
Mutants in *hisP* are most simply obtained by selection of
resistance to the inhibitory histidine analogue α-hydrazino
imidazole propionic acid (HIPA) which is transported by
the high-affinity histidine permease; there are several
other methods of *hisP* selection available. The fifth line
in Table 1 describes the effect of a *hisP* mutation in an
otherwise wild-type background: the uptake of histidine
is lower (rate: 0.005) than in a *hisJ* mutant, because neither the J-P nor the K-P system (of which more will be said
later) are functioning. However, the J protein level is
unaffected. Similarly, the sixth line describes the pheno-
type of a *dhuA hisP* mutant: the level of histidine transport is as low as that of a simple *hisP* mutant, despite the
fact that the *dhuA hisP* strain still contains the *dhuA* mutation which by itself confers high levels of transport and
still has a J protein level identical to that of the parent
strain *dhuA*.

In conclusion, these results clearly indicate that the
J protein is necessary, but not sufficient to carry on histidine transport, and that the product of the *hisP* gene is
also necessary; the product of the *hisP* gene is a protein,
as shown by the isolation of amber mutants in that gene (6).
At present, the working hypothesis is that the P protein is

a membrane component and efforts are aimed at its isolation and characterization.

To my knowledge no other amino acid transport system has been analyzed in a similar way as to indicate the presence of two separate components, both necessary for transport. The available evidence does not indicate how the combination of these components (two at least) constitutes the transport system. The next section is a discussion of a few possibilities concerning this question.

INTERACTION BETWEEN THE J AND THE P PROTEINS

The J and the P proteins might perform their function either by interacting directly with each other or by working in a separate and sequential fashion. The distinction between these two possibilities has not yet been made definitively by means of either genetic or physiological experiments. The following argument however lends some indirect support to the protein-protein interaction hypothesis.

The elevation in J protein which accompanies the elevation in transport in a *dhuA* mutant suggested that this protein might be the rate-limiting step in transport (e.g. of D-histidine, the uptake of which is completely dependent upon the J protein). However, strains with wild type levels of the J protein are *completely* unable to grow on D-histidine, while the *dhuA* mutation allows growth on D-histidine at rates approximately equal to those on L-histidine (17). If the only reason for the inability of the wild type to grow on D-histidine were the limiting amount of J protein, one would expect that with approximately a five-fold elevation of the J protein observed in *dhuA*-containing strains, the bacteria would only be able to grow five times faster. Therefore, there is no proportionality between the rate of growth on D-histidine and the amount of J protein. Since an increased affinity of the J protein for D-histidine cannot be invoked as the explanation for the greatly improved growth rate of a *dhuA* mutant (4,18) we favor the following hypothesis. Since the mutation responsible for the J protein elevation has occurred at a probable regulatory site (*dhuA*), it might have elevated also another component which is the limiting step in transport. The *hisP* gene is located very near the *dhuA* and *hisJ* genes, and its product, the P protein, might be under control of the *dhuA* site and thus also be elevated in a *dhuA* mutant. A protein-protein interaction between the J and the P proteins could give a more than proportional increase in permease function for the J-P system in a *dhuA* mutant.

In this respect it should be pointed out that kinetic data obtained with the temperature-sensitive *hisJ* revertant (5) might support the interaction hypothesis. In fact, while the level of the binding protein in this strain grown at 30°, as measured *in vitro*, is approximately the same as the parent level (i.e. elevated), this strain does not grow on D-histidine at 30°, and its transport of L-histidine through the J-P permease is only 60% of the parent level. It seems therefore that the temperature-sensitivity of the J protein in this strain, as assayed *in vitro* by histidine-binding, does not necessarily account for the *in vivo* temperature-sensitivity of growth on D-histidine. If an interaction between the J protein and the P protein (or any other transport protein) is required for transport, the *in vivo* temperature-sensitivity may be due to an alteration in this interaction, rather than to an alteration in the binding of substrate to the J protein.

It should also be pointed out that the *dhuA* mutant analyzed has a lower Km (6.6 x 10^{-9} M) for L-histidine uptake than the wild type (2.6 x 10^{-8} M (4)). We previously postulated (4) that the measurement of the lower Km might be due to the increased preponderance, in the *dhuA* mutant, of the J-P permease, which has a higher affinity for histidine than the other permease, K-P. However, if several permease components have been elevated simultaneously in a *dhuA* mutant, then the rate-limiting step in transport might be a different one, therefore resulting in a different apparent affinity. On the other hand, if several transport components interact with each other, it is possible that the altered kinetics of uptake in the *dhuA* mutant is due to the elevation of one or more of the components.

Evidence consistent with the idea that the P protein is also elevated in *dhuA* mutants is discussed in the next section.

THE P PROTEIN IS MULTIFUNCTIONAL

From Table 1 it can be seen that a mutation in either *hisJ* or *hisP* lowers the level of histidine transport, but a *hisP* mutation has a more drastic effect than a *hisJ* mutation.

This suggests that in a *hisJ* mutation there is a residual transport mechanism, which is also dependent on an intact P protein, because it is lacking in *hisP* mutants. This system has been temporarily named the K-P permease; it could actually represent the sum of several permeases, each of which however is dependent on the P protein for function. The other strong evidence that the P protein functions in both of these transport systems comes from the isolation of HIPA-resistant mutants. The fact that they all are in *hisP* can be explained by assuming that HIPA is transported by more than one system and that the common component for these systems is the P protein (Figure 3). In this case the only type of HIPA-resistant mutants which can be obtained by a one-step mutation is in the common component *hisP*.

The P protein therefore is needed for at least two permeases: J-P and K-P. Evidence is appearing now (Govons Küstü and Ferro-Luzzi Ames, unpublished results, (20)) that other transport systems might be also utilizing the P protein. Uptake of arginine is dependent upon the P protein, when arginine is used as the sole source of nitrogen, because *hisP* mutants are unable to utilize it. Moreover, either *dhuA* or *dhuA hisJ* mutants grow better than the parent on arginine as nitrogen source. This suggests that the *dhuA* mutation has elevated the level of a component limiting for the growth on arginine, and that the J protein is not that component. The P protein is a possible candidate. Also the uptake of histidinol is partially dependent upon the P protein, because *hisP* mutants grow more slowly on histidinol as a histidine source than the HisP$^+$ parent.

Speculation on what the function of the P protein might be and about its multifunctionality are premature. However, the P protein does not appear to have a universal function, meaning by that that transport through at least some of the other amino acid transport systems is unaltered in *hisP* mutants. The system best studied under these conditions is the *aroP* permease, which is not affected by a *hisP* mutation. Also the proline permease does not seem to be dependent upon the P protein, because no resistance to the proline analogue, azetidine carboxylic acid, appears in *hisP* mutants. (This analogue is thought to be transported by the proline permease, and preliminary results indicate that mutants resistant to it are affected in proline transport (Ferro-Luzzi Ames, unpublished observations)).

In support of this is also the fact that no single mutation giving resistance to both HIPA and a second unrelated analogue (azetidine carboxylic acid or azaserine) has been obtained.

Thus, the P protein seems to have a certain amount of specificity, making it unlikely that it has an essential function in the transport of all amino acids (an energy-supplying step?), while it is essential for the transport of some of them.

CONCLUSIONS

The identification of the J binding protein as a component of histidine transport is the first step towards an understanding of the mechanism by which histidine is transported. The isolation and characterization of the P protein is the next obvious aim followed in this laboratory. The biochemical approach such as solubilization of membrances and analysis by a variety of methods, is being accompanied by further genetic and physiological studies which have often cleared the path for biochemistry. At the same time, a function for the minor histidine-binding proteins is being sought and the problem of multifunctionality of the P protein is being explored.

ACKNOWLEDGMENTS

My research in this field has been made particularly pleasant by the multiple interactions I have had with my friends and collaborators: Bruce N. Ames, Sydney Govons Küstü and Julia Lever.

REFERENCES

1. G. Ferro-Luzzi Ames, Arch. Biochem. Biophys. 104, 1 (1964).
2. G. Ferro-Luzzi Ames, in S. Fleischer, L. Packer and R. W. Eastabrook (Editors), Methods in Enzymology (in press), Academic Press, New York.
3. M. Brenner and B. N. Ames, in H. J. Vogel (Editor), Metabolic Regulation, page 349. Academic Press, New York (1971).
4. G. Ferro-Luzzi Ames and J. Lever, Proc. Nat. Acad. Sci. U.S.A. 66, 1096 (1970).

5. G. Ferro-Luzzi Ames and J. Lever, J. Biol. Chem., submitted for publication (1972).
6. G. Ferro-Luzzi Ames and J. R. Roth, J. Bacteriol. 96, 1742 (1968).
7. K. D. Brown, J. Bact. 104, 177 (1970).
8. W. W. Kay, J. Biol. Chem. 246, 7373 (1971).
9. J. H. Weiner and L. A. Heppel, J. Biol. Chem. 246, 6933 (1971).
10. B. P. Rosen, J. Biol. Chem. 246, 3653 (1971).
11. C. E. Furlong and J. H. Weiner, Biochem. Biophys. Res. Comm. 38, 1076 (1970).
12. J. Guardiola and M. Iaccarino, J. Bacteriol. 108, 1034 (1971).
13. H. M. Kalckar, Science 174, 557 (1971).
14. I. Pastan and R. Perlman, Science 169, 339 (1970).
15. S. Shifrin, B. N. Ames and G. Ferro-Luzzi Ames, J. Biol. Chem. 241, 3424 (1966).
16. L. A. Heppel, Science 156, 1451 (1967).
17. K. Krajewska-Grynkiewicz, W. Walczak and T. Klopotowski, J. Bacteriol. 105, 28 (1971).
18. J. E. Lever, J. Biol. Chem., submitted for publication (1972).
19. B. P. Rosen and F. D. Vasington, J. Biol. Chem. 246, 5351 (1971).
20. G. Ferro-Luzzi Ames, S. Govons and J. E. Lever, Fed. Proc. in press (1972).

TABLE 1

Genotype	Histidine transport[a]	J protein level[b]	HIPA-sensitivity	D-histidine growth[c]
wild type	0.25	1	sensitive	-
dhuA	0.43	5	supersensitive	+
dhuA hisJ	0.02	0.06	sensitive	-
dhuA hisJ revertant	t.s.[d]	t.s.	sensitive	+t.s.
hisP	0.005	1	resistant	-
dhuA hisP	0.005	5	resistant	-

[a]Rate (in μmoles per g dry weight per minute, at 10^{-8} M L-histidine) measured by the growing-cells assay. At this histidine concentration uptake is due essentially to the high-affinity histidine permease.

[b]Values normalized to wild type level.

[c]Assayed in the corresponding histidine-requiring derivatives.

[d]t.s. means that the presence of that property is dependent upon the temperature of assay.

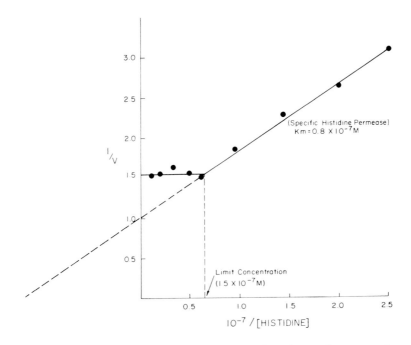

Fig. 1. Lineweaver-Burk plot for histidine uptake, measured by the growing-cells assay (G. Ferro-Luzzi Ames, Arch. Biochem. Biophys. 104, 1 (1964)).

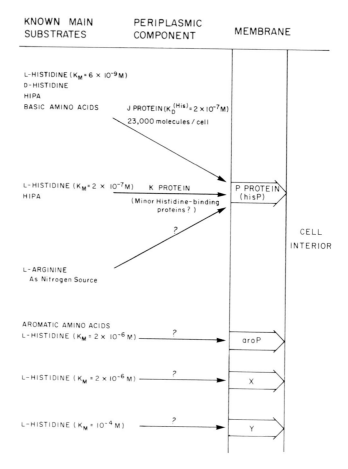

Fig. 2. Schematic representation of the multiple histidine permeases.

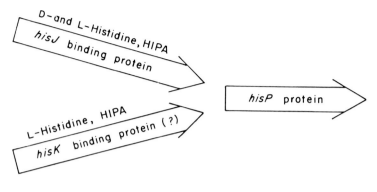

Fig. 3. Schematic representation of transport through the high-affinity transport system (G. Ferro-Luzzi Ames and J. Lever, Proc. Nat. Acad. Sci. U.S.A. 66, 1096 (1970)).

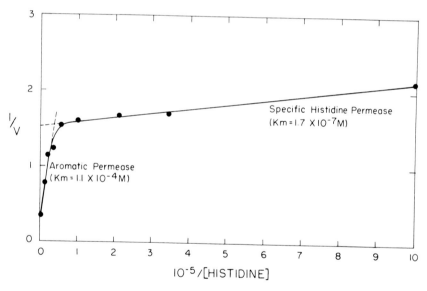

Fig. 4. Lineweaver-Burk plot for histidine uptake in starved cells (G. Ferro-Luzzi Ames, Arch. Biochem. Biophys. 104, 1 (1964)). The Km of the aromatic permease is higher when assayed by this assay than with the growing-cells assay.

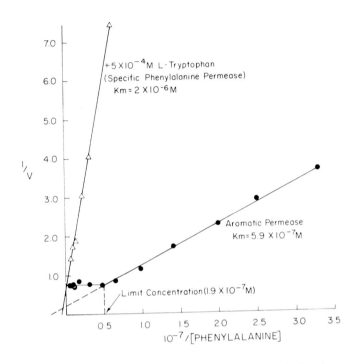

Fig. 5. Lineweaver-Burk plot for phenylalanine uptake, measured by the growing-cells assay. In the absence of tryptophan (o) the curve corresponds to the aromatic permease. In the presence of 5×10^{-4} M tryptophan (Δ) the curve obtained is due to the specific phenylalanine permease (G. Ferro-Luzzi Ames, Arch. Biochem. Biophys. 104, 1 (1964)).

PROBES OF CARBOXYL ACTIVATION AS A COMMON FEATURE OF

ENERGY-TRANSDUCING PROCESSES IN MEMBRANES[†]

P. D. Boyer, R. L. Cross[*], A. S. Dahms and T. Kanazawa

Molecular Biology Institute and
Department of Chemistry
University of California
Los Angeles, California

ABSTRACT. Energy transducing processes in membranes, such as oxidative phosphorylation, active transport of ions and metabolites driven by oxidative or ATP-derived energy, and energy-linked reduction may involve common "high-energy" compounds or states. Selected aspects are presented of research in progress on possible carboxyl group activation in mitochondrial membranes, Na^+,K^+-ATPase of kidney and nerve, Ca^+,Mg^{++}-ATPase of muscle sarcoplasmic reticulum, and bacterial membranes.

[†]Supported in part by Grant GM 11094 of the Institute of General Medical Sciences, U. S. Public Health Service and Contract AT(04-3)-34, proj. 102 of the U. S. Atomic Energy Commission.

[*]Jane Coffin Childs Memorial Fund Postdoctoral Fellow

INTRODUCTION

As this audience well recognizes, our knowledge of the molecular mechanism of energy transduction processes in membranes is, to put it mildly, unsatisfying. The possibility of common energized compounds or states in oxidative phosphorylation, active transport, energy-linked reduction, and even in muscle contraction, makes the problems even more tantalizing. It is not my privilege in this short communication to report any apparent or incipient solutions to the problem. I felt it may be of interest, nonetheless, to summarize the present status of researches within my laboratory group exploring the possibility that activation of carboxyl groups may be a common and fundamental process in membrane energy transducing processes.

The concept of carboxyl activation induced by conformational changes was presented at an East Coast symposium in 1964 (1). The concept, although intriguing, remained experimentally dormant for several years. But in the past two years, there has been increased activity and experimentation within my group. Several factors account for this. One is the data of Dr. Richard Cross, currently a postdoctoral fellow with my group, demonstrating that an uncharacterized phosphoprotein of mitochondria, which appears to be an acyl phosphate, may be related to oxidative phosphorylation (2). A second factor has been the recognition of the presence of protein bound acyl phosphates in membranes and in isolated ATPases concerned with Na^+ and Ca^{++} transport. Common features with mitochondrial systems have been indicated by recent observations in our laboratory of Dahms (3) and Kanazawa (4) of oxygen exchange reactions somewhat akin to those of the mitochondrial membranes capable of oxidative phosphorylation. Another indication of possible carboxyl group function has come from the demonstrations of Cattell *et al.* (see 5) on the inhibition of mitochondrial energy linked processes by low concentrations of dicyclohexylcarbodiimide. Also, the continued lack of satisfying explanations for energy transducing processes favors exploration of possible carboxyl activation inasmuch as there appear to be experimentally plausible although sometimes somewhat difficult approaches available.

The principal experimental data that I will deal with in this paper will concern the metabolic characteristics of the unidentified mitochondrial phosphoprotein, and the oxygen exchange reactions and phosphoprotein formation characteristics of the Na^+,K^+ and the Ca^{++},Mg^{++}-ATPases, with brief mention of some probes of the anaerobic bacterial membrane ATPase with dicyclohexylcarbodiimide, and mention of some experimental plans with myosin.

RESULTS AND DISCUSSION

Some properties of a mitochondrial protein phosphate of unknown function. A full report of the factors governing the formation, metabolic characteristics, and chemical properties of the mitochondrial phosphoprotein is in preparation. The present procedure used for detection of the mitochondrial component is indicated in Table 1. Essentially, the assay depends on measurement of the amount of $^{32}P_i$ incorporated within a few seconds into mitochondrial protein which is stable to precipitation by cold perchloric or trichloroacetic acid. The "E~P" fraction represents that portion of the ^{32}P-label that disappears during a 4 second exposure to 2,4-dinitrophenol (DNP). Another parameter of the assay, not indicated in Table 1, is the alkali lability of the DNP-sensitive phosphoprotein. This alkali lability, and other hydrolytic and chemical characteristics are the principal factors indicating that the substance is an acyl phosphate.

Many of its properties suggest that the phosphorylated protein may function as an intermediate in oxidative phosphorylation. However, at this stage although the substance is readily detectable in liver and kidney mitochondria, it can only be questionably demonstrated in heart mitochondria. This suggests caution in interpretation of its function. In this regard, it is instructive that mitochondrial phosphohistidine when first observed was felt to be a good candidate for an intermediate of oxidative phosphorylation, but subsequently was found to participate in the substrate level phosphorylation catalyzed by mitochondria (6).

Some more recent studies appear to localize the E~P to the inner membrane of mitochondria. Considerations in this regard are indicated in Fig. 1. Here the hypothetical phosphorylated intermediate of oxidative phosphorylation is indicated as X~P and localized in the inner membrane. It is also possible that our phosphorylated protein is not labeled directly from inorganic phosphate, but arises by phosphoryl transfer from ATP to give a "Y~P" in the inter-membrane space, or a "Z~P" in the matrix space, as indicated in the figure.

Characteristics of carboxyatractyloside inhibition make it unlikely that the DNP-sensitive phosphorylated compound is a Y~P type protein. As shown in Table 2, carboxyatractyloside does not prevent formation of E~P from inorganic phosphate, or the discharge of E~P by DNP. Carboxyatractyloside is known to inhibit translocation of ATP between the matrix and intermembrane space. Carboxy-atractyloside would thus be expected to prevent or de-crease formation of any Y~P type intermembrane protein from $^{32}P_i$ via AT^{32}P; no such inhibition is noted.

Evidence indicating location of E~P in the inner membrane comes from separation of the respiratory enzyme components from proteins soluble in dilute acetic acid. Since E~P is maximally stable near pH 2, it is possible to use the procedure of Zahler <u>et al</u>. (7) to remove matrix proteins, most of Racker's factor F_1, and other proteins soluble in dilute acetic acid, leaving the insoluble mitochondrial membranes behind. Results of such a fract-ionation applied to mitochondria containing radioactive E~P are shown in Table 3. The specific activity of the E~P in the fraction containing the inner membranes is over 20 times that in the soluble proteins. Other data that time does not permit to discuss here also point to an inner membrane location of this interesting substance. Although the inner membrane is the site of oxidative phosphorylation, other activities, both known and un-known, may be implicated. Further study is obviously necessary to define the function of Cross' phosphoprotein.

<u>Characteristics of oxygen exchange and enzyme phos-phate formation with Na$^+$, K$^+$ and Ca^{++}, Mg$^+$-ATPases</u>. Obser-vations of Stephen Dahms and Tohru Kanazawa have demon-

strated that under appropriate conditions the Na^+,K^+-ATPase of kidney and of the electroplax organ of the eel, and the Ca^{++},Mg^{++}-ATPase of the sarcoplasmic reticulum have the capacity to catalyze a relatively rapid exchange of H_2O oxygens with inorganic phosphate, as well as show a smaller ATP-induced extra oxygen exchange. Table 4 gives data for the Na^+,K^+-ATPase showing the extent of E-P formation and exchange in the presence and absence of ouabain. The striking feature of the data of Table 4 is that ouabain nearly completely inhibits the exchange reaction, yet promotes the formation of the membrane-bound acyl phosphate in nearly stoichiometric amounts. The most plausible explanation for the origin of the exchange is that it represents rapid formation and hydrolytic cleavage of the acyl phosphate formed from P_i. Formation of the acyl phosphate from P_i in the presence of ouabain is a most unusual reaction when one considers the apparent free energy of hydrolysis of acyl phosphates at neutral pH is minus 15 kilocalories. In some manner ouabain binding induces conformational changes, or perhaps even covalent bond change, that greatly shifts the equilibrium in the favor of phosphorylation of the carboxyl group. But even though more E-P is formed in the presence of ouabain, the rate of interchange with inorganic phosphate is markedly slowed. It appears pertinent that changes of the type produced by ouabain binding can suffice to drive the formation of the acyl phosphate.

With sarcoplasmic reticulum ATPase, a strikingly rapid, Mg^{++}-dependent exchange of P_i oxygens with water oxygens occurs, as noted in Table 5. This exchange is strongly inhibited by the presence of Ca^{++} ions. A similar but less pronounced inhibition by Na^+ ions occurs with the Na^+,K^+-ATPase (3). Thus with both the Na^+,K^+ and the Ca^{++},Mg^{++}-ATPases the presence of the ions actively transported decrease the capacity for the exchange, probably representing a decreased rate or extent of formation of acyl phosphate from P_i.

The results in Table 5 show that a small but significant fraction of the enzyme is phosphorylated in presence of inorganic phosphate and Mg^{++}. The preparation in presence of ATP and Ca^{++} would show nearly 40 times as much phosphorylated protein. This implies an exceptionally rapid turnover of the small amount of E-P formed from P_i, equivalent to about 20,000 per minute.

The nature of the carboxyl activation for enzyme-phosphate formation. The two most likely means for formation of the phosphoryl enzyme from inorganic phosphate are (a) the phosphorolysis of a pre-existing covalent bond or (b) the displacement by a carboxyl oxygen of a hydroxyl group from the entering P_i. As noted in Fig. 2, these two modes of phosphoenzyme formation would give a differing distribution of oxygen-18 originally in the P_i. Thus if the carboxyl oxygen carries out a displacement reaction, the C-O-P bridge oxygen would obviously be furnished by the carboxyl group. If phosphorolysis occurs, the P_i would furnish the bridge oxygen.

A means of distinguishing between the two possibilities is provided by use of appropriate chemical cleavages of the protein acyl phosphate. As illustrated in Fig. 2, if phosphorolysis were responsible for formation of the acyl phosphate, more ^{18}O would appear in the P_i isolated after C-O cleavage than after P-O cleavage. The experimentally observed result (Fig. 2 and Table 6) is that all of the incorporated ^{18}O from P_i can be accounted for in the phosphoryl oxygens, with none in the bridge oxygen, demonstrating that the carboxyl group makes a direct displacement.

The presence or absence of K^+ during the hydrolysis does not change the apparent incorporation pattern of the ^{18}O, as indicated in Table 7. The fraction of the ^{18}O incorporated that can be accounted for in the phosphoryl group is within 10 per cent of the total ^{18}O present Because of the extreme limitations of sample size and sensitivity, this is about the range of the present experimental assessment. If the C-O-P bridge oxygen were labeled, no more than 75 per cent of the total ^{18}O would be accountable for in the phosphoryl group.

In the exchange accompanying ATP cleavage, the distributions of ^{18}O from water might be as indicated in Fig. 3. The experimentally observed results (Fig. 3 and Table 7) show all the detectable ^{18}O is present in the phosphoryl group and not in the bridge oxygen group. Although some preliminary data had indicated the possi-

bility of oxygen incorporation into the carboxyl groups
(8) the present more sensitive experiments are consistent
with all oxygen exchanges in the Na^+,K^+-ATPase occurring
through reversal of direct displacement of P_i oxygens by
the carboxylate function. Whether a similar pattern holds
for the Ca^{++},Mg^{++}-ATPase is currently under investigation.

Bacterial membrane inhibition by dicyclohexylcarbodiimide. As noted in the fine studies of Abrams and
co-workers (9), the membrane-bound ATPase of Streptococcus
faecalis can be inhibited by combination of dicyclohexylcarbodiimide (DCCD) with a membrane component, distinct from the ATPase per se. Based on the approaches used
by Beechey and colleagues with mitochondria (5), Eugene
Dinovo, in his Ph.D. studies in our group, has devised
conditions where remarkably sensitive inhibition of the
ATPase by DCCD can be obtained. The important feature is
that an amount of DCCD that is roughly equivalent to 1 or
2 molecules per ATPase will suffice for inhibition of the
ATPase. ATP protects against the inhibition. Such high
specificity suggests covalent modification of a specific
group at a critical site of the membrane protein that
interacts with the ATPase. Such a non-ATPase protein
component is obviously a prime candidate for the energy
transduction from the ATPase cleavage. The sensitivity to
the carbodiimide is consistent with the presence of an
essential carboxyl group, but other groups, such as the
sulfhydryl, will likewise react with the carbodiimide.
Dinovo has attempted specific labeling of the fraction by
using the displacement reaction with radioactive glycine
methyl ester, as introduced by Hoare and Koshland (10).
He has been able to show presence of a small protein
fraction whose labeling can be specifically prevented by
presence of ATP during exposure to the DCCD. Labeling by
such a displacement reaction is characteristic of a carboxyl adduct with the DCCD.

The inhibition of the bacterial ATPases may be quite
pertinent to the nature of the high energy compound or
state produced by oxygen uptake. Various recent findings
in our and other laboratories suggest that a common
energized compound or state is produced in bacterial membranes with use of energy derived either from oxidations
or ATP cleavage.

433

Some projected experiments with myosin. Our current
data on carboxyl activation and some other intriguing
data in the literature suggest that it may be of value to
explore the possibility that ATP cleavage by the myosin
molecule activates a carboxyl group located elsewhere than
the active site, in a manner such that a covalent bond is
formed by the activated carboxyl structure. Observations
reported several years ago by Tonomura and colleagues (11)
showed that myosin when exposed to p-nitrothiophenol for
a period of hours at 0° during ATP cleavage will form a
covalent thioester bond between a protein carboxyl group
and the p-nitrothiophenol. The ATPase capacity still
remains, although the interactions with actin are modified.
The interesting possibility which might be checked with
appropriate ^{18}O labeling techniques, is that the carboxyl
group may be induced to form an internal or possibly
external acyl-S linkage or acyl-X linkage concomitant with
the ATP cleavage. This carboxyl activation, spatially
distinct from the site of ATP hydrolysis, could represent
a link in the energy-transducing process of muscle. Results will not be forthcoming quickly, however, because
adequate methodology has to be developed for isolation
and detection of ^{18}O in the specific carboxyl group.

CONCLUSION

A specific and definable chemical change, either
involving covalent bond formation or a marked change in
reactivity of certain groups, may play a vital role in
energy transducing processes of membranes. Our experiments
have yielded several approaches for probing the possibility
of carboxyl activation as a central process. Future
perspective will likely show that these approaches are
inept or circuitous. But at some future time, some
approach is going to reveal essential facets of membrane
activation not yet understood. The motivation for experimentation thus remains.

REFERENCES

1. Boyer, P. D., in "Oxidases and Related Redox Systems". Vols. 1 & 2, Eds. T. E. King, H. S. Mason, M. Morrison, John Wiley & Sons, Inc., 1965, p. 994.

2. Cross, R. L. and Boyer, P. D., Fed. Proc., 30, 1127 (1971).

3. Dahms, A. S., Fed. Proc., 30, 680 (1971).

4. Kanazawa, T. and Boyer, P. D., Submitted for publication, (1971).

5. Cattell, K. J., Knight, I. G., Lindop, C. R., and Beechey, R. B., Biochem. J., 117, 1011 (1970).

6. Mitchell, R. A., Butler, L. G., and Boyer, P. D., Biochem. Biophys. Research Communs., 16, 545 (1964).

7. Zahler, W. L., Saito, A., and Fleischer, S., Biochem. Biophys. Research Communs., 32, 512 (1968).

8. Boyer, P. D., Klein, W. L., and Dahms, A. S., "International Colloquium on Bioenergetics", Pugnochioso, Italy, 1970, in press.

9. Harold, F. M., Baarda, J. R., Baron, C., and Abrams, A. J. Biol. Chem., 244, 2261 (1969).

10. Hoare, D. G., and Koshland, D. E., Jr., J. Biol. Chem., 242, 2447 (1967).

11. Kinoshita, N., Kubo, S., Onishi, H., and Tonomura, Y., J. Biochem., 65, 285 (1969).

Table 1

PROCEDURE FOR DETECTION OF MITOCHONDRIAL E~P[a]

Time (sec)	Sample	2,4-Dinitrophenol Discharged Control
0	Mitochondria	Mitochondria
60	$\longrightarrow$ $^{32}P_i$	$\longrightarrow$ $^{32}P_i$
70	$\longrightarrow$ H_2O	$\longrightarrow$ DNP
74	$\longrightarrow$ Acid stop	$\longrightarrow$ Acid stop

[a] E~P $\rightleftharpoons$ Protein-bound ^{32}P (Sample - Control)

Rotenone-inhibited mitochondria equivalent to between 5 and 20 mg protein per ml of 0.25 M Sucrose, 20 mM Tris, 10 mM KCl, and 2 mM $MgCl_2$, pH 7.4 are brought to about 21°. Additions are made in small volumes; final concentrations are 7.5 mM Succinate (0 sec), 0.1 mM $^{32}P_i$ containing 10^6 to 10^8 (60 sec), and 200 μM 2,4-dinitrophenol or an equivalent amount of water (70 sec). The reaction is stopped at 74 seconds by the addition of 4 volumes of 0.375 M PCA, 12.5 mM P_i. The protein precipitate is washed free of unreacted $^{32}P_i$ and counted by liquid scintillation.

Table 2

INABILITY OF CARBOXYATRACTYLOSIDE TO PREVENT
E~P FORMATION AND DINITROPHENOL DISCHARGE

Description	Carboxy-atractyloside	Total Protein-bound ^{32}P	E~^{32}P
		cpm	cpm
Sample	+	3770	
			1640 ± 200
DNP-discharged Control	+	2130	
Sample	−	4380	
			2020 ± 170
DNP-discharged Control	−	2360	

Mitochondria equivalent to 10 mg of protein in 1 ml of
the Sucrose-Tris-K^+-Mg^{2+} medium were treated as described
in Table 1 except that where noted 100 mmoles of carboxy-
atractyloside was added two minutes before succinate.
The CPM listed are averages of 4 replications. The error
estimates are at the 67% confidence level.

Table 3

PRESENCE OF E~P IN RESPIRATORY MEMBRANE FRACTION

Fraction	% in Fraction		E~^{32}P per mg Protein
	Protein	E~P	
			cpm
Soluble Proteins (Colorless) F$_1$, Matrix Proteins, etc. Soluble in Acetic Acid	78	14	600
Insoluble Proteins (Red-Brown) Respiratory Enzymes, Inner Membranes	22	86	12,900

Mitochondria equivalent to 25 mg of protein in 2 ml of the Sucrose-Tris-K$^+$-Mg^{2+} medium were treated as described in Table 1 except that the reaction was stopped by addition of 3 ml of 2.5% acetic acid, 2 mM H$_3$PO$_4$. The insoluble and soluble protein fractions were separated as described by Zahler et al. (7), washed free of unreacted ^{32}P$_i$ and counted.

Table 4

$^{32}P_i$ INCORPORATION INTO E-P AND $P_i \rightleftarrows$ HOH EXCHANGE
OBSERVED WITH Na^+,K^+-ATPase
IN THE PRESENCE OR ABSENCE OF OUABAIN

Ouabain added	E-P	$P_i \rightleftarrows$ HOH Exchange
mM	μmoles/g protein	oxygens exchanged per P_i present
1.0	0.269	< 0.02
0	0.058	0.75

$^{32}P_i$ incorporation into E-P: the reaction mixture contained 50 mM glycylglycine-imidazole (pH 7.2), 0.1 mM
$KH_2$$^{32}PO_4$ (10 μC/μmole), 5 mM $MgCl_2$, and 1.5 mg electroplax
Na^+,K^+-ATPase in a total volume of 1.0 ml. The reaction
was quenched after 15 minutes by the addition of 40 ml
4% perchloric acid containing 20 mM P_i. The acid-washed
protein-bound phosphate was determined. The above values
are corrected for $^{32}P_i$ incorporation into denatured
protein in the absence of ouabain.
$P_i \rightleftarrows$ HOH exchange: a mixture containing 50 mM glycyl-
glycine (pH 7.2), 40 mM KH_2PO_4, 0.25 mg electroplax
Na^+,K^+-ATPase, 5 mM $MgCl_2$, and 1.09 a.p.e. $H_2$$^{18}O$ in a
total volume of 1.0 ml was incubated for 1 hour at 37°C.
P_i was isolated after deproteinization with perchloric
acid, and its ^{18}O content determined.

Table 5

E-P FORMATION AND EXCHANGE CAPACITIES OF
SARCOPLASMIC RETICULUM IN PRESENCE OF Mg^{++} OR Ca^{++}

Exp.	Additions			$P_i \rightleftarrows$ HOH	E-P
	$MgCl_2$	$CaCl_2$	EGTA	exchange	
	mM	mM	mM	g atoms/sec /10^6 g protein	mole/10^7 g protein
1	5	0	5		0.91
	5	0.1	0		0.02
2	5	0	5	22	
	5	0.5	0.5[a]	0.6	

[a]Gives 14.5 µM free Ca^{++}

For $P_i \rightleftarrows$ HOH exchanges, 1 ml reaction, $H^{18}OH$, 100 mM KCl,
40 mM P_i, 100 mM Tris-HCl (pH 7.0), 0.5 mg sarcoplasmic
reticulum protein, 15°, 30 min.
For E-P formation, as above, but 5 mM $^{32}P_i$, 3 mg sarcoplasmic reticulum protein, 30 sec.

Table 6

DIRECT LOSS OF P_i OXYGEN ACCOMPANYING MEMBRANE
ACYL PHOSPHATE FORMATION IN Na^+,K^+-ATPase

Position of acyl phosphate bond cleavage	Observed atom percent excess ^{18}O	Expected atom percent excess if acyl phosphate bridge oxygen were furnished by P_i
C–O	0.053 ± .002	0.073
P–O	0.053[a]± .006	0.053

[a]Equivalent to 0.073 a.p.e. if calculated for the 3 oxygens of the phosphoryl group.

The reaction mixture contained 7.21 mM $^{32}P_i^{18}O$ (73 atom per cent excess ^{18}O), 5 mM $MgCl_2$, 50 mM imidazole-glycyl-glycine (pH 7.2), 1 mM ouabain, and 1.35 g porcine outer medullar Na^+,K^+-ATPase (specific activity = 13.1 μmoles per hour per mg protein) in a final volume of 60 ml at 4°C. Enzyme and ouabain were added at zero time and 10 seconds, respectively. The reaction was terminated at 40 seconds by the addition of 200 ml 4% perchloric acid containing 20 mM P_i. Inorganic phosphate was isolated following hydrolysis of the acid-washed, protein-bound acyl phosphate under conditions selective for P–O or C–O cleavage (Dahms and Boyer, unpublished) and the ^{18}O content measured. Values tabulated above indicate averages of triplicate determinations.

Table 7

INCORPORATION OF ^{18}O INTO THE PHOSPHORYL OXYGENS OF
MEMBRANE ACYL PHOSPHATE DURING ATP HYDROLYSIS

Exp. #	Additions	Position of acyl bond cleavage	Atom percent excess ^{18}O	Phosphoryl or phosphate oxygens exchanged per acyl phosphate phosphoryl oxygen	phosphate oxygen
1	$-K^+$	P–O	0.016	1.00	
		C–O	0.014		0.92
2	$-K^+$	P–O	0.014	0.93	
		C–O	0.013		1.02
3^a	$+K^+$	P–O	0.053	1.02	
		C–O	0.048		1.09

^{a}Twice as much protein and ATP were used.
The reaction mixture contained 3 mM ATP-γ-^{32}P (17 μC/
μmole), 6 mM $MgCl_2$, 140 mM NaCl, $\pm$ 20 mM KCl, 7.1 mM P_i,
50 mM imidazole-glycylglycine (pH 7.3), 300 mg electroplax Na^+,K^+-ATPase (specific activity = 309 μmoles per
hour per mg protein) in a total volume of 8.0 ml $H_2^{18}O$
(9.64 atom percent ^{18}O) at 15°. The reaction was initiated upon addition of ATP and quenched after 15 seconds
by the rapid addition of 300 μl 92% trichloroacetic acid
containing 0.1 M P_i. Inorganic phosphate was isolated
following hydrolysis of the acid-washed, protein-bound
acyl phosphate under conditions selective for P–O or C–O
cleavage (Dahms and Boyer, unpublished) and the ^{18}O
content measured.

INTRAMITOCHONDRIAL LOCATION OF E~P

MATRIX SPACE	INNER MEMBRANE	INTERMEMBRANE SPACE	OUTER MEMBRANE	OUTSIDE SPACE

ELECTRON TRANS.

DNP

X~I
OLIGOMYCIN

Z~P   X~P

ADP   Pᵢ

ATP

ATP → ATP
ADP ← ADP   Y~P

CARBOXY-ATRACTYLOSIDE

ADP

Fig. 1 Diagrammatic sketch of possible location of mitochondrial phosphoproteins.

RETENTION OF ¹⁸O FROM Pᵢ WHEN ENZYME + Pᵢ ⟶ E-P

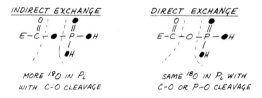

INDIRECT EXCHANGE

MORE ¹⁸O IN Pᵢ
WITH C-O CLEAVAGE

DIRECT EXCHANGE

SAME ¹⁸O IN Pᵢ WITH
C-O OR P-O CLEAVAGE

OBSERVED ATOM PERCENT EXCESS ¹⁸O:

BRIDGE OXYGEN	PHOSPHORYL OXYGEN
$C-O-P$	$-PO_3$
< 0.003	$0.073 \pm .006$

Fig. 2 Experimental approach for determining ^{18}O incorporation patterns from P_i-^{18}O during $P_i \rightleftarrows HOH$ exchange by Na^+,K^+-ATPase.

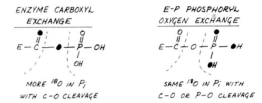

UPTAKE OF ^{18}O FROM WATER WHEN ENZYME + ATP ⟶ E-P

OBSERVED ATOM PERCENT EXCESS ^{18}O:

BRIDGE OXYGEN	PHOSPHORYL OXYGEN
$C-O-P$	$-PO_3$
<0.003	$0.053 \pm .005$

Fig. 3 Experimental approach for determining ^{18}O incorporation patterns from $H^{18}OH$ during ATP hydrolysis by Na^+, K^+-ATPase (see data of table 7, exp. 3).

THE RELATIONSHIP OF BACTERIAL MEMBRANE ORIENTATION TO OXIDATIVE PHOSPHORYLATION AND ACTIVE TRANSPORT*

A.F. Brodie, H. Hirata, A. Asano, N.S. Cohen, T.R. Hinds, H.N. Aithal and V.K. Kalra

Department of Biochemistry
University of Southern California School of Medicine
Los Angeles, California 90033

ABSTRACT. Membrane vesicles differing in size and orientation differ in their ability to carry out coupled phosphorylation and active transport. Ghost preparations were capable of oxidation whereas phosphorylation was cryptic. The ghost membranes were oriented as in vivo. The ETP appear to contain two types of particles, the bulk of which are oriented inside out. Transport of proline occurred in the ghosts and ETP, and required Na^+. With ETP, in which SDH is on the outside, the addition of succinate following proline uptake resulted in an efflux of proline, whereas with the ghosts the efflux did not occur. Irradiation (360nm) resulted in a loss of oxidation and transport. Lapachol, which bypasses a segment of the respiratory chain, restored both activities. Heat treatment produced conformational changes in the membranes of ETP but not of ghosts.

INTRODUCTION

Various types of membrane preparations of bacterial origin have been resolved which differ in their capacity to transport amino acids and to couple phosphorylation to the oxidation of substrate. Oxidative phosphorylation and

---

*This paper is dedicated to the late Dr. Shinji Ishikawa whose association and scientific contributions have influenced our group.

active transport are membrane-related phenomena which exhibit a requirement for structural organization and spatial
orientation. In a number of respects the bioenergetic
mechanisms for the two processes differ; however, both
active transport and oxidative phosphorylation in aerobic
organisms require substrate oxidation. Nevertheless, it is
not known whether they share a common respiratory chain or
whether the respiratory pathways differ. In the present
communication an attempt will be made to characterize these
processes with membrane structures which differ in degree
of integrity and in vectorial orientation.

METHODS

Types of membrane structures

A number of bacterial systems have been used to study
membrane related phenomena. In most instances little is
known about the nature of the membrane, electron transport
pathways or the bioenergetic processes. In contrast, the
membrane vesicles from Mycobacterium phlei have been characterized with regard to the nature of the respiratory
carriers (1,2), the sequence of electron transport (1) and
the coupling process (3). In addition, both the protoplast
ghosts and the smaller membrane vesicles derived from the
ghost preparations or by sonication of whole cells have
been shown to transport proline against a concentration
gradient (4).

Sonic oscillation of whole cells followed by differential centrifugation (144,000 x g) results in the formation of membrane vesicles which are referred to as regular
particles or the electron transport particles (ETP). Membrane vesicles may be obtained from sonication of either
whole cells or protoplast ghosts, and appear to be similar
in morphology and in their ability to carry out oxidative
phosphorylation or active transport of amino acids. These
ETP preparations appear to consist of a heterogeneous
population of membrane vesicles (5) which differ in size
and membrane orientation, (Table 1). Thus what has been
referred to as the electron transport particles contains a
number of different types of membrane vesicles.

The ETP obtained by sonic oscillation of whole cells
or protoplast ghosts range in size from 80 to 120 nm in
diameter. Negative staining revealed that the bulk of the
membrane structures contain repeating units or spherical

446

bodies (90 - 120 Å) protruding from the membrane attached
to a stalk (23 Å). These vesicular structures are capable
of oxidative phosphorylation and are morphologically identical to those observed with submitochondrial particles.

The membrane vesicles in the ETP fraction can be
resolved by sucrose density gradient centrifugation in the
absence of Mg^{++} ions (5,6). The heaviest fraction appears
as a red pellet and contains the complete electron transport pathway for succinate and NAD^+-linked oxidation. However, this fraction fails to couple phosphorylation to the
oxidation of either substrate and is referred to as depleted -ETP (DETP). Morphologically the DETP appear as
membrane vesicles without the repeating unit or stalk. Restoration of phosphorylation, and reappearance of the repeating units occur upon the addition of one of the light
fractions (Fraction IV) from the sucrose gradient, to the
DETP in the presence of Mg^{++} ions (Table 2). The repeating
units appear to be a "particle-bound" coupling factor (5)
and contain about 95% of the latent ATPase observed in the
ETP.

The bulk of the ETP and DETP preparation represents
membrane vesicles oriented inside out; however, it should be
noted that these preparations are contaminated with membrane vesicles oriented right side out (Fig. 1). Antibodies
have been prepared to the repeating units (Fraction IV)
which precipitate membrane vesicles oriented inside out
due to the presence of repeating units. Following low
speed centrifugation the supernatant contains membrane
vesicles which are composed almost entirely of membranes
oriented right side out. As will be seen these vesicles
contain the respiratory chain and are similar in enzymatic
activities to the ETP or DETP preparations.

The lack of respiratory control and low levels of
phosphorylation observed with most bacterial systems have
been attributed to the harsh procedures used to disrupt
bacterial cells. Nevertheless, protoplasts obtained by
mild disruption of M. lysodeikticus fail to exhibit respiratory control and exhibit low levels of phosphorylation (7).
Attempts to obtain protoplasts of M. phlei by treatment
with lysozyme, penicillin or following removal of waxes and
lipoidal materials from the cell wall have been unsuccess-
ful. However, incubation of the whole cells in glycine and
sucrose for 3 hours followed by lysozyme treatment according to the method of Mizuguchi and Tokunaga (8) has result-

ed in the formation of protoplasts of M. phlei. Ghosts of
the protoplasts are obtained by bursting the protoplasts in
2 mM $MgCl_2$.

The isolated ghosts appear to be about the same size
as the whole cells and appear to be intact. Although the
protoplast ghosts were capable of oxidation, their ability
to couple phosphorylation to oxidation was found to be
cryptic (9). It is also of interest that the different
types of membrane vesicles and the ghost preparation contain similar ratios of cytochromes b, c, and a + a_3. Sonic
oscillation of the ghost fraction results in the formation
of smaller membrane vesicles which appear to be similar to
the ETP obtained by sonic oscillation of whole cells.

The existence of different types of membrane structures provides a means of examining phenomena which require structural integrity and strict spatial orientation
for function. The ability of membrane structures to carry
out oxidative phosphorylation and active transport is taken
as an index of structural integrity of the membrane. However, it should be noted that these two activities are
separable; for example the membrane vesicles referred to as
DETP are capable of oxidation and active transport but fail
to couple phosphorylation to the oxidation of substrate. In
contrast, membrane vesicles which have been stored carry
out oxidative phosphorylation but lose the ability to carry
out active transport of proline.

RESULTS

Comparison of protoplast ghosts to ETP

Protoplast ghosts from M. phlei were found to contain
small amounts of the soluble enzymes, (adenylate kinase,
β-hydroxybutyrate dehydrogenase and fumarase) found in the
intact cell (10). In contrast, the content of cytochromes
b, c_1+c and a+a_3, latent ATPase, and phospholipid was
similar to that of the membrane vesicles (ETP) obtained by
direct sonication of intact cells. Malate-vitamin K reductase is associated with the ghost fraction but is solubilized
during sonication of the ghosts or intact cells.

Although the oxidative activities of the ghosts with
all substrates tested were similar or slightly lower than
those observed with the ETP, the ghost preparations failed
to couple phosphorylation to any of the substrates tested
(Table 3). Phosphorylation with the protoplast ghosts was

observed following sonic oscillation or upon addition of a
soluble protein component from the supernatant (9). The
lack of phosphorylation with the ghost fraction was attributed to the lack of permeability of the bacterial membrane
to exogenous cofactors used to assay phosphorylation. Preincubation of the ghost preparation with high concentrations
of the components necessary for phosphorylation for 60
minutes at 0°, followed by elevation of the temperature to
30°, increased the P/O ratio. However, the formation of ATP
appeared to occur inside the ghosts and only a small amount
of the energy-rich phosphate bond was transferred to exogenous nucleotides.

The vectorial orientation of the ghosts and ETP preparations was determined by examination of the distribution
of latent ATPase following sucrose density gradient centrifugation in the absence of Mg^{++} ions and by studying the
effect of uncoupling agents whose inhibitory effects on
mitochondria or submitochondrial particles have been shown
to be dependent on the orientation of the membrane. Higashi
et. al., (5) have shown that the repeating unit which exhibits latent ATPase can be removed from the membrane by
sucrose density gradient centrifugation in the absence of
Mg^{++} ions. However, sucrose density gradient centrifugation
of the ghost fraction did not dissociate the latent ATPase
from the membrane. In addition, other proteins which are
released from the ETP were not released from the ghosts by
this treatment.

A comparison of the effects of uncoupling agents related to ion transport on oxidative phosphorylation of the
ghosts (preincubation method) with that of the ETP revealed
differences between these membrane structures (Table 4).
With the ghost preparations, valinomycin in the presence of
K^+ ions, dimethyldibenzylammonium chloride (DDA$^+$) and carbonyl cyanide m-chlorophenylhydrazone (CCP) uncoupled phosphorylation whereas valinomycin in the presence of Na^+ ions,
nigericin and tetraphenylboron (TPB) had little or no uncoupling effect. In contrast, the ETP preparation, like
submitochondrial particles (11-13), was uncoupled by valinomycin but required ammonium ions or nigericin and K^+ ions.
The ETP were insensitive toward DDA$^+$.

The ETP preparation may consist of a mixture of vesicles, some with membranes oriented as in the intact cell
(ETP_o) and others in which the membrane orientation is inside out (ETP_i). Phosphorylation was observed only with

membrane vesicles oriented inside out. The removal of coupling factors from ETP (5), the inhibition of phosphorylation by specific antibodies to the coupling factors (14), the effect of uncoupling agents and electron microscopic observation of structures similar to the repeating units of the mitochondrial inner membrane (15) support the conclusion that the bulk of the ETP preparation consists of membranes in which the inner surface is on the outside of the vesicle. In contrast, similar types of evidence indicates that the protoplast ghosts are oriented right side out as in the intact cell.

It is of interest that the addition of a soluble protein factor to the ghost preparation elicits a response similar to that observed with inside out vesicles with respect to oxidative phosphorylation (9). The mechanism of action of the soluble factor is not known; however, it appears to eliminate the permeability barrier and facilitate the diffusion of small molecules required for phosphorylation.

Active transport of proline by ghosts and ETP

Active transport is a characteristic of biological membranes and has been studied in bacterial and mammalian systems. However, little is known about the relationship between the transport and the energy yielding processes. Respiration dependent transport of proline has been demonstrated in the ETP from M. phlei (4,16). The uptake of proline proceeded against a concentration gradient with succinate, generated NADH, exogenous NADH or ascorbate-TPD as substrate. The latter two electron donors were the most effective for the transport of proline and for oxidation but they were the least efficient for oxidative phosphorylation. The transport of proline appears to be independent of oxidative phosphorylation since it proceeded in the absence of coupling factors or phosphate and was not inhibited by arsenate.

Kaback and his collaborators (17,18) have developed a subcellular membrane transport system using isolated membrane preparations from E. coli. They have suggested that the transport of amino acids and sugars is coupled primarily with D-lactic dehydrogenases and a carrier protein. The carrier protein appears to be one of the electron transport mediators (19,20). A requirement for oxidative energy has been suggested by several investigators (21,22); however, little is known about the bioenergetic process in these microorganisms.

The protoplast ghosts from M. phlei appear to have an orientation similar to that described for isolated membrane vesicles from E. coli. The ghost preparations from M. phlei have been found to concentrate proline against a gradient with various electron donors. The uptake of proline by the ghosts and ETP is compared in Table 5. On the assumption that the ^{14}C-inulin impermeable and ^{3}H$_2$O permeable space (approximately 7 μl/mg protein) equals the intravesicular space, the concentration of proline in the ghosts is about 1.7 mM with succinate or ascorbate-TPD. Compared to the medium (9 μM, initial) the ghosts established a gradient of 200-fold. In contrast, the ETP vesicles established a concentration gradient of greater than 30-fold with ascorbate-TPD as the electron donor. It is of interest that with membrane vesicles oriented inside out the transport of proline is low with succinate and best with ascorbate-TPD. The succinate dependent transport or proline differs with different types of membranes and is dependent on the vectorial orientation of the structure.

Active transport of proline was inhibited in the ETP by respiratory inhibitors, uncoupling agents or anaerobiosis (Table 6). A similar pattern of inhibition was observed with the ghost preparation. However, the insensitivity of transport to high concentrations of arsenate and DCCD, both of which are known to inhibit the energy transfer reaction of oxidative phosphorylation, suggests that neither the coupling of high energy phosphate, ATP, nor high energy phosphorylated intermediates (23) are involved in the bioenergetics of the transport process. The data suggest that active transport is inextricably associated with oxidation.

Furthermore, as shown in Table 7, DETP which are depleted of their membrane-bound coupling factor (BCF$_4$) exhibit no phosphorylative activity but are capable of active transport to the level observed with ETP. Although one can conclude that the formation of ATP coupled to the oxidation of substrate is not necessary for transport, oxidation via the respiratory chain is obligatory for active transport. However, the inhibition of proline uptake by uncoupling agents (mCCP or PCP) and the ionophoretic antibiotic valinomycin (Table 6) does not rule out the participation of a high energy intermediate or high energy state created by oxidation.

The transport of proline by ETP was found to be

dependent on the presence of sodium or lithium ions, (Table 8). Kinetic studies of the effect of varying concentrations of sodium ions on proline transport indicate that the system has the same K_m for proline but that the V_{max}. differs. Thus it would appear that sodium is necessary for the transport of proline in M. phlei but that this was not a sodium co-transport system. A sodium co-transport system for sugar and amino acids has been observed in mammalian cells (24) and for sugar transport in Salmonella typhimurium (25).

Orientation of the membrane and its relationship to proline transport

Since different membrane preparations differ in membrane orientation it was of interest to investigate amino acid transport in preparations which exhibit vectorial orientation. Sonication* of the ghost preparation (Fig. 2) was found to result in a loss of proline transport (88%) with succinate as an electron donor whereas with ascorbate-TPD the transport decreased 50%. Even if one considers the change in intravesicular volume before and after sonication (7 μl and 1.7 μl per mg protein respectively) the loss in transport activity with succinate should be reflected by a similar loss of proline transport with ascorbate-TPD. These results are in agreement with those shown in Table 5 in which the succinate dependent transport of proline was low in the ETP but was the same as ascorbate-TPD in the ghosts.

The membrane-bound coupling factor (BCF$_4$) exhibits latent ATPase activity and is one of the enzymes bound to the inner surface of the cytoplasmic membrane. Thus, it can be used as a marker enzyme to detect differences in membrane orientation. In addition, antibodies have been prepared to the purified latent ATPase, and as shown in Fig. 2, sonication of ghosts results in an increase in the latent ATPase activity which was susceptible to the antibodies. Thus, it would appear that sonication of ghosts resulted in the turning over of the membrane structure. If one assumes that the direction of active transport is fixed from the in vivo outside of the membrane to the inside, it is difficult to explain the observation in Fig. 2, since the inside out vesicles should not carry out active transport.

In order to explain the differences observed with

---

*It should be noted that 4 minutes of sonication of whole cells is used to obtain ETP since longer treatment results in a loss of oxidative phosphorylation.

ghost and ETP preparations the following possibilities were considered: (a) the sonicated vesicles consist of a mixed population of inside out and right side out membrane vesicles; (b) that ascorbate-TPD can drive transport from both directions whereas succinate is unidirectional; (c) the third coupling site, which derives electrons from ascorbate-TPD, remains structurally in the same configuration in the inverted membrane vesicles and in the ghosts.

In order to examine the suggested hypotheses, proline transport with the ETP preparations was tested with both substrates (Table 9). Succinate plus ascorbate-TPD resulted in a significant decrease in the steady state level of proline in the ETP. Since succinic dehydrogenase may be located on the outer surface of the ETP it was of interest to determine the effect of malate-vitamin K reductase (MKR), an enzyme solubilized from ETP during sonication. As seen in Table 9 the addition of soluble MKR also caused a decrease in the steady state level of proline which was established with ascorbate-TPD.

A steady state level of proline accumulation was established by the addition of ascorbate-TPD; however, the further addition of succinate resulted in an efflux of the accumulated proline. Furthermore, the rate of proline efflux upon addition of succinate was similar to the rate of uptake of proline with ascorbate-TPD (Fig. 3a). It is of interest, however, that the efflux with succinate does not occur with ghost preparations. Instead, the addition of succinate resulted in a further uptake of proline which then returned to the same level as that observed before the addition of succinate (Fig. 3b).

The efflux of proline from the ETP after the addition of succinate was inhibited by malonate but not by NHQNO (Fig. 4). Since malonate is a competitive inhibitor of succinic dehydrogenase and NHQNO inhibits succinoxidase activity by blocking electron flow from cytochrome b to c (23), it seems likely that succinic dehydrogenase plays a role in the efflux process. In M. phlei ascorbate-TPD enters the respiratory chain at cytochrome c beyond the NHQNO block (24). This inhibitor does not affect oxidation or transport with ascorbate-TPD.

In preliminary experiments it was found that succinic dehydrogenase activity in the ghost preparations was very low; however, by the addition of non-ionic detergent (Triton X-100) the activity was increased several fold. On the other hand, ETP exhibited very high succinic dehydrogenase activity without any treatment. These results suggest that the succinic dehydrogenase is located on the inner surface of cytoplasmic membrane, similar to that observed in the inner membrane of mitochondria (27).

It is essential to know the nature of the respiratory chain utilized for the transport process. Irradiation with light at 360 nm has been shown in M. phlei to destroy the menaquinone (MK_9 (II-H)) and result in the inhibition of oxidation (28). Restoration of oxidative phosphorylation was found to be specifically dependent on the natural quinone or closely related homologues. However, other quinones were found which restore oxidation only and bypass electrons from the coupled respiratory pathway (29). Irradiation of the ETP was found to inhibit respiration and the active transport of proline; of particular interest was the finding that proline transport can be partially restored (38%) by the addition of lapachol (2 hydroxy-3-methyl-2-butenyl 1,4 naphthoquinone) a quinone which bypasses a major section of the respiratory pathway (Table 10).

Conformational changes in membrane structure

ETP from M. phlei exhibited an increased level of phosphorylation when subjected to heat treatment at 50° for 10 minutes (30). In contrast ETP subjected to freezing (-75° for 10 minutes) followed by slow thawing (frozen ETP) showed a decreased level of phosphorylation (31). This loss in activity was reversed by heat treatment of the frozen ETP (Table 11). Since freezing and thawing is known to result in changes in membrane structures leading to altered function (31-37), the reversible effect of heat treatment may represent structural reorganization of the membrane (31).

The fluorochrome, 1-anilino 8-naphthalene sulfonate (ANS) has been used to detect conformational changes in membranes and proteins (38-44). As in mammalian mitochondria (38-40,44), the ETP of M. phlei displayed both non-energized and energized fluorescence of ANS (NE_f and E_f respectively) with succinate as substrate (45). Heat treatment of the membranes resulted in an enhancement of NE_f and E_f whereas a decrease in E_f was observed following freezing the ETP (Fig. 5). When frozen ETP were heated to 50° for 10 minutes the particles exhibited the same level of E_f as that elicited by heat treated ETP; however, the level of phosphorylation was similar to that of untreated ETP but did not attain the level observed in heat treated ETP (Table 11). Thus a direct correlation between the level of phosphory- lation and E_f could not be established. There was a signif- icant decrease in the dissociation constant (K_D) of the heat treated membrane-dye complex with no change in the number of binding sites in the energized state (Table 11). It was also observed that in the energized state there was a 3-fold increase in the relative quantum yield (42) in heat treated ETP as compared to untreated ETP. These re- sults indicate that the observed changes in fluorescence were due to changes in the membrane structure.

The increase in the E_f following heat treatment of the membrane vesicles was also obtained with other substrates such as NADH or ascorbate-PMS (Table 12). In contrast to submitochondrial particles (39), ATP and Mg^{++} ions caused a decrease in fluorescence of ETP. Latent ATPase activity in M. phlei is unmasked by trypsin treatment, and this treatment also increases oxidative phosphorylation (47). ATP, in the presence of Mg^{++}, induced an increase in fluore- scence in trypsin treated ETP (Table 13).

Depleted ETP lack the ability to couple phosphorylation but have been shown to carry out amino acid transport (Table 7). The same level of E_f was observed in both DETP and un- treated ETP (46). Thus particulate-bound coupling factor(s) was not required for the substrate induced E_f. The addition of proline to ETP or DETP in the E_f state failed to induce any change in fluorescence. It should be noted that these results do not preclude the possibility that some conform- ational change occurs during active transport.

DISCUSSION

Membrane preparations differing in size and vectorial orientation have been obtained from M. phlei. These struct- ures differ in their ability to carry out oxidative phos- phorylation and active transport. Protoplast ghost prepara- tions were capable of oxidation but coupled phosphorylation was found to be cryptic. The effects of uncoupling agents, the distribution of latent ATPase following sucrose density gradient centrifugation in the absence of Mg^{++} ions, and morphological studies have indicated that the membranes of the ghost preparations were oriented as in the intact cell. However, the ETP preparation appeared to be composed of at least two classes of particles the bulk of which are oriented inside out.

The substrate dependent active transport of proline has been shown to occur in both the ghost and ETP preparations. Proline transport required Na^+ or Li^+ ions but did not appear to involve a co-transport process. Stock and Roseman (25) have shown that in Salmonella typhimurium the transport of sugars was found to involve a co-transport of Na^+ ions. A similar co-transport phenomenon has been observed in animal cells (24).

The dependence of active transport on the orientation of the membrane vesicles was indicated by the succinate dependent proline efflux observed in the ETP but not in the ghost preparation. Assuming that succinic dehydrogenase is located on the cytoplasmic side of the membrane whereas the cytochrome oxidase complex is accessible from either side as in mitochondria (27), the following hypothesis was formulated: the driving force for transport is controlled by the location of certain enzymes, such as succinic dehydrogenase or cytochrome oxidase (Fig. 6).

Kaback et. al., (17-20) with a system from E. coli have indicated that oxidative phosphorylation is not involved in active transport. However, little is known about the nature of the respiratory carriers the sequence of electron transport, and the nature of coupled phosphorylation in this microorganism (48). For example, there is no evidence for the involvement of nonheme iron and a quinone on the oxygen side of cytochrome b. In addition a number of oxidative enzymes of the respiratory chain are inhibited by sulfhydryl agents. Thus the inhibition of transport by sulfhydryl agents is difficult to interpret.

The studies with M. phlei indicated that whereas oxidative phosphorylation is not required for proline transport, the participation of a high energy intermediate or high energy state created by oxidation is not ruled out. Irradiation with light (360nm) resulted in a loss of oxidation and of transport of proline. Lapachol, a naphthoquinone known to bypass a segment of the respiratory chain was found to restore both oxidation and transport. It would be expected that if there were a correlation between the high energy intermediate of oxidative phosphorylation and the transport process, substrates such as generated NADH and succinate which utilize 3 and 2 phosphorylative sites respectively would provide more energy for transport than ascorbate-TPD. It would appear likely that the energy used for active transport may be derived from a chemical gradient established by ions or proton movement.

ACKNOWLEDGMENTS

This work supported by the National Science Foundation Grant #GB6257X, USPHS National Institutes of Health Grant #AI05637, and the Hastings Foundation of the University of Southern California School of Medicine. This is the 59th paper in a series dealing with oxidative phosphorylation in fractionated bacterial systems. We gratefully acknowledge the technical assistance of Miss Patricia Brodle and Mrs. Keiko Kikekawa.

457

REFERENCES

1. Asano, A. and Brodie, A.F., J. Biol. Chem., 239, 4280 (1964).
2. Brodie, A.F. and Gutnick, D., in "Electron and Coupled Energy Transfer in Biological Systems" (T.E. King and M. Klingenberg, eds.), Marcel Dekker Inc., New York, Vol. 1, part B, p. 599 (1972).
3. Asano, A. and Brodie, A.F., J. Biol. Chem., 240, 4002 (1965).
4. Hirata, H., Asano, A. and Brodie, A.F., Biochem. Biophys. Res. Commun., 44, 368 (1971).
5. Higashi, T., Bogin, E. and Brodie, A.F., J. Biol. Chem., 244, 500 (1969).
6. Bogin, E., Higashi, T. and Brodie, A.F., Biochem. Biophys. Res. Commun., 38, 478 (1970).
7. Ishikawa, S. and Lehninger, A., J. Biol. Chem., 237, 240 (1962).
8. Mizuguchi, Y. and Tokunaga, T., J. Bacteriol., 104, 1020 (1970).
9. Asano, A., Hirata, H. and Brodie, A.F., Biochem. Biophys. Res. Commun., 46, 1340 (1972).
10. Asano, A., Cohen, N., Baker, R.F. and Brodie, A.F., unpublished observations.
11. Montal, M., Chance, B., Lee, C.P. and Azzi, A., Biochem. Biophys. Res. Commun., 34, 104 (1969).
12. Cockrell, R.S. and Racker, E., Biochem. Biophys. Res. Commun., 35, 414 (1969).
13. Smith, E.H. and Beyer, R.E., Arch. Biochem. Biophys., 122, 614 (1967).
14. Bogin, E., Higashi, T. and Brodie, A.F., Arch. Biochem. Biophys., 136, 337 (1970).
15. Asano, A., Cohen, N.S. and Brodie, A.F., Fed. Proc., 30, 1285 Abs. (1971).
16. Hirata, H. and Brodie, A.F., Fed. Proc., in press (1972).
17. Kaback, H.R. and Milner, L.S., Proc. Natl. Acad. Sci. U.S., 66, 1008 (1970).
18. Barnes, Jr., E.M. and Kaback, H.R., Proc. Natl. Acad. Sci. U.S., 66, 1190 (1970).
19. Barnes, Jr., E.M. and Kaback, H.R., J. Biol. Chem., 246, 5518 (1971).
20. Kaback, H.R. and Barnes, Jr., E.M., J. Biol. Chem., 246, 5523 (1971).

21. Klein, W.L., Dahns,A.S. and Boyer, P.D., Fed. Proc., 29, 341 (1970).
22. Konings, W.N.and Freese, E., Fed. Eur. Biochem. Soc. Lett., 14, 65 (1971).
23. Brodie, A.F. and Adelson, J., Science, 149, 265 (1965).
24. Schultz, S.G. and Curran, P.F., Physiol. Rev., 50, 637 (1970).
25. Stock, J. and Roseman, S., Biochem. Biophys. Res. Commun., 44, 132 (1971).
26. Orme, T.W., Revsin, B. and Brodie, A.F., Arch. Biochem. Biophys., 134, 172 (1969).
27. Lee, C.P., Johansson, B. and King, T.E. in "Probes of Structure and Function of Macromolecules and Membranes" (B. Chance, C.P. Lee and J. K. Blasie, ed.), Academic Press, New York and London, Vol. 1, p. 401 (1971).
28. Brodie, A.F. and Ballantine, J., J. Biol. Chem., 235, 226 (1960).
29. Brodie, A.F. and Ballantine, J., J. Biol. Chem., 235, 232 (1960).
30. Bogin, E., Higashi, T. and Brodie, A.F., Proc. Natl. Acad. Sci. U.S., 67, 1 (1970).
31. Aithal, H.N., Kalra, V.K. and Brodie, A.F., Biochem. Biophys. Res. Commun., 43, 550 (1971).
32. Chilson, O.P., Costello, L.A. and Kaplan, N.O., Fed. Proc., 24, S55 (1965).
33. Aithal, H.N., Ramasarma, T., Biochem. J., 115, 77 (1969).
34. Fischbein, W.N. and Stowell, R.E., Cryobiology, 4, 283 (1968).
35. Camerino, P.W., and King, T.E., Biochim. Biophys. Acta, 96, 19 (1965).
36. Walton, K.G., Kervina, M., Fleischer, S. and Don, D.S., J. Bioenergetics, 1, 3 (1970).
37. Heber, U., Tyankova, L.and Santarius, K.A., Biochim. Biophys. Acta, 241, 578 (1971).
38. Azzi, A. and Vainio, H., in "Probes of Structure and Function of Macromolecules and Membranes" (B. Chance, C.P. Lee and J.K. Blasie, eds.), Academic Press, New York, Vol. 1, p. 209 (1971).
39. Chance, B., Proc. Natl. Acad. Sci. U.S., 67, 560 (1970).
40. Datta, A. and Penefsky, H.S., J. Biol. Chem., 245, 1537 (1970).
41. Vanderkooi, J. and Martonosi, A., in "Probes of Structure and Function of Macromolecules and Membranes" (B. Chance, C.P. Lee and J.K. Blasie, eds.), Academic Press,

New York, Vol. I, p. 293 (1971).

42. Brocklehurst, J.R. and Radda, G.K., in "Probes in Structure and Function of Macromolecules and Membranes" (B. Chance, C.P. Lee and J.K. Blasie, eds.), Academic Press, New York, Vol. II, p. 59, (1971).

43. Stryer, L., J. Mol. Biol., 13, 482 (1965).

44. Nordenbrand, K. and Ernster, L., Europ. J. Biochem., 18, 258 (1971).

45. Kalra, V.K., Aithal, H.N. and Brodie, A.F., Biochem. Biophys. Res. Commun., 46, 979 (1972).

46. Aithal, H.N., Kalra, V.K. and Brodie, A.F. unpublished observations.

47. Bogin, E., Higashi, T. and Brodie, A.F., Biochem. Biophys. Res. Commun., 41, 995 (1970).

48. Kashket, E.R. and Brodie, A.F., J. Biol. Chem., 238, 2564 (1963).

TABLE 1

Summary of Different Types of Membrane Fractions from M. phlei

and Their Properties

	Membrane Preparation	Appearance EMG (negative staining)	Cytochrome[***]	Oxidation[*]	Phosphorylation[*]	Amino Acid Transport (Active)	Latent ATPase
Types	ETP		b, c_1+c, a+a$_3$	++	++	proline ++	++++
	DETP		b, c_1+c, a+a$_3$	++	0	proline ++	0
	DETP + Fraction IV		b, c_1+c, a+a$_3$	++	++	proline ++	++++
	Ghosts		b, c_1+c, a+a$_3$	++	cryptic	proline ++	++++
	Sonicated ghost vesicles		b, c_1+c, a+a$_3$	++	++	proline ++	++++
Orientation Of Membrane	ETP outer		b, c_1+c, a+a$_3$	++	not tested	proline	not tested
	ETP inner		b, c_1+c, a+a$_3$	++	++	proline	++++
	DETP inner					not tested	
Physical Treatment	Heat-Treated ETP		b, c_1+c, a+a$_3$	++	++++	proline ++	++++
	Freeze-Treatment ETP	distorted membrane fragments	b, c_1+c, a+a$_3$	+-	+-	not tested	++
	Freeze-heat treated ETP		b, c_1+c, a+a$_3$	++	+++	not tested	++++

[*] Either succinate or NAD$^+$-linked substrates.

O indicates particle with thickened membrane may be These structures may be considered as a small ghost preparation.

[**] Although the cytochrome content in the different preparations differ, the ratio of one cytochrome to the other is the same in all the membranes except ETP_O.

461

TABLE 2

Sucrose Density Gradient Centrifugation of the ETP

Fraction	Oxidation µatoms	Phosphorylation µmoles	P/O ratio
ETP*	10.9	8.2	0.75
I (DETP)	10.6	0.0	0.0
IV	1.5	0.7	---
DETP + IV	10.4	7.5	0.72

Data from Higashi et. al., (5).   *ETP before centrifugation.

TABLE 3

A Comparison of Oxidative Phosphorylation in Ghosts and ETP*

Substrate	Ghosts		ETP	
	ΔO µatoms	ΔP_i µmoles	ΔO µatoms	ΔP_i µmoles
Generated NADH	7.6	0.0	10.0	9.5
Succinate	7.2	0.3	7.4	6.4
	Sonicated Ghosts			
Generated NADH	7.8	9.4	----	---

*Unpublished observations of Asano et. al., (10).   The system contained 10 mg of protein and the reaction was carried out for 5 minutes.

TABLE 4

Effect of Uncoupling Agents*

Agent	Conc.	Ghosts (P/O)	ETP (P/O)
None	----	0.7	2.34
Valinomycin	3.3 µM	0.0	1.95
Nigericin	1.3 µM	0.6	1.96
Valinomycin	3.3 µM		
+ NH$_4$	5 mM	---	0.73
TPB	67 µM	0.4	0.24
DDA	16.7 mM	0.0	2.21
CCP	2 µM	0.2	----

*Unpublished observations of Asano et. al., (10). Generated NADH was used as a substrate.

TABLE 5

Transport of Proline by Ghosts and ETP

	Ghosts (pmoles/mg protein)	ETP (pmoles/mg protein)
None	870	6.5
Succinate	12,770	18.7
Generated NADH	2,100	47.8
NADH	8,850	129.5
Ascorbate-TPD	12,030	225.0

The reaction mixture (0.1 ml) contained 5 x 10^{-2} M potassium phosphate buffer (pH 7.0), 10^{-2} M MgCl$_2$, substrate, 9 x 10^{-8} M L-proline-U-^{14}C (200 mCi/mmole) and ETP (0.21 mg) or ghosts (0.07 mg). Chloramphenicol (10 µg) was also added in the experiment with ghosts. The reaction was started by the addition of substrate and ^{14}C-proline. The amount of substrate used was 2.5 µmoles Na-succinate, 0.4 µmoles ethanol, 0.01 µmole semicarbazide, 0.1 µmole NAD$^+$, 5 µg alcohol dehydrogenase, 1 µmole NADH, 1.7 µmoles Na-ascorbate, 0.015 µmoles TPD.

TABLE 6

Effect of Inhibitors on Proline Transport by ETP*

Additions	Conc.	Proline Uptake pmoles/mg protein	%
None		120.5	100
Anaerobic		19.2	15.9
m-ClCCP	5×10^{-5}M	11.8	9.7
DCCD	5×10^{-5}M	140.0	116.0
None		49.2	100
Cyanide	10^{-2}M	27.4	55.8
Pentachlorophenol	10^{-4}M	14.6	30.0
None		42.6	100
Irradiation (360nm)		7.4	17.5
None		57.4	100
Arsenate	5×10^{-2}M	58.2	101.0
None (2% ethanol)		72.3	100
Valinomycin	2×10^{-5}M	5.7	7.9

Generated NADH was used as a substrate. *Data from Hirata, et. al., (4).

TABLE 7

Proline Uptake by Coupling Factor Depleted ETP

	0 uptake (μatom)	Pi esterified (μmole)	P/0	Proline uptake cpm/mg protein
ETP	9.7	11.2	1.15	2,645*
DETP	4.6	0	0	8,950
DETP + BCF$_4$	8.3	13.1	1.54	7,220

*The depleted ETP comprise approximately 30% of the ETP preparation. From Hirata, et. al., (4).

TABLE 8

Effect of Salts on Proline Transport by ETP

Additions	Proline Uptake pmoles/mg protein
None	37.2
NaCl	577.0
KCl	51.5
LiCl	442.4
RbCl	36.3
Choline-Cl	45.2
CsCl	32.7
NH_4Cl	38.5
Na_2CO_3	563.5

Ascorbate-TPD (Tris neutralized) was used as a substrate. The concentration of each monovalent cation was 20 mM (final).

TABLE 9

Effect of Combination of Substrates on Proline Transport by ETP

Substrate	Proline Uptake pmoles/mg protein	%
Ascorbate-TPD	418.2	100
Ascorbate-TPD + Succinate	169.0	40.4
Ascorbate-TPD + Malate*	249.5	59.7

*Partially purified malate-vitamin K reductase (96 µg) and FAD (25 µM) were also added.

TABLE 10

Restoration of Proline Transport by Lapachol
Following Irradiation (360nm)

Preparation	Proline uptake pmoles/mg protein	O_2 uptake mμatoms/min
ETP	134.4	49.1
ETP Irradiated	11.4	8.4
ETP Irradiated + Lapachol	51.8	35.0*

The ETP were irradiated with light at 360 nm for 40 minutes
and the reaction was carried out for 15 minutes with generated NADH as the electron donor. *Data taken from a
different experiment.

TABLE 11

Effect of Different Treatment on Oxidative Phosphorylation and ANS Response in Energized and Nonenergized Membranes of Electron Transport Particles of M. phlei

	Oxygen uptake (μatoms)	ΔP_i (μmoles)	P/O	Increase in fluorescence (arbitrary units)		Binding Sites (nanomoles/mg prot)		Dissociation constant (K_D) ($\times 10^{-5}$ M)	
				NE_f	E_f	NE_f	E_f	NE_f	E_f
ETP	5.80	6.53	1.12	27	9	23	44	16.7	10.0
Frozen ETP	5.53	4.82	0.87	24	6	22	37	17.0	12.5
Heated ETP	6.78	10.66	1.57	38	33	26	46	15.0	2.7
Frozen and Heated ETP	6.71	6.80	1.19	33	31	25	44	15.5	4.5

Data from Kalra et. al., (45).

467

TABLE 12

Changes in ANS Fluorescence Induced
by NADH and Ascorbate-PMS

| Treatment | Increase in fluorescence (arbitrary units) | | | |
| | Exogenous NADH $(405 \rightarrow 580nm)$ | | Ascorbate-PMS $(358 \rightarrow 480nm)$ | |
	NE_f	E_f	NE_f	E_f
None	26	8	40	7
Heat-treated	32	25	42	40

13.5 mM NADH or a mixture of 10 mM ascorbate + 10 µM PMS
were added to obtain E_f.

TABLE 13

Effect of ATP on Energized Fluorescence
in Trypsinized ETP from <u>M. phlei</u>

| | Changes in fluorescence (arbitrary units) | | |
| | NE_f | E_f | |
		Succinate	ATP
Untreated ETP	32	18	-8
Trypsinized ETP	26	6	18

The reaction mixture also contained 2 mM $MgCl_2$. 1.5 mM ATP
was added to obtain E_f.

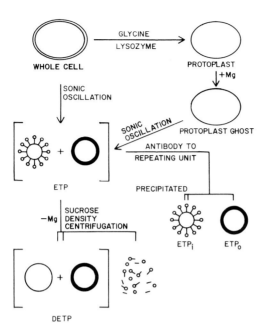

Fig. 1

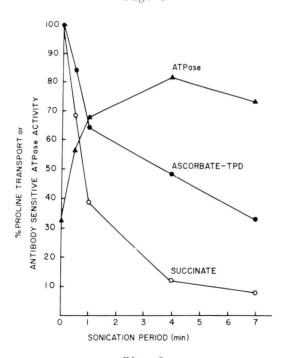

Fig. 2

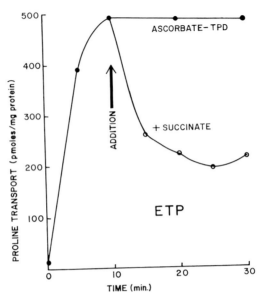

Fig. 3 A

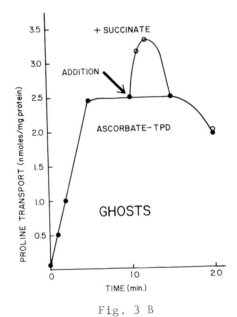

Fig. 3 B

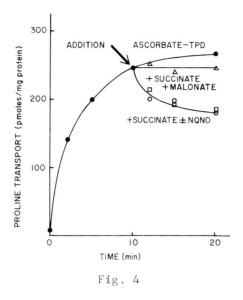

Fig. 4

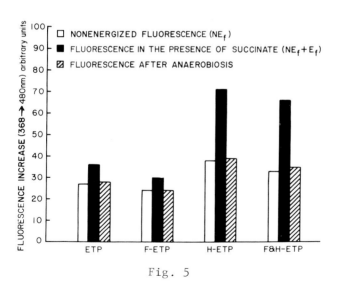

Fig. 5

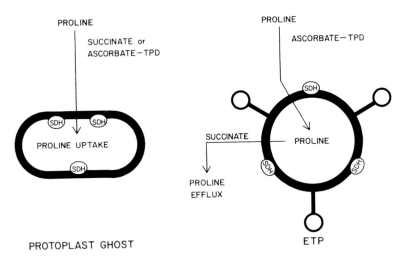

Fig. 6

TRANSPORT MECHANISMS IN ISOLATED BACTERIAL
CYTOPLASMIC MEMBRANE VESICLES*

H. R. Kaback
The Roche Institute of Molecular Biology
Nutley, New Jersey

ABSTRACT. The P-enolpyruvate-P-transferase system catalyzes transport by "vectorial phosphorylation". In
Escherichia coli, the system is involved in the transport
of relatively few sugars.

The transport of 15 amino acids, β-galactosides,
galactose, arabinose, glucuronic acid, gluconic acid, and
glucose-6-P in E. coli membrane vesicles is coupled to a
membrane-bound, flavin linked D-lactic dehydrogenase.
Valinomycin-induced K^+ and Rb^+ transport is also coupled to
this dehydrogenase.

The D-lactic dehydrogenase-coupled transport systems
are dependent upon electron transfer; however, they are
independent of the P-enolpyruvate-P-transferase system,
oxidative phosphorylation, or the generation of proton or
potential gradients. The site of energy-coupling is localized to a segment of the respiratory chain which is above
80% of the membrane-bound flavoproteins but below cytochrome b_1, and the activity of the "carriers" reflects the
redox state of the respiratory chain at this site. A conceptual model is presented in which the "carriers" are
depicted as electron transfer intermediates undergoing
reversible conformational changes during reduction and oxidation of a critical disulfide moiety within the molecules.

Using ascorbate and phenazine methosulfate, membrane
vesicles prepared from a variety of bacteria catalyze
active transport. With the exception of the primary dehydrogenases coupled to transport, all other aspects of the
systems which have been studied in detail are similar to

*A similar manuscript has been submitted to the Fourth
Annual Biochemistry Winter Symposium, Miami, Florida.

those described for E. coli.

Some of the amino acid "carriers" have been solubilized and partially purified from E. coli membrane vesicles.

INTRODUCTION

In 1960, it was first reported that a sub-cellular preparation from E. coli catalyzed the uptake of glycine (1). Subsequent studies confirmed the initial observation and demonstrated that cell-free preparations of membrane vesicles, essentially devoid of cytoplasmic constituents, catalyzed the uptake of glycine (2,3), its ultimate conversion to phosphatidylethanolamine (3), and the concentrative uptake of proline (4,5). These studies established a model system which allowed an examination of the biochemistry of transport and the relationship of the mechanisms involved to the cell membrane

METHODS

The methods used to prepare bacterial membrane vesicles, their morphology, homogeneity, purity, composition, and physical properties, as well as the methods used to assay transport have been described (6-10).

RESULTS AND DISCUSSION

The P-enolpyruvate-P-transferase System

A bacterial phosphotransferase system was described in 1964 (11) which catalyzes the transfer of phosphate from P-enolpyruvate to various carbohydrates according to the following reactions:

$$\text{Enzyme I, Mg}^{++}$$

$$\text{P-enolpyruvate + HPr} \rightleftharpoons \text{pyruvate + P-HPr} \qquad (1)$$

$$\text{Enzyme II, Mg}^{++}$$
$$\text{(Factor III)}$$

$$\text{P-HPr + sugar} \longrightarrow \text{sugar-P + HPr} \qquad (2)$$

Enzyme I

$$\text{P-enolpyruvate} + \text{sugar} \xrightarrow{\hspace{2cm}} \text{sugar-P} + \text{pyruvate} \hspace{1cm} (3)$$

HPr, Mg^{++}
Enzyme II,
(Factor III)

Subsequently biochemical and genetic evidence was presented indicating that this system might be involved in bacterial carbohydrate metabolism and/or transport. For an extensive review of this work, the reader is directed to reference (12).

Studies utilizing membrane vesicles demonstrate that the P-transferase system catalyzes the vectorial phosphorylation of glucose and related monosaccharides in E. coli, S. typhimurium and B. subtilis (7,12-15). By this means, sugar is transported into the vesicles as sugar-P without the mediation of an intramembranal free sugar pool. The most direct evidence for this mechanism is derived from double isotope experiments in which the intravesicular pool is pre-loaded with ^{14}C-glucose under conditions in which there is no phosphorylation of the sugar. After removal of the external isotope, the pre-loaded vesicles are exposed to ^{3}H-glucose in the presence of P-enolpyruvate. The vesicles show an almost absolute preference for ^{3}H-glucose added to the outside simultaneously with P-enolpyruvate. Other lines of evidence which substantiate this interpretation have been discussed in detail (7,12,13), and will not be covered here.

Membrane vesicles take up little glucose-P when incubated with or without P-enolpyruvate (7,12,13). Moreover, the diffusion-limited rates of appearance of glucose-6-P or free glucose into the intramembranal pool are similar. In light of the previous observations, this finding implies that external sugar reaches a catalytic site within the membrane and is translocated as a result of phosphorylation. Also, these experiments suggest that the P-transferase system does not function as a trap since there would be little advantage in trapping sugar as sugar-P in order to decrease its outward diffusion through the membrane.

In S. aureus, the P-transferase system catalyzes the transport of many carbohydrates (16-20). In most other organisms, however, this does not appear to be the case. In E. coli, for instance, this system is involved in the transport of relatively few sugars, i.e., glucose, mannose, fructose, and mannitol (21). It is not involved in amino acid transport or in most inducible sugar transport systems in E. coli, as will be discussed subsequently, nor is it apparently involved in glucose transport in obligate aerobes (22).

A conceptual model which has been proposed as a possible mechanism for vectorial phosphorylation (7,12), as well as experiments related to the regulation of the P-transferase system by sugar-P's (7,8,12) and to the role of phosphatidylglycerol in P-transferase activity have been presented elsewhere (7,9,12) and will not be discussed.

Coupling of a Membrane-Bound D-Lactic Dehydrogenase (D-LDH) to Amino Acid and Sugar Transport in E. coli Membrane Vesicles

Amino Acid Transport--Recent studies (14,23,24) have defined the energetics of amino acid transport in detail. The addition of D-lactate to suspensions of vesicles dramatically stimulates proline uptake with a 20- to 30-fold increase over baseline levels. Only succinate, L-lactate, D,L-α-hydroxybutyrate, and NADH replace D-lactate to any extent whatsoever, and each is much less effective than D-lactate. Furthermore, NAD in the presence of D-lactate causes no additional stimulation of proline transport.

D-lactate or succinate are converted stoichiometrically to pyruvate or fumarate, respectively (12,23-25), and neither pyruvate nor fumarate has any effect on proline transport. These results indicate that the concentrative uptake of proline involves electron transfer, and more specifically, that a membrane-bound lactic dehydrogenase with a high degree of specificity towards D-lactate is tightly coupled to proline transport.

Since the rate and extent of conversion of D-lactate to pyruvate is much greater than can be accounted for by proline transport alone, the effect of D-lactate on the transport of other amino acids was investigated. The results of these experiments (10,14,23) demonstrate that the con-

version of D-lactate to pyruvate markedly stimulates the
initial rates of uptake and the steady-state levels of
accumulation of proline, glutamic acid, aspartic acid,
tryptophan, serine, glycine, alanine, lysine, phenylalanine,
tyrosine, cysteine, leucine, isoleucine, valine, and histidine. The transport of glutamine, arginine, cystine,
methionine, and ornithine is stimulated only marginally.
Virtually all of the radioactivity taken up by the vesicles
is recovered as the unchanged amino acid (23). Furthermore, the steady-state concentration of each of the amino
acids taken up is many times higher than that of the
medium. Primarily D-lactate and, to a much lesser extent,
succinate, L-lactate, D,L-α-hydroxybutyrate, and NADH are
the only energy sources which initiate the uptake of any
of the amino acids. However, the relative effects of these
compounds on a particular amino acid transport system
varies (14). Generally, when effective, succinate is only
one-third to one-half as active as D-lactate, and L-lactate
only one-tenth to one-fifth as effective.

When vesicles are prepared from cells grown on either
glycerol or enriched media, a membrane-bound α-glycerol-P
dehydrogenase is induced which is coupled to amino acid
transport about as effectively as succinic dehydrogenase
(26,27). Similar observations have also been made for
formate dehydrogenase (26).

Competition experiments indicate that there are at
least 9 amino acid transport systems, each one of which is
specific for a structurally related group of amino acids
(10). Genetic evidence, in some cases, corroborates the
assignment of a particular group of amino acids to one
transport system.

β-Galactoside Transport--Although the β-galactoside
transport system in E. coli has been examined in detail,
the mechanism of coupling of metabolic energy to active
galactoside transport was poorly understood. A role for
ATP in lactose transport in E. coli has been suggested
(28); however, studies on anaerobic TMG uptake (29) indicate that uncouplers of oxidative phosphorylation block
TMG accumulation but do not alter ATP levels. Fox and
Kennedy (30) demonstrated the existence of a "permease"
protein (the M protein) which is a product of the y gene

(31), but data indicating that the P-enolpyruvate-P-trans-ferase system catalyzed TMG uptake in E. coli (32) raised the possibility that the M protein might be an inactivated Enzyme II. This topic has been discussed in detail in recent reviews (7,12).

Since much of the interest in this laboratory over the past few years was directed towards the role of the P-transferase system in sugar transport, and since all attempts to implicate this system in the transport of gal-actosides were uniformly negative, the effect of D-lactate on the uptake of β-galactosides by E. coli vesicles was investigated (25). Conversion of D-lactate to pyruvate in membrane vesicles prepared from cells containing a func-tional y gene markedly stimulates the initial rate of transport of β-galactosides, and in a short time, the vesicles accumulate these sugars to concentrations at least 100-times higher than the medium. All of the galactosides accumulated are recovered as the unchanged substrates. As demonstrated for amino acid transport, only D-lactate, and to a lesser extent, D,L-α-hydroxybutyrate, succinate, and L-lactate increase lactose transport above endogenous levels. NADH does not stimulate lactose transport. α-Glycerol-P and formate also stimulate lactose transport in membranes prepared from cells grown on appropriate media (26,27).

Vesicles prepared from E. coli GN-2 (33), a mutant lack-ing Enzyme I of the P-transferase system, but constitutive for lac, transport β-galactosides in the presence of D-lactate but are completely unable to vectorially phos-phorylate α-methylglucoside (13). D-Lactate does not stimulate α-methylglucoside uptake by membrane vesicles containing an intact P-transferase system. Moreover, P-enolpyruvate does not stimulate lactose, IPTG or TMG up-take, nor is lactose-P or TMG-P detected. Finally, mem-branes which transport β-galactosides fail to exhibit phosphatase activity towards TMG-P, and the addition of lactose to vesicles incubated with ^{32}P-enolpyruvate does not accelerate the appearance of ^{32}P$_i$, as might be expected if a lactose-P P-hydrolase were involved (12). β-galacto-side transport in E. coli clearly does not involve the P-enolpyruvate-P-transferase system.

Coupling of Other Sugar Transport Systems to D-LDH--The transport systems for galactose (34), arabinose, glucuronic acid, gluconic acid, and glucose-6-P are coupled to D-LDH in a manner identical to that described for β-galactosides (24,35). Transport of these sugars by vesicles requires induction of the parent cells, does not involve the P-enol-pyruvate-P-transferase system, and is inhibited by the same conditions which affect amino acid and β-galactoside transport (see below).

D-Galactose transport is exhibited by lac y⁻ vesicles prepared from induced E. coli ML 3 and ML 35, and E. coli ML 32400 (36) and W3092cy⁻ (37) which transport galactose constitutively. Of particular interest is the observation that "galactose binding protein" is totally absent from the vesicles, as is a high affinity, low V_{max} galactose transport system present in the whole cells. These findings, in addition ot the observation that the vesicles do not transport β-methylgalactoside, indicate that the galactose transport system retained by the vesicles is the so-called "gal permease" system (38). Moreover, it is obvious that this system does not require the galactose binding protein for activity.

Activity of Membrane Vesicles Compared to Whole Cells-- Although previous data suggest that the rates of sugar and amino acid transport by membrane vesicles is 10 to 20% of that found in whole cells, recent experiments show that vesicles lose significant activity due to excessive manipu-lation during preparation. Vesicles prepared using gentle conditions (i.e., avoiding vigorous homogenization in tight-fitting homogenizers) and assayed for lactose or pro-line transport prior to freezing have specific activities 3 to 5 times higher than whole cells (comparing initial rates of uptake per mg protein). However, quantitative comparisons between vesicles and whole cells are extremely difficult to interpret, especially when the transport activity manifested by whole cells may be the composite of more than one system (e.g., galactose transport).

Source of D-lactate in E. coli--E. coli has two dis-tinct D-LDH's--a soluble, pyridine nucleotide-dependent enzyme which catalyzes the conversion of pyruvate to D-lac-tate (39,40), and a membrane-bound, flavin-linked enzyme which catalyzes the reverse reaction (26,41). Apparently, the soluble enzyme produces D-lactate which may then be utilized by the membrane-bound enzyme to drive many transport systems and perhaps other cellular processes.

Substrate Oxidation by Membrane Vesicles

There is no relationship between the oxidase activity of vesicles towards various substrates and their ability to stimulate transport (26). With vesicles prepared from succinate-grown cells, succinate is oxidized much faster than D-lactate and NADH is oxidized approximately as fast, yet D-lactate is markedly more effective as an electron donor for transport.

D-Lactate-dependent transport by vesicles is inhibited by anoxia. Moreover, the electron transfer inhibitors cyanide, 2-heptyl-4-hydroxyquinoline-N-oxide (HOQNO), amytal and the specific D-lactic dehydrogenase inhibitor oxamic acid effectively block transport (26). Inhibition of D-lactate oxidation by these compounds is similar to the inhibition of transport (10,24), and these inhibitors also effectively block succinate and NADH oxidation. Investigations of the respiratory chain of $\underline{E}$. $\underline{coli}$ (42) have identified the amytal-sensitive site as a flavoprotein between D-LDH and cytochrome b_1. HOQNO acts between cytochrome b_1 and cytochrome a_2, perhaps at a quinone-containing component, and cyanide blocks cytochrome a_2. Thus, each of the dehydrogenases studied is coupled to oxygen via a membrane-bound respiratory chain.

The D-lactate-dependent transport systems are not significantly inhibited by high concentrations of arsenate or oligomycin (23-25). Moreover, vesicles have been prepared with arsenate buffer rather than phosphate buffer throughout the procedure, and subsequently transport was assayed in arsenate buffer in the absence of inorganic phosphate. The results were the same--little or no inhibition of transport. It is apparent from these studies that the effect of D-lactate is not mediated by the production of stable high-energy phosphate compounds. This conclusion is supported by many observations, among which are the absence of transport in the presence of ATP under conditions in which ATP is demonstrably accessible to reactive

sites within the membrane (43), the failure of ADP or other
nucleoside-diphosphates to stimulate transport in the presence of D-lactate or other electron donors, and the observation that the membrane preparations do not carry out oxidative phosphorylation (44).

Transport and D-lactate oxidation by the vesicles are
inhibited by the sulfhydryl reagents N-ethylmaleimide (NEM)
and p-chloromercuribenzoate (PCMB) (25,26,35). These
reagents will be discussed in greater detail below.

DNP, carbonyl cyanide m-chlorophenylhydrazone (CCCP),
and azide do not significantly affect D-lactate oxidation,
despite profound inhibition of transport (24-26). This
finding is not surprising since most bacterial electron
transfer systems are not subject to respiratory control.

The observations, taken as a whole, indicate that the
specificity of the transport systems for D-LDH cannot be
accounted for solely on the basis of its presence in the
vesicles (to the exclusion of other dehydrogenases), and
furthermore, that the coupling of D-LDH to transport involves the flow of electrons through a respiratory chain to
oxygen as the terminal electron acceptor.

Site of Energy-Coupling Between D-LDH and Transport

Difference spectra between D-lactate-, succinate-,
NADH-, L-lactate-, or dithionite-reduced samples and oxidized samples are indistinguishable (24,26). Furthermore,
difference spectra between D-lactate-reduced and NADH-,
succinate-, L-lactate- and dithionite-reduced samples show
no absorption bands. Since the rate of reduction of cytochrome b_1 by these substrates is directly proportional to
their rates of oxidation, the data indicate that each dehydrogenase is coupled to the same cytochrome chain. Thus,
the site of energy-coupling between D-LDH and transport
must lie between the primary dehydrogenase and cytochrome
b_1, the first cytochrome in the E. coli respiratory chain.

Direct evidence for this hypothesis is provided by
experiments in which the effects of NEM and PCMB on D-lactate and NADH oxidation were investigated (26). Both of
these sulfhydryl reagents markedly inhibit D-lactate

oxidation at concentration which block transport. Moreover,
PCMB inhibition is reversed by dithiothreitol, providing
further evidence for sulfhydryl involvement in D-lactate
oxidation. The effect of these thiol reagents on D-lactate
oxidation is not expressed at the level of the primary dehydrogenase. Neither the D-lactate:dichlorophenolindo-
phenol (DCI) reductase activity of the intact vesicles nor
that of the solubilized, partially purified preparation of
this enzyme (26) is sensitive to NEM or PCMB. It is extremely important that NADH oxidation is not sensitive to
NEM or PCMB. Thus, neither the primary dehydrogenase nor
the cytochrome system contains the reactive sulfhydryl
group(s), and the site of inhibition of D-lactate oxidation
by NEM and PCMB must lie between D-LDH and the cytochromes.

An abbreviated schematic representation of the sequence
of events thought to occur is presented in Fig. 1.

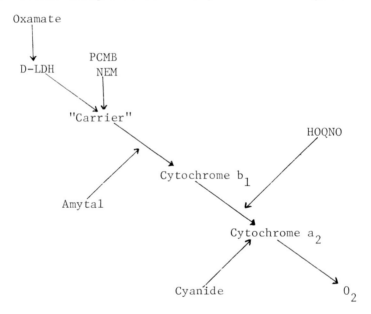

Fig. 1. Electron transfer pathway from D-LDH to oxygen
showing sites of inhibition of electron transfer inhibitors.

Electrons from D-lactate flow through the "carriers" or
something closely aligned with the "carriers" before enter-

ing the cytochrome chain, after which they ultimately reduce oxygen to water. Each of the electron transfer inhibitors shown inhibits solute transport by interrupting the flow of electrons above (oxamate), at (PCMB, NEM), or below (amytal, HOQNO, cyanide, or anoxia) the site of energy-coupling.

Mechanism of Energy-Coupling of D-LDH to Transport

The findings presented above, especially when considered in conjunction with those to be discussed, indicate that the "carrier" components of these transport systems reflect the redox state of the respiratory chain between D-LDH and cytochrome b_1. Although it cannot be concluded definitively that the "carriers" are electron transfer intermediates, most of the available evidence is consistent with this probably oversimplified interpretation.

D-LDH activity and the initial rates of transport respond identically to temperature; furthermore, both phenomena have the same activation energy of 8400 calories per mole (35). Thus, the behavior of the "carriers" could determine the activation energy for D-lactate oxidation. Some support for this suggestion is provided by experiments in which the D-lactate:DCI reductase activity of membrane vesicles was studied as a function of temperature. This reaction exhibits a markedly different temperature profile from that of D-lactate oxidation and has an activation energy of approximately 30,000 calories per mole.

Evidence has been presented (35) which demonstrates that the steady-state levels of solute accumulation at temperatures ranging from 0° to 53° represent equilibrium states in which there is a balance between influx and efflux. Moreover, efflux induced at 45° is a saturable, "carrier"-mediated phenomenon with a much lower affinity (i.e., higher K_m) than the influx system, but a very similar maximum velocity (35). Anoxia, cyanide, HOQNO, and DNP also induce the efflux of solute and the kinetics of cyanide-induced efflux manifest the same apparent K_m as temperature-induced efflux and the same maximum velocity as the influx process. Thus, efflux occurs at rates which are comparable to those of influx, but much higher concentrations of substrate are required to saturate the "carrier" on the inside of the membrane. Furthermore, the rate of efflux responds to temperature in a manner that is essentially the inverse of the response of the steady-state level of lactose accumulation to temperature (35).

The β-galactoside transport system in E. coli is inhibited by sulfhydryl reagents (45-47), and the ability of two substrates of this system to protect against sulfhydryl inactivation led to the identification and purification of M protein, the product of the y gene (30,31,48,49). However, very little evidence has been presented which has any bearing on the mechanistic role of sulfhydryl groups in galactoside transport. Evidence discussed above demonstrates that D-lactate-induced respiration is inhibited by PCMB and NEM, and that the site(s) of action of these compounds is between D-LDH and cytochrome b_1, i.e., at the site of energy-coupling. Virtually every D-LDH-coupled system is inhibited by PCMB and NEM, and PCMB inhibition is reversed by dithiothreitol (35). Moreover, each "carrier"-mediated aspect of transport, even those that are independent of D-lactate oxidation, is inhibited by PCMB and NEM; and in each case, the inhibition by PCMB is reversed by dithiothreitol (35). D-lactate oxidation and "carrier" activity are both dependent upon functional sulfhydryl groups.

Electron transfer inhibitors whose sites of action are well documented (cf., Fig. 1) were studied with respect to their effect on the ability of the vesicles to retain accumulated solute (35). Each inhibitor used blocks D-lactate oxidation and the initial rate of influx by at least 70 to 80% (10,24,26). Only anoxia and those inhibitors which block electron transfer after the site of energy-coupling (i.e., anoxia, cyanide, HOQNO, and amytal) cause efflux. Thus, reduction of a portion of the electron transfer chain between D-LDH and cytochrome b_1 is responsible for efflux. Strikingly, oxamate which inhibits electron transfer before the site of energy-coupling does not induce efflux, nor does PCMB or NEM. Thus the redox state of the respiratory chain at the site of energy-coupling determines the rate of efflux.

The effect of anoxia and the same inhibitors on the time course of uptake is consistent with this interpretation (35). Since the removal of oxygen or the presence of electron transfer inhibitors which inhibit after the site of energy-coupling result in reduction of the energy-coupling site, membranes incubated under these conditions exhibit profound inhibition of uptake throughout the time course of the experiment. On the other hand, inhibitors which work before or at the site of energy coupling inhibit the reduction of the energy-coupling site, and vesicles incubated under these conditions exhibit markedly diminished initial rates of uptake but eventually accumulate significant quantities of solute.

These experimental findings are consistent with the conceptual working model presented in Fig. 2.

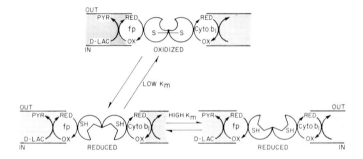

Fig. 2. Conceptual working model for D-lactic dehydrogenase-coupled transport systems. D-LAC, D-lactate; PYR, pyruvate; fp, flavoprotein; cyto b₁, cytochrome b_1; ox, oxidized; red, reduced. OUT signifies the outside surface of the membrane; IN signifies the inside surface. The spheres located between fp and cyto b₁ represent the

"carrier"; ⌣ , a high affinity binding site and ◯ , a low affinity binding site. The remainder of the cytochrome chain from cytochrome b_1 to oxygen has been omitted (35).

The "carriers" are depicted as electron transfer intermediates which undergo reversible oxidation-reduction. In the oxidized state, the "carrier" has a high affinity site for ligand which it binds on the exterior surface of the membrane. Electrons coming ultimately from D-lactate through one or possibly more flavoproteins reduce a critical disulfide in the "carrier" resulting in a conformational change. Concomitant with this conformational change, the affinity of the "carrier" for ligand is markedly reduced, and ligand is released on the interior surface of the membrane. The reduced "sulfhydryl" form of the "carrier" is then oxidized by cytochrome b_1 and the electrons flow through the cytochrome chain to reduce molecular oxygen to water. The reduced form of the "carrier" can also "vibrate" and catalyze a low affinity, "carrier"-mediated, non-energy dependent transport of ligand across the membrane.

The proposed mechanism implies that there are functionally heterogeneous intermediates between different D-LDH molecules and cytochrome b_1. For each transport system, there should be an intermediate which has a binding site that is specific for a particular transport substrate. Supportive evidence for this prediction is provided by studies of lactose transport in the presence of structurally unrelated substrates which are also transported by D-LDH-coupled systems (35). Little or no inhibition of either the initial rate of lactose uptake or the steady-state level of lactose accumulation is observed. Moreover, the sum of the V_{max} values of the known D-LDH-coupled transport systems in a particular membrane preparation is equal to or less than the V_{max} of D-LDH in the same membrane preparation.

Although no direct evidence has been presented which demonstrates unequivocally that the "carriers" undergo oxidation-reduction, this formulation is consistent with most of the experimental observations and is certainly the simplest conception possible. D-lactate oxidation and the

D-lactate-dependent and -independent aspects of transport are inhibited by PCMB and NEM. Furthermore, PCMB inhibition is reversed by dithiothreitol. Since the site of energy-coupling lies between the primary dehydrogenase and cytochrome b_1, and the site(s) of inhibition by sulfhydryl reagents is also between D-LDH and cytochrome b_1, the proposed mechanism is supported by more than simplicity of conception. However, the model does not account for the action of "uncoupling agents" (i.e., DNP, CCCP) or azide on the system. These reagents abolish transport without inhibition of D-lactate oxidation (24-26).

The observations, and the conceptual model to which they have led, do not conflict with studies in whole cells (50-52). Furthermore, the proposed mechanism would explain some apparent inconsistencies. Only TDG and melibiose protect the galactoside transport system against inhibition by NEM (48). Since many other galactoside analogues are transported via this system, it was proposed that M protein has two binding sites and that TDG and melibiose bind to one site only. However, no evidence has been presented to substantiate this suggestion. According to the mechanism proposed here, M protein has two sites, but one site is involved in electron transfer and the other in binding substrate. Since presumably all galactosides bind to M protein by virtue of their galactose moieties, it does not seem unlikely that TDG and melibiose, due to their size (TDG) or shape (melibiose is an α-galactoside), sterically protect a sulfhydryl group which is not in the binding site.

The proposed mechanism could also explain so-called "energy-uncoupled" galactoside transport mutants (53,54). These mutants have increased amounts of M protein, but do not catalyze the concentrative uptake of galactosides as well as the parent. Uptake of TMG by membrane vesicles prepared from one of these mutants (53) is stimulated by D-lactate, but only about one third as well as the parent preparation. D-Lactate-dependent transport of proline and P-transferase-mediated uptake of α-methylglucoside, on the other hand, are identical. Possibly the defect in these mutants is related to electron transfer properties of M protein.

Coupling of Ascorbate-Phenazine Methosulfate (PMS) to Transport

Konings, Barnes, and Kaback (55) demonstrated that an artificial electron donor system--ascorbate-PMS-- can be coupled to transport in vesicle preparations. The initial rate of lactose transport in E. coli membrane vesicles is stimulated about 3-times more effectively by ascorbate-PMS than D-lactate, the best physiological electron donor. This finding is consistent with the observation that reduced PMS is oxidized much faster than D-lactate by the vesicles. However, comparative stimulation of various transport systems by ascorbate-PMS or D-lactate differs. The significance of this observation will be discussed below.

The effect of ascorbate-PMS on transport is inhibited by anoxia, and by cyanide, HOQNO, PCMB, NEM and amytal in a manner identical to that described for D-lactate. Significantly, oxamic acid has no effect, since this inhibitor acts at the level of D-LDH. Thus, ascorbate-PMS reduces the respiratory chain of the vesicles at a redox level below that of cytochrome b_1. This conclusion is substantiated by spectrophotometric data. Moreover, ascorbate-PMS reduces only 21% of the membrane-bound flavoprotein, whereas D-lactate reduces 63%.

Since PCMB and NEM inhibit transport in the presence of ascorbate-PMS, the postulated sulfhydryl component(s) of the respiratory chain which is(are) common to both transport and D-lactate oxidation must lie within the segment of the respiratory chain reduced by ascorbate-PMS.

Amytal inhibits the effect of ascorbate-PMS on transport, suggesting that there are flavoproteins between the "carriers" and cytochrome b_1 or possibly that the "carriers: are flavoproteins themselves. No evidence has been presented which can resolve these possibilities. Since the methods used in these experiments do not distinguish between flavoprotein and non-heme iron (42), it is also possible that non-heme iron protein is involved. Recent experiments carried out with the iron chelator cyclic dihydroxybenzoyl serine triplex (cyclic DBS_3) demonstrate that this compound inhibits transport with D-lactate or ascorbate-PMS as electron donors. Moreover, cyclic DBS_3 inhibits D-lactate oxidation and the initial rate of transport, but does not cause efflux of solute accumulated in the intramembranal pool. It is also noteworthy that Bragg (56) has shown that ascorbate-PMS reduces non-heme iron in respiratory particles from E. coli. These observations place non-heme iron near the site of ascorbate-PMS reduction either before or at the site of energy-coupling. Regarding the proposed mechanism (cf. Fig. 2), it is interesting that the relative effects of ascorbate-PMS or D-lactate differ depending upon the transport system studied. Preliminary observations indicate that although most transport systems are stimulated more by ascorbate-PMS than by D-lactate, some are stimulated more by D-lactate, and others are stimulated about equally by both. Possibly, there are small differences in the redox potentials of various "carriers" such that a few are not completely reduced by ascorbate-PMS.

A Kinetic Model for the Redox Transport Mechanism

The mechanistic model presented in Fig. 2 may be expressed kinetically (10) and Dr. F. J. Lombardi has derived the following differential equation to describe the proposed mechanism:

$$v_{net} = v_{in} - v_{out}$$

$$v_{net} = \left[\frac{V_{max}^{\uparrow} S^o}{K_m^{\uparrow} + S^o} \right] (1-\alpha) - \left[\frac{\dfrac{V_{max}^{eff}}{K_m^i} (S^i - S^o)}{1 + \dfrac{S^o}{K_m^o} + \dfrac{S^i}{K_m^i}} \right] \alpha$$

where v=velocity

$V_{max}^{\uparrow}$ =maximum velicity of energy-dependent active uptake.

$K_m^{\uparrow}$ =apparent Michaelis constant for energy-dependent active uptake.

S^o=external substrate concentration.

α=percentage of carriers in the oxidized form under steady-state conditions.

V_{max}^{eff}=maximum velocity of efflux.

K_m^i=apparent Michaelis constant for efflux.

S^i=internal substrate concentration.

K_{max}^o=apparent Michaelis constant for energy-dependent influx.

It should be emphasized that each term in this equation can be measured experimentally.

Using values from experiments carried out at various temperatures with the equation shown, a computer was programmed to generate uptake curves as a function of time. The computer-generated curves are indistinguishable from experimental data (35). Although this type of treatment does not discriminate between mechanisms in which the "carrier" is or is not an obligatory electron transfer intermediate, it provides strong evidence for the general type of redox mechanism presented.

Solubilization and Partial Purification of "Carriers"

Recently, Dr. Adrienne S. Gordon solubilized and partially purified a fraction from vesicles which apparently contains many of the amino acid "carriers" (57). Membrane vesicles are extracted with non-ionic detergents, and the extract is subjected to gel filtration in the presence of detergent. Three 280 nm absorbance peaks are eluted from Sephadex G-100 columns: 1) excluded protein (peak I); 2) included protein of relatively high molecular weight (peak II); and 3) included protein of low molecular weight (peak III).

Material in each of the three protein-containing fractions binds proline. However, peak III exhibits a much higher specific activity than the two other fractions. Proline is not altered chemically by each fraction nor by the unfractionated extract, no detectable D-LDH activity is associated with peak III, and succinic dehydrogenase and NADH dehydrogenase activities are not detected in the column effluent.

Judging from the UV absorption spectrum of peak III, this fraction is composed predominantly of protein. Moreover, when extracts from ^{32}P-labeled membranes are chromatographed, no significant radioactivity appears in peak III. Thus, very little, if any, phospholipid is associated with this fraction.

The proline binding activity of peak III is highly specific. Only proline itself inhibits the binding of the radioactive amino acid. In contrast, binding is not affected by a mixture of structurally-unrelated amino acids.

Proline-binding by peak III is inhibited by NEM and PCMB; and PCMB inhibition is reversed by dithiothreitol. These results are analogous to the behavior of the transport system (35,57). In contrast to inhibition of binding by sulfhydryl reagents, electron transfer and general metabolic inhibitors such as amytal, cyclic DBS_3, HOQNO, CCCP, and DNP do not inhibit.

Peak III also exhibits binding activity for lysine, serine, tyrosine, and glycine. The amino acids, like proline, are transported by systems with low apparent K_m's suggesting that the "carriers" have high affinities. Binding activity for each of these amino acids is inhibited reversibly by PCMB; moreover, binding of these amino acids is not altered by the presence of structurally unrelated amino acids. Binding of amino acids other than those mentioned has not been tested.

These solubilized membrane components are distinctly different from the "binding proteins" isolated by cold osmotic shock (58-60). The latter are water soluble and localized in the periplasmic space, whereas these components are associated with cytoplasmic membrane and are soluble only in detergent. In addition, as discussed previously, the vesicles contain no "galactose binding protein". Finally, the soluble binding proteins obtained by osmotic shock are insensitive to sulfhydryl reagents and contain no cysteine.

Although the "carrier" components of other D-LDH-coupled transport systems may also be present in peak III, many of them would be difficult to detect by binding assays. It is unlikely, for instance, that M protein could be detected, since the β-galactoside transport system has a relatively high K_m (10,25,35).

General Importance of Dehydrogenase-Coupled Transport

The use of ascorbate-PMS has extended the study of res-piration-coupled transport to membrane vesicles prepared from a wide variety of bacteria (55). This artificial sys-tem markedly stimulates the transport of amino acids by membrane vesicles prepared from E. coli, S. typhimurium, Ps. putida, P. mirabilis, B. megaterium, B. subtilis, M. denitrificans, and S. aureus.

Although the physiological electron donor(s) for the transport systems in many organisms is(are) not known at present, the following observations are significant:

1) Short, White, and Kaback (61) have demonstrated that the transport of 16 amino acids by membrane vesicles prepared from S. aureus is coupled exclusively to a mem-brane-bound α-glycerol-P dehydrogenase.

2) Konings and Freese have shown that the concentra-tive uptake of a variety of amino acids, in addition to L-serine (62), by B. subtilis membranes is coupled pri-marily to α-glycerol-P and NADH dehydrogenase, and also, to some extent, to L-lactic dehydrogenase.

3) Membrane vesicles from M. denitrificans accumulate glycine, alanine, glutamine, and asparagine in the pre-sence of D-lactate.

These and other observations demonstrate that active transport systems which are basically similar to the D-LDH-coupled systems in E. coli are present in a variety of other organisms. The coupling of particular dehydrogenases to transport may be important with regard to the ecology of various bacterial species.

Valinomycin-Induced Rb^+ Transport

Recently, Bhattacharyya, Epstein, and Silver (63) reported that addition of valinomycin to E. coli membrane vesicles results in the accumulation of K^+ or Rb^+ by a temperature- and energy-dependent process. Moreover, these workers demonstrated that vesicles prepared from mutants altered in K^+ transport show defects manifested by the intact cells. Since the vesicles establish what appears to be a proton gradient (64), it seemed likely that the extrusion of protons might be the driving force for this transport system and possibly others.

Dr. F. J. Lombardi of this laboratory and Dr. John P. Reeves of Rutgers University have shown that valinomycin-induced Rb^+ uptake is analogous in nearly all respects to the respiration-linked transport of sugars and amino acids. D-lactate and ascorbate-PMS are the most effective electron donors in E. coli and M. denitrificans vesicles, while α-glycerol-P and ascorbate-PMS are most effective in membranes from S. aureus. In E. coli vesicles, both D-lactate-dependent Rb^+ uptake and D-lactate oxidation are blocked by anoxia, oxamate, amytal, HOQNO, and cyanide. Studies of oxygen utilization and spectrophotometric evidence indicate that the energy-coupling site for Rb^+ transport in E. coli membranes is located between D-LDH and cytochrome b_1. Agents which block electron transport beyond this site (anoxia, amytal, HOQNO, and cyanide) cause rapid efflux of accumulated Rb^+, whereas oxamate produces no efflux. These results provide strong evidence that valinomycin does not simply enhance the passive flux of Rb^+ across the membrane since the vesicles maintain a large Rb^+ concentration gradient in the presence of oxamate despite inhibition of D-lactate oxidation and Rb^+ influx.

Proton and Potential Gradients

Chemiosmotic coupling has been suggested by Mitchell (65-67) as a mechanism for oxidative phosphorylation in mitochondria, and Harold (68) and West (69) have applied this theory to active transport in bacteria. Although the

493

pH of lightly buffered membrane suspensions decreases on addition of D-lactate and other electron donors (64), a proton gradient is probably <u>not</u> established under most conditions. The evidence for this conclusion is as follows:

1) Vesicles do not take up the lipid-soluble weak acid 5,5-dimehtyloxazolidine-2,4-dione (DMO) in the presence of D-lactate or other electron donors. DMO is a weak acid, metabolically inert, which diffuses passively across many biological membranes. The permeability coefficient of the uncharged acid is generally much greater than that of the anion. The distribution of DMO is therefore a function of pH, and these experiments indicate that the intravesicular pH is not alkaline with respect to the medium.

2) The addition of transport substrates (i.e., lactose plus a mixture of amino acids) has no effect on the rate or absolute amount of acidification.

3) Membrane vesicles treated with phospholipase A-B (70) such that they retain the catalytic activities associated with transport (i.e., P-enolpyruvate-dependent phosphorylation of α-methylglucoside and D-lactate oxidation) but are unable to retain transported solute (7,10,16,18,19,28) exhibit similar pH changes. Moreover, so-called "proton conductors" such as DNP and CCCP have the same effect in normal and phospholipase-treated vesicles. Since these preparations are devoid of a diffusion barrier, it is highly likely that the observed pH changes can be due to proton gradients.

4) Vesicles containing high intramembranal concentrations of Na^+ incubated in the presence of D-lactate exhibit no change in the rate of acidification on addition of valinomycin and K^+. Moreover, when such vesicles are loaded with ^{22}Na, the addition of valinomycin and D-lactate in the presence of K^+ results in the movement of ^{22}Na out of the vesicles. On the other hand, when the vesicles contain low Na^+ concentrations, addition of valinomycin and K^+ doubles the rate and extent of acidification over that observed with D-lactate alone. Finally, membranes incubated under the latter conditions (i.e., vesicles containing low internal Na^+, incubated in the presence of valinomycin, K^+, and D-lactate) accumulate DMO. In the presence of D-lactate

alone, as mentioned above, DMO is not accumulated. These results indicate that the extrusion of Na^+ or protons from the vesicles is the result of (rather than the cause of) the active accumulation of Rb^+ or K^+.

Regarding potential gradients, a number of experimental approaches were utilized:

1) The transport of each ionic species in the reaction mixtures was measured in the presence of D-lactate or ascorbate-PMS, and the vesicles do not accumulate magnesium, sulfate, phosphate, potassium (in the absence of valinomycin), sodium, chloride, or pyruvate.

2) An intensive study of 1-anilino-8-naphthalene sulfonic acid (ANS) fluorescence was initiated since the behavior of this fluorescent probe has been postulated to reflect membrane potential changes in mitochondria (71). When D-lactate is added to membrane vesicles in the presence of ANS, there is a rapid decrease in fluorescence which is maintained until the reaction mixture becomes anaerobic at which time there is a large increase in fluorescence. The rate at which the initial fluorescence decrease progresses is most marked with D-lactate, and much slower with succinate or L-lactate. Electron transfer inhibitors whose site of action is after the site of energy-coupling (i.e., cyanide and HOQNO) reverse the initial decrease in fluorescence, whereas oxamate, NEM, and PCMB which inhibit before or at the site of energy-coupling cause a further decrease in fluorescence. Moreover, dithiothreitol reverses the effect of PCMB. These effects are obviously analogous to certain aspects of transport. As demonstrated with the pH measurements, the behavior of phospholipase-treated membranes is similar to that of untreated membranes. Thus, it is unlikely that ANS reflects membrane potentials in this system.

Evidence that the fluorescence changes described are due to structural changes in the membrane has recently been obtained from studies of ANS fluorescence in the energy transfer mode (72). In this case, tryptophan residues in the membrane proteins are excited and the fluorescence emission of ANS is recorded. It can be demonstrated directly that the ANS fluorescence under these conditions is due to absorption from tryptophan by the decrease in tryptophan emission. Moreover, the ratio of the 290 nm to 370 nm excitation maxima of ANS is increased when energy transfer is the mode of excitation. In any case, the effects of various electron donors and electron transfer inhibitors on ANS fluorescence in the energy transfer mode are similar to the findings described above. Of considerable interest is the observation that the effect of CCCP is greater in the energy transfer mode than in the fluorescent mode. These studies provide strong support for the contention that conformational changes in components of the membrane are intimately related to the mechanism of active transport.

Genetic Studies

Dr. Jen-shiang Hong, who recently joined this laboratory has isolated and characterized a number of primary dehydrogenase mutants. Primary D-LDH mutants were selected based on their ability to grow on pyruvate but not D-lactate. The whole cells are able to transport, as might be expected since other dehydrogenases can drive transport. Membrane vesicles prepared from these mutants, however, exhibit unexpected properties. In the presence of ascorbate-PMS, mutant vesicles transport at the same rate and to the same extent as wild type vesicles. D-Lactate, as expected, does not stimulate transport by the mutant vesicles. Surprisingly, L-lactate or succinate, depending on growth conditions, stimulates transport in mutant vesicles as well as D-lactate in the wild type. Even more striking is the observation that the rate of oxidation of L-lactate or succinate by mutant vesicles is similar to that observed in wild type vesicles. In other words, in the absence of D-LDH, coupling of L-lactate or succinate dehydrogenase to transport is enhanced. In vesicles prepared from a mutant defective in both D-lactic and succinic dehydrogenases, addition of L-lactate stimulates transport as well as D-lactate in wild type vesicles. Hopefully, genetic manipulations combined with enzymologic and physical techniques currently in use will extend the study of membrane transport to the molecular level.

REFERENCES

1. Kaback, H. R., Fed. Proc., 19, 130 (1960).

2. Kaback, H. R. and Kostellow, A. B., *J. Biol. Chem.*, 243, 1384 (1968).

3. Kaback, H. R. and Stadtman, E. R., *J. Biol. Chem.*, 243, 1390 (1968).

4. Kaback, H. R. and Stadtman, E. R., *Proc. Nat. Acad. Sci., U.S.A.*, 55, 920 (1966).

5. Kaback, H. R. and Deuel, T. F., *Arch. Biochem. Biophys.* 132, 118 (1969).

6. Kaback, H. R., in: *Methods in Enzymology*, Vol. XXII, ed. W. B. Jakoby (Academic Press, Inc., New York, 1971) p. 99.

7. Kaback, H. R., in: *Current Topics in Membranes and Transport*, Vol. I, eds. A. Kleinzeller and F. Bronner (Academic Press, Inc., New York, 1970) p. 35.

8. Kaback, H. R., *Proc. Nat. Acad. Sci. U.S.A.*, 63, 724 (1969).

9. Milner, L. S. and Kaback, H. R., *Proc. Nat. Acad. Sci. U.S.A.*, 65, 683 (1970).

10. Kaback, H. R., *Biochim. Biophys. Acta*, in press.

11. Kundig, W., Ghosh, S., and Roseman, S., *Proc. Nat. Acad. Sci. U.S.A.*, 52, 1067 (1964).

12. Kaback, H. R., *Ann. Rev. Biochem.*, 39, 561 (1970).

13. Kaback, H. R., *J. Biol. Chem.*, 243, 3711 (1968).

14. Kaback, H. R., Barnes, E. M., Jr., and Milner, L. S., in: *Recent Adv. in Microbiol.*, eds. A. Pérez-Miravete and D. Paláez (Liberia Internacional, S.A., Mexico City, Mexico, 1971) p. 171.

15. Kaback, H. R., in: *The Molecular Basis of Membrane Transport*, ed. D. C. Tosteson (Prentice-Hall, Englewood Cliffs, New Jersey, 1969) p. 421.

16. Egan, J. B. and Morse, M. L., Biochim. Biophys. Acta, 97, 310 (1965).

17. Egan, J. B. and Morse, M. L., Biochim. Biophys. Acta, 109, 172 (1965).

18. Egan, J. B. and Morse, M. L., Biochim. Biophys. Acta, 112, 63 (1966).

19. Henstenberg, W., Egan, J. B., and Morse, M. L., J. Biol. Chem., 243, 1881 (1968).

20. Laue, D. and MacDonald, R. E., Biochim. Biophys. Acta, 165, 410 (1968).

21. Tanaka, S., Lerner, S. A., and Lin, E. C. C., J. Bacteriol., 93, 64 (1967).

22. Romano, A. H., Eberhard, S. J., Kingle, S. L., and McDowell, T. D., J. Bacteriol., 104, 808 (1970).

23. Kaback, H. R., and Milner, L. S., Proc. Nat. Acad. Sci. U.S.A., 66, 1008 (1970).

24. Kaback, H. R., Barnes, E. M., Jr., Gordon, A. S., Lombardi, F. J., and Kerwar, G. K., in First Meeting of the Pan American Biochemical Society, Caracas, Venezuela, 1971, in press.

25. Barnes, E. M., Jr. and Kaback, H. R., Proc. Nat. Acad. Sci. U.S.A., 66, 1190 (1970).

26. Barnes, E. M., Jr. and Kaback, H. R., J. Biol. Chem., 246, 5518 (1971).

27. Dietz, G. W., Fed. Proc., 30, 52 (1971).

28. Scarborough, G. A., Rumley, M. K., and Kennedy, E. P., Proc. Nat. Acad. Sci. U.S.A., 60, 951 (1968).

29. Pavlasova, E. and Harold, F. M., J. Bacteriol., 98, 198 (1969).

30. Fox, C. F. and Kennedy, E. P., Proc. Nat. Acad. Sci. U.S.A., 54, 891 (1965).

31. Fox, C. F., Carter, J. R., and Kennedy, E. P., Proc. Nat. Acad. Sci. U.S.A., 57, 698 (1967).

32. Kundig, W., Kundig, F. D., Andersen, B. E., and Roseman, S., J. Biol. Chem., 241, 3243 (1966).

33. Tanaka, S., Fraenkel, P. G., and Lin, E. C. C., Biochem. Biophys. Res. Commun., 27, 53 (1967).

34. Kerwar, G. K., Gordon, A. S., and Kaback, H. R., J. Biol. Chem., 247, 291 (1972).

35. Kaback, H. R. and Barnes, E. M., Jr., J. Biol. Chem., 246, 5523 (1971).

36. Horecker, B. L., Thomas, J., and Monod, J., J. Biol. Chem., 235, 1580 (1960).

37. Wu, H. C. P., J. Mol. Biol., 24, 213 (1967).

38. Ganesan, A. K. and Rotman, B., J. Mol. Biol., 16, 42 (1966).

39. Tarmy, E. M. and Kaplan, N. O., J. Biol. Chem., 243, 2579 (1968).

40. Tarmy, E. M. and Kaplan, N. O., J. Biol. Chem., 243, 2587 (1968).

41. Kline, E. S. and Mahler, H. R., Ann. N.Y. Acad. Sci., 119, 105 (1965).

42. Cox, G. B., Newton, N. A., Gibson, F., Snoswell, A. M., and Hamilton, J. A., Biochem. J., 117, 551 (1970).

43. Weissbach, H., Thomas, E. L., and Kaback, H. R., Arch. Biochem. Biophys., 147, 249 (1971).

44. Klein, W. L., Dhams, A. S., and Boyer, P. D., Fed. Proc., 29, 341 (1970).

45. Kepes, A. and Cohen, G. N., in: The Bacteria, Vol. IV, ed. I. C. Gunsalis and R. Stanier (Academic Press, New York, 1962) p. 179.

46. Kepes, A., in: Current Topics in Membranes and Transport, Vol. I, eds. A. Kleinzeller and F. Bronner (Academic Press, New York, 1970) p. 101.

47. Kepes, A., Biochim. Biophys. Acta, 40, 70 (1960).

48. Carter, J. R., Fox, C. F., and Kennedy, E. P., Proc. Nat. Acad. Sci. U.S.A., 60, 725 (1968).

49. Jones, T. H. D. and Kennedy, E. P., J. Biol. Chem., 244, 5981 (1969).

50. Schachter, D. and Mindlin, A. J., J. Biol. Chem., 244, 1808 (1969).

51. Manno, J. A. and Schachter, D., J. Biol. Chem., 245, 1217 (1970).

52. Koch, A. L., J. Mol. Biol., 59, 447 (1971).

53. Wong, P. T. S., Kashket, E. R., and Wilson, T. H., Proc. Nat. Acad. Sci. U.S.A., 65, 63 (1970).

54. Wilson, T. H., Kusch, M., and Kashket, E. R., Biochem. Biophys. Res. Commun., 40, 1409 (1970).

55. Konings, W. N., Barnes, E. M., Jr., and Kaback, H. R., J. Biol. Chem., 246, 5857 (1971).

56. Bragg, P. D., Canad. J. Biochem., 48, 777 (1970).

57. Gordon, A. S., Lombardi, F. J., and Kaback, H. R., Proc. Nat. Acad. Sci. U.S.A., 69, 358 (1972).

58. Heppel, L. A., Science, 156, 1451 (1967).

59. Pardee, A. B., Science, 162, 632 (1968).

60. Neu, H. C. and Heppel, L. A., J. Biol. Chem., 240, 3685 (1965).